Recent Advances in Wireless Power Transfer

Recent Advances in Wireless Power Transfer

Editor: Shawn Guy

MURPHY & MOORE
www.murphy-moorepublishing.com

Published by Murphy & Moore Publishing
1 Rockefeller Plaza,
New York City, NY 10020, USA
www.murphy-moorepublishing.com

Recent Advances in Wireless Power Transfer
Edited by Shawn Guy

Cataloging-in-Publication Data

Recent advances in wireless power transfer / edited by Shawn Guy.
p. cm.
Includes bibliographical references and index.
ISBN 978-1-63987-478-1
1. Wireless power transmission. 2. Electric power transmission.
3. Wireless communication systems--Power supply. I. Guy, Shawn.
TK3088 .R43 2022
621.319--dc23

Contents

Preface

The world is advancing at a fast pace like never before. Therefore, the need is to keep up with the latest developments. This book was an idea that came to fruition when the specialists in the area realized the need to coordinate together and document essential themes in the subject. That's when I was requested to be the editor. Editing this book has been an honour as it brings together diverse authors researching on different streams of the field. The book collates essential materials contributed by veterans in the area which can be utilized by students and researchers alike.

Wireless power transfer (WPT) is the transmission of electrical energy without wires using electromagnetic fields. It is also called electromagnetic power transfer. In a WPT system, a transmitter device driven by electric power from a power source generates a time-varying electromagnetic field. This field transfers electricity to a receiver device through space. The receiver device then extracts the power from the electromagnetic field and delivers it to an electrical load. WPT technology can eliminate the use of wires and batteries, and thus increase the mobility, convenience and safety of an electronic device. It is useful in operating electrical equipment when connecting wires is inconvenient, dangerous or not possible. WPT mainly falls into two categories, near-field or non-radiative technique and far field or radiative technique. The ever growing need of advanced technology is the reason that has fueled the research in the field of wireless power transfer in recent times. This book strives to provide a fair idea about this discipline and to help develop a better understanding of the latest advances within this field. It aims to equip students and experts with the advanced topics and upcoming concepts in this area.

Each chapter is a sole-standing publication that reflects each author's interpretation. Thus, the book displays a multi-facetted picture of our current understanding of application, resources and aspects of the field. I would like to thank the contributors of this book and my family for their endless support.

Editor

Performance Analysis for NOMA Relaying System in Next-Generation Networks with RF Energy Harvesting

Dac-Binh Ha and Jai P. Agrawal

Abstract

In this chapter, we investigate the performance of the non-orthogonal multiple access (NOMA) relaying network with radio-frequency (RF) power transfer. Specifically, this considered system consists of one RF power supply station, one source, one energy-constrained relay, and multiple energy-constrained NOMA users. The better user and relay can help the source to forward the message to worse user by using the energy harvested from the power station. The triple-phase harvest-transmit-forward transmission protocol is proposed for this considered system. The exact closed-form expressions of outage probability and throughput for each link and whole system are derived by using the statistical characteristics of signal-to-noise ratio (SNR) and signal-to-interference-plus-noise ratio (SINR) of transmission links. In order to understand more detail about the behavior of this considered system, the numerical results are provided according to the system key parameters, such as the transmit power, number of users, time switching ratio, and power allocation coefficients. The simulation results are also provided to confirm the correctness of our analysis.

Keywords: outage probability, non-orthogonal multiple access, relaying, radio frequency, wireless energy harvesting

Introduction

Radio-frequency (RF) energy harvesting (EH), abbreviated as RF-EH, enabled wireless power transfer (WPT) is an emerging and promising approach to supply everlasting and cost-effective energy to low-power electronic devices, that is, sensor nodes, low-power cellphone, wireless control devices, and so on [1–3]. This approach is expected to have abundant applications in next-generation wireless networks, such as Internet-of-Things (IoT) wireless networks or 5G networks, which can be a platform for varied fields, for example, manufacturing [4, 5], smart agriculture [5], and smart city [5]. Specifically, the IoT-based wireless sensor networks for smart agriculture usually consist of a large number of battery-powered wireless sensor nodes for data collecting, data processing, and data transmission. Therefore, these energy-constrained devices need to be replaced or recharged periodically; this leads the lifetime of network limited. The

RF energy harvesting can prolong the lifetime of these networks using the energy harvested from the RF sources (e.g., base station, TV/radio broadcast station, microwave station, satellite earth station, etc.). Compared with other resources available for the EH, the RF power is easy to be converted. Therefore, RF energy harvesting is the solution to enhance the system energy efficiency in the energy-constrained networks, including network lifetime; reduction of carbon footprint, without requiring battery replacement; easy and fast deployment in complicated or toxic environments; etc. In the past few years, there have been a number of works on RF energy harvesting communications, and the main works focused on the development of energy harvesting models, protocols, transmission schemes, and security in communication systems [1–3]. In practice, RF-EH can be operated in a time switching (TS) scheme in which the receiver uses a portion of time duration for energy harvesting and the remaining time for information receiving or a power splitting (PS) scheme in which the received signal power is divided into two parts for energy harvesting and information receiving, separately [6].

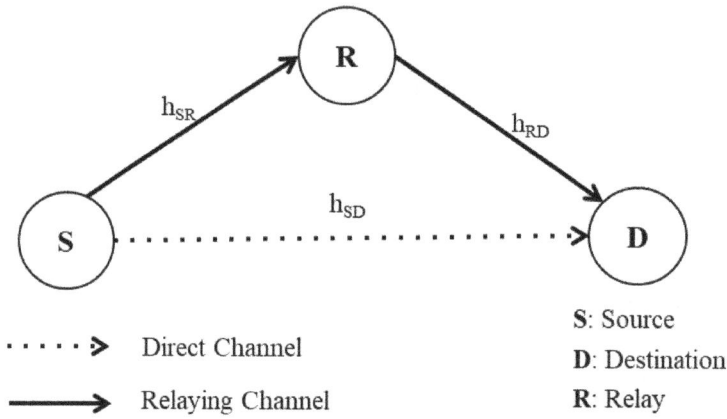

Figure 1. *A system model of cooperative network.*

Relaying communication technique can mitigate the wireless channel fading and improve the reliability of wireless links by exploiting the spatial diversity gains inherent in multiple user environments [7]. This can be achieved by using collaboration of relay nodes to form virtual multiple input multiple output (MIMO) without the need of multiple antennas at each node. **Figure 1** depicts a system model of cooperative network. We can observe from this figure that the destination D can receive two signals from direct link and relaying link. It means that D has more opportunities to decode its own message; thus the performance of this system can be improved. There are two schemes of relaying technique: amplify-and-forward (AF) or decode-and-forward (DF). In AF relaying scheme, the relay simply sends a scaled copy of the received noisy signal to the destination, while in DF relaying scheme, the relay transmits a re-encoded copy to the destination, if the relay can successfully decode the transmitted message. In wireless relaying networks (e.g., energy-constrained wireless sensor networks), the relay nodes (e.g., cluster head nodes) are often subject to space limitation to equip a large battery for long lifetime using [8]. Thus, RF energy harvesting technique has been applied for this type of relay nodes to not only improve the throughput and reliability by exploiting the virtual spatial diversity but also promise everlasting network lifetime without requiring battery replacement. Due to the new imposed time-varying energy constraints, several technical issues, such as relaying protocols, power allocation, energy-information tradeoff, relay selection, cooperative spectrum sensing and sharing, security, etc., have been investigated for various relaying network models [6, 9, 10]. The challenges in these works become more complicated because the harvested energy varies according to channel fading, and the energy usage overtime needs to make a tradeoff between energy harvesting time and information processing time.

The next-generation networks (5G and beyond) are supported with very high data rate, ultralow latency, massive connections, and very high mobility to satisfy the fast-increasing users and demands. To fulfill these targets, the relaying and non- orthogonal multiple access (NOMA) techniques are proposed to extend the coverage of network, improve the performance, achieve high spectral efficiency, and support dense networks [11]. In NOMA scheme, the source superposes all messages before transmitting them to users as **Figure 2**. In this figure, we can see that the near receiver (or better user) uses successive interference cancelation (SIC) to obtain the far user's message first (due to it is allocated with more transmit power) and subtracts this component from the received signal to obtain its own message. Compared to conventional orthogonal multiple access (OMA), for example, frequency division multiple access (FDMA), time division multiple access (TDMA), and code division multiple access (CDMA), NOMA simultaneously serves multiple user equipment on the same resource blocks by splitting users into power domain [11]; therefore it can improve spectral efficiency of wireless network.

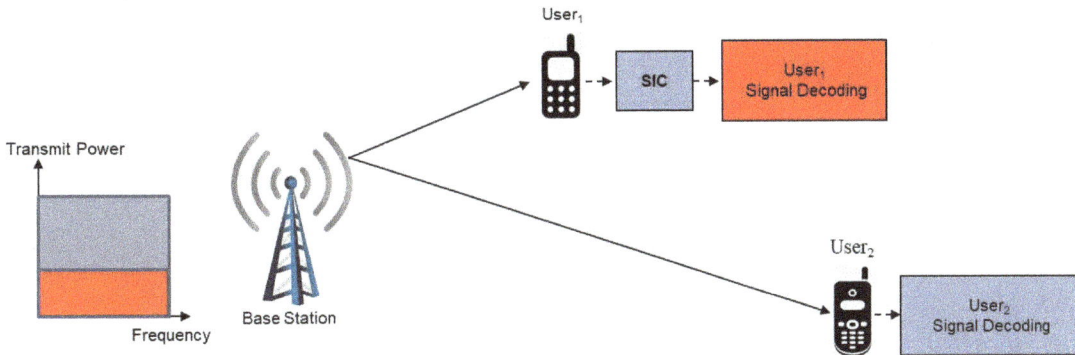

Figure 2. *Illustration of NOMA.*

The above three techniques (i.e., RF-EH, relaying, NOMA) can be integrated into next-generation networks. However, there are many related issues that need to be addressed before these techniques can be deployed in next-generation networks, such as the network architecture, power allocation, relaying scheme selection, the combination between NOMA and other multiple access methods, fixed or dynamic user pairing/clustering, optimal user allocation and beamforming in NOMA MIMO systems, the impact of imperfect CSI, and joint optimization of diverse aspects of NOMA (spectrum efficiency, energy efficiency, security) [11]. In recent years, a number of works investigated some related issues such as performance of energy harvesting DF/AF relaying, cooperative, cognitive, and MIMO NOMA networks [12–15]. In addition, the work of [16] studied the secrecy performance of MIMO NOMA system over Nakagami-m channels with transmit antenna selection protocol. However, almost in these works the information sources are assumed that it can transmit RF energy and information by using TS or PS scheme.

Different from the above works, in this chapter we investigate the cooperative NOMA network in which the power station and information source (e.g., base station) are separated and the energy-constrained user nodes collaborate with the energy-constrained relay nodes to help source forward the information to destinations. The main contributions of this chapter are as follows:

- The triple-phase harvest-transmit-forward transmission protocol is proposed for this considered system.

- The exact closed-form expressions of outage probability and throughput for each link and whole system are derived by using the statistical characteristics of signal-to-noise ratio (SNR) and signal-to-interference-plus-noise ratio (SINR) of transmission links.

- In terms of outage probability, the numerical results are provided according to the system key parameters, such as the transmit power, number of users, time switching ratio, and power allocation coefficients to look insight this considered system.

System and channel model description

Figure 3 depicts the system model for RF-EH NOMA relaying network, in which the power station (P) intends to transfer energy to energy-constrained relay (R) and energy-constrained destination nodes (D); information source (S) intends transmit the information to destinations by the help of relay node R.

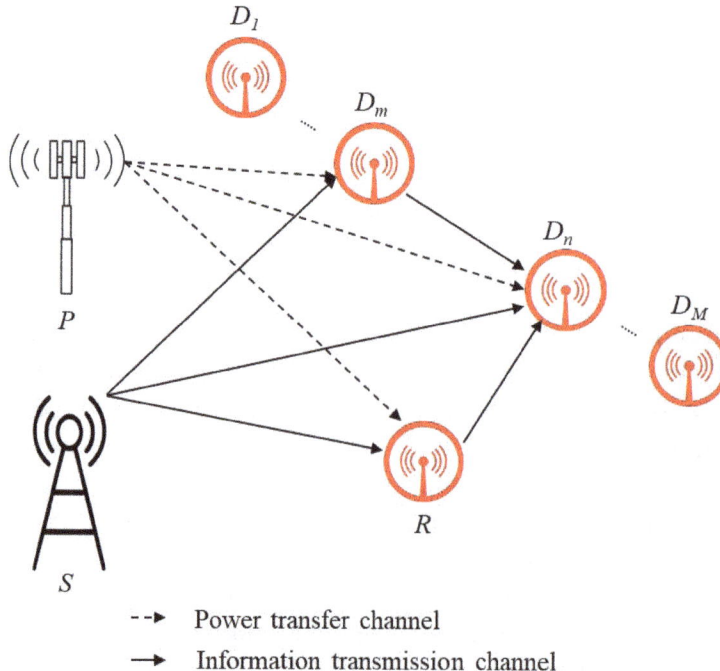

Power transfer channel
Information transmission channel

Figure 3. *System model for RF-EH NOMA relaying network.*

Notation: Denote P, S, R, and D as power station, information source, relay, and destination, respectively. $|h_{SDm}|^2$ and $|h_{SDn}|^2$ are denoted as the ordered channel gains of the mth user and the

nth user, respectively. Denote $|h_{PR}|^2$, $|h_{PDm}|^2$, $|h_{SR}|^2$, $|h_{RDn}|^2$, and $|h_{mn}|^2$ as the channel gains of the links P—R, P—D_m, S—R, R—D_n, and D_m—D_n, respectively. Denote d_{PR}, d_{PDm}, d_{SR}, d_{RDn}, d_{SDm}, and d_{mn} as the Euclidean distances of P—R, P—D_m, S—R, S—D_m, R—D_n, and D_m—D_n, respectively. Symbol θ is denoted as the path loss exponent. Let $X_1 = |h_{PDm}|^2$, $Y_1 = |h_{PR}|^2$, $X_2 = |h_{SDm}|^2$, $Y_2 = |h_{SR}|^2$, $X_3 = |h_{mn}|^2$, and $Y_3 = |h_{RDn}|^2$.

In this system, NOMA scheme is applied for M destination users in pair division manner, such as $\{D_m, D_n\}$ with $m < n$ [17]. Without loss of generality, we assume that all the channel power gains between S and D_i ($1 \leq i \leq M$) follow the following order: $|h_{SD1}|^2 \geq \ldots \geq |h_{SDm}|^2 \geq |h_{SDn}|^2 \geq \ldots \geq |h_{SDM}|^2$.

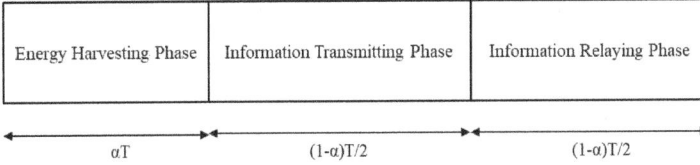

Energy Harvesting Phase	Information Transmitting Phase	Information Relaying Phase
αT	$(1-\alpha)T/2$	$(1-\alpha)T/2$

Figure 4. *The triple-phase protocol for RF-EH NOMA relaying network.*

The scenario of this considered system is investigated as follows:

- Due to the severe shadowing environment, the worse node (i.e., D_n) cannot detect message signal transmitted from S. Thus, the better node (i.e., D_m) or relay node R is selected to help S forwarding the message signal to worse node.

- All the transceivers are equipped by single antenna and operate in half duplex mode.

- All wireless links are assumed to undergo independent frequency nonselective Rayleigh block fading and additive white Gaussian noise (AWGN).

- The power gains of all links are modeled by random variables with zero mean and the same variance σ^2, that is, rvCN($0,\sigma^2$).

In this work, we propose a triple-phase harvest-transmit-forward transmission protocol for this RF-EH NOMA relaying system as shown in **Figure 4**:

1. In the first phase (power transfer phase): P transfers RF energy to the users with power P_0 in the time αT ($0 \leq \alpha \leq 1$, time switching ratio; T, block time for each transmission).

2. In the second phase (information transmitting phase): S uses power P_S to transmit superimposed message signal

$$x = \sqrt{a_m}s_m + \sqrt{a_n}s_n \tag{1}$$

to user pair (D_m, D_n) in the time of $(1-\alpha)T/2$, where s_m and s_n are the message for the mth user D_m and the nth user D_n, respectively, and a_m and a_n are the power allocation coefficients which satisfied the conditions: $0 < a_m < a_n$ and $a_m + a_n = 1$ by following the NOMA scheme. By applying NOMA, D_m uses SIC to detect message s_n and subtracts this component from the received signal to obtain its own message s_m.

3. In the third phase (information relaying phase): in this phase, D_m re-encodes and forwards s_n to D_n in the remaining time of $(1-\alpha)T/2$ with the energy harvested from P. At the same time, relay decodes x and forwards x to D_n.

Finally, D_n combines two received signals, that is, the relaying signals from D_m and R, to decode its own message by using selection combining (SC) scheme. For more detailed purpose, we continue to present the transmission of this protocol for RF-EH NOMA relaying system in mathematical manner.

Power transfer phase

In this phase, the energy of D_m and R harvested from P in the time of αT can be respectively expressed as

$$E_1 = \frac{\eta P_0 |h_{PD_m}|^2 \alpha T}{d_{PD_m}^\theta},$$ (2)

$$E_2 = \frac{\eta P_0 |h_{PR}|^2 \alpha T}{d_{PR}^\theta},$$ (3)

where η is the energy conversion efficiency ($0 \le \eta \le 1$).

Information transmitting phase

In this duration of $(1-\alpha)T/2$, the source S broadcasts superimposed message signal x as Eq. (1) to the user pair and relay. The received signal at D_m is written as

$$y_{SD_m} = \sqrt{\frac{P_S}{d_{SD_m}^\theta}}(\sqrt{a_m}s_m + \sqrt{a_n}s_n)h_{SD_m} + n_{SD_m},$$ (4)

where $n_{SDm} \sim CN(0,\sigma^2)$ is AWGN.

Similarly, the received signal at R is expressed as

$$y_{SR} = \sqrt{\frac{P_S}{d_{SR}^\theta}}(\sqrt{a_m}s_m + \sqrt{a_n}s_n)h_{SR} + n_{SR},$$ (5)

where $n_{SR} \sim CN(0,\sigma^2)$ are AWGN.

Applying NOMA, D_m uses SIC to detect message s_n and subtracts this component from the received signal to obtain its own message s_m. Therefore, the instantaneous SINR at D_m to detect s_m and s_n transmitted from S can be respectively given by

$$\gamma_{SD_m}^{s_n} = \frac{a_n \gamma_S |h_{SD_m}|^2}{a_m \gamma_S |h_{SD_m}|^2 + d_{SD_m}^\theta} = \frac{b_2 X_2}{b_1 X_2 + 1},$$ (6)

$$\gamma_{SD_m}^{s_m} = \frac{a_m \gamma_S |h_{SD_m}|^2}{d_{SD_m}^\theta} = b_1 X_2,$$ (7)

where $\gamma_S = \dfrac{P_S}{\sigma^2}, b_1 = \dfrac{a_m \gamma_S}{d_{SD_m}^\theta}, \quad b_2 = \dfrac{a_n \gamma_S}{d_{SD_m}^\theta}.$

And in the meanwhile the relay applying DF scheme first decodes its received signal from S to obtain superimposed message x and then re-encodes and forwards it to the destination. Therefore, in this phase the instantaneous SNR at R to detect x transmitted from S can be given by

$$\gamma_{SR}^x = \frac{\gamma_S |h_{SR}|^2}{d_{SR}^\theta} = b_3 Y_2, \tag{8}$$

where $b_3 = \dfrac{\gamma_S}{d_{SR}^\theta}.$

Information relaying phase

In this phase, D_m and R spend the harvested energy E_1 and E_2, respectively, as Eqs. (2) and (3) to forward received signals to D_n in duration of (1-α)T/2. Notice that we ignore the processing power required by the transmit/receive circuitry of D_m and R. Therefore, the transmit power of D_m and R is respectively given by

$$P_1 = \frac{2\eta\alpha P_0 |h_{PD_m}|^2}{(1-\alpha)d_{PD_m}^\theta}, \tag{9}$$

$$P_2 = \frac{2\eta\alpha P_0 |h_{PR}|^2}{(1-\alpha)d_{PR}^\theta}. \tag{10}$$

The received signals at D_n that are transmitted from D_m and R are, respectively, expressed as

$$y_{D_m D_n} = \sqrt{\frac{P_1}{d_{mn}^\theta}} s_n h_{mn} + n_{mn}, \tag{11}$$

where $n_{mn} \sim CN(0,\sigma^2)$, and

$$y_{RD_n} = \sqrt{\frac{P_2}{d_{RD_n}^\theta}} (\sqrt{a_m} s_m + \sqrt{a_n} s_n) h_{RD_n} + n_{RD_n}, \tag{12}$$

where $n_{RD_n} \sim CN(0,\sigma^2)$.

Because D_n applies SC scheme, the instantaneous SNR/SINR at D_n to detect s_n transmitted from S can be given by

$$\gamma_{D_n}^{s_n} = \max \left\{ \frac{2\eta\alpha\gamma_0 |h_{PD_m}|^2 |h_{mn}|^2}{(1-\alpha)d_{PD_m}^\theta d_{mn}^\theta}, \frac{2\eta\alpha a_n \gamma_0}{(1-\alpha)d_{PR}^\theta} \frac{|h_{PR}|^2 |h_{RD_n}|^2}{\left(a_m \gamma_0 |h_{RD_n}|^2 + d_{RD_n}^\theta\right)} \right\}$$
$$= \max \left\{ c_1 X_1 X_3, \frac{c_2 Y_1 Y_3}{(c_3 Y_3 + 1)} \right\}, \tag{13}$$

$$\text{where } \gamma_0 = \frac{P_0}{\sigma^2}, c_1 = \frac{2\eta\alpha\gamma_0}{(1-\alpha)d_{PD_m}^\theta d_{mn}^\theta}, \quad c_2 = \frac{2\eta\alpha a_n\gamma_0}{(1-\alpha)d_{PR}^\theta d_{RD_n}^\theta}, c_3 = \frac{a_m\gamma_0}{d_{RD_n}^\theta}.$$

The independent and identically distributed (IID) Rayleigh channel gains $(X_1, Y_1, X_2, Y_2, X_3, Y_3)$ follow exponential distributions with parameters $\lambda_1, \lambda_2, \lambda_3, \lambda_4, \lambda_5$, and λ_6, respectively. According to [18], the cumulative distribution function (CDF) and the probability density function (PDF) of ordered random variable X_2 are respectively written as follows

$$F_{X_2}(x) = \frac{M!}{(M-m)!(m-1)!} \sum_{k=0}^{m-1} \frac{C_k^{m-1}(-1)^k}{M-m+k+1} \left[1 - e^{\frac{-x(M-m+k+1)}{\lambda_3}}\right], \tag{14}$$

$$f_{X_2}(x) = \frac{M!}{(M-m)!(m-1)!} \frac{1}{\lambda_3} \sum_{k=0}^{m-1} C_k^{m-1}(-1)^k e^{\frac{-x(M-m+k+1)}{\lambda_3}}. \tag{15}$$

Because all links undergo Rayleigh fading, the PDF and CDF of random variable $V \in \{X_1, Y_1, Y_2, X_3, Y_3\}$ have the following forms:

$$f_V(x) = \frac{1}{\lambda} e^{-\frac{x}{\lambda}}, \tag{16}$$

$$F_V(x) = 1 - e^{-\frac{x}{\lambda}}, \tag{17}$$

where $\lambda \in \{\lambda_1, \lambda_2, \lambda_4, \lambda_5, \lambda_6\}$.

Performance analysis

Outage probability

Outage probability is an important performance metric for system designers [12]. It is generally used to characterize a wireless communication system and defined as the probability that the instantaneous end-to-end SNR (γ_{e2e}) falls below the predetermined threshold γ_{th} ($\gamma_{th} = 2^\Omega-1$, where Ω is fixed transmission rate at the source), given by

$$P_{out} = \Pr(\gamma_{e2e} < \gamma_{th}): \tag{18}$$

In this considered system, the outage event at the destination occurs when D_m cannot detect successfully s_n or s_m or when D_m can detect successfully s_n and s_m, but an outage occurs in information relaying phase. Accordingly, outage probability of this RF-EH NOMA relaying system is written as

$$P_{out} = \Pr\left(\gamma_{SD_m}^{s_n} < \gamma_{th}\right) + \Pr\left(\gamma_{SD_m}^{s_n} > \gamma_{th}, \gamma_{SD_m}^{s_m} < \gamma_{th}\right)$$
$$+ \Pr\left(\gamma_{SD_m}^{s_n} > \gamma_{th}, \gamma_{SD_m}^{s_m} > \gamma_{th}\right) \Pr\left(\gamma_{D_n}^{s_n} < \gamma_{th}\right). \tag{19}$$

Notice that because the messages are transmitted in the duration of (1-α)T/2, thus γth is calculated by γ_{th} = 2²Ω/(1-α) - 1, where Ω is fixed source transmission rate.

Substituting Eqs. (6), (7), and (13) into Eq. (19), we obtain the following equation:

$$P_{out} = \Pr\left(\frac{b_2 X_2}{b_1 X_2 + 1} < \gamma_{th}\right) + \Pr\left(\frac{b_2 X_2}{b_1 X_2 + 1} > \gamma_{th}, \ b_1 X_2 < \gamma_{th}\right)$$

$$+ \Pr\left(\frac{b_2 X_2}{b_1 X_2 + 1} > \gamma_{th}, \ b_1 X_2 > \gamma_{th}\right)\Pr(c_1 X_1 X_3 < \gamma_{th}) \qquad (20)$$

$$\times \left[\Pr(b_3 Y_2 < \gamma_{th}) + (1 - \Pr(b_3 Y_2 < \gamma_{th}))\Pr\left(\frac{c_2 Y_1 Y_3}{c_3 Y_3 + 1} < \gamma_{th}\right)\right]$$

$$= I_1 + I_2 + I_3 I_4 [I_5 + (1 - I_5) I_6],$$

where

$$I_1 = \Pr\left(\frac{b_2 X_2}{b_1 X_2 + 1} < \gamma_{th}\right), \qquad (21)$$

$$I_2 = \Pr\left(\frac{b_2 X_2}{b_1 X_2 + 1} > \gamma_{th}, \ b_1 X_2 < \gamma_{th}\right), \qquad (22)$$

$$I_3 = \Pr\left(\frac{b_2 X_2}{b_1 X_2 + 1} > \gamma_{th}, \ b_1 X_2 > \gamma_{th}\right), \qquad (23)$$

$$I_4 = \Pr(c_1 X_1 X_3 < \gamma_{th}), \qquad (24)$$

$$I_5 = \Pr(b_3 Y_2 < \gamma_{th}), \qquad (25)$$

$$I_6 = \Pr\left(\frac{c_2 Y_1 Y_3}{c_3 Y_3 + 1} < \gamma_{th}\right). \qquad (20)$$

By the help of Eqs. (14)–(17), we obtain the exact closed-form expressions of I_1, I_2, I_3, I_4, I_5, and I_6, respectively, as follows:

$$I_1 = \begin{cases} 1, & \gamma_{th} > \dfrac{a_n}{a_m} \\[2ex] F_{X_2}\left(\dfrac{\gamma_{th}}{b_2 - b_1 \gamma_{th}}\right), & \gamma_{th} < \dfrac{a_n}{a_m} \end{cases}$$

$$= \begin{cases} 1, & \gamma_{th} > \dfrac{a_n}{a_m}. \\[2ex] \dfrac{M!}{(M-m)!(m-1)!} \displaystyle\sum_{k=0}^{m-1} \dfrac{C_k^{m-1}(-1)^k}{M-m+k+1}\left[1 - e^{\frac{-(M-m+k+1)\gamma_{th}}{\lambda_3(b_2 - b_1\gamma_{th})}}\right], & \gamma_{th} < \dfrac{a_n}{a_m}. \end{cases} \qquad (27)$$

$$I_2 = \begin{cases} F_{X_2}\left(\dfrac{\gamma_{th}}{b_1}\right) - F_{X_2}\left(\dfrac{\gamma_{th}}{b_2 - b_1 \gamma_{th}}\right), & \gamma_{th} < \dfrac{a_n}{a_m} - 1 \\[2ex] 0, & \gamma_{th} > \dfrac{a_n}{a_m} - 1 \end{cases}$$

$$= \begin{cases} \dfrac{M!}{(M-m)!(m-1)!} \displaystyle\sum_{k=0}^{m-1} \dfrac{C_k^{m-1}(-1)^k}{M-m+k+1}\left[e^{\frac{-(M-m+k+1)\gamma_{th}}{\lambda_3(b_2 - b_1\gamma_{th})}} - e^{\frac{-(M-m+k+1)\gamma_{th}}{\lambda_3 b_1}}\right], & \gamma_{th} < \dfrac{a_n}{a_m} - 1. \\[2ex] 0, & \gamma_{th} > \dfrac{a_n}{a_m} - 1. \end{cases} \qquad (28)$$

$$I_3 = \begin{cases} 1 - F_{X_2}\left(\dfrac{\gamma_{th}}{b_1}\right), & \gamma_{th} < \dfrac{a_n}{a_m} - 1 \\[3mm] 1 - F_{X_2}\left(\dfrac{\gamma_{th}}{b_2 - b_1\gamma_{th}}\right), & \dfrac{a_n}{a_m} - 1 < \gamma_{th} < \dfrac{a_n}{a_m} \\[3mm] 0, & \gamma_{th} > \dfrac{a_n}{a_m} \end{cases}$$

$$= \begin{cases} 1 - \dfrac{M!}{(M-m)!(m-1)!}\displaystyle\sum_{k=0}^{m-1}\dfrac{C_k^{m-1}(-1)^k}{M-m+k+1}\left[1 - e^{-\frac{(M-m+k+1)\gamma_{th}}{\lambda_3 b_1}}\right], & \gamma_{th} < \dfrac{a_n}{a_m} - 1. \\[4mm] 1 - \dfrac{M!}{(M-m)!(m-1)!}\displaystyle\sum_{k=0}^{m-1}\dfrac{C_k^{m-1}(-1)^k}{M-m+k+1}\left[1 - e^{-\frac{(M-m+k+1)\gamma_{th}}{\lambda_3(b_2-b_1\gamma_{th})}}\right], & \dfrac{a_n}{a_m} - 1 < \gamma_{th} < \dfrac{a_n}{a_m}. \\[4mm] 0, & \gamma_{th} > \dfrac{a_n}{a_m}. \end{cases} \tag{29}$$

$$I_4 = \Pr\left(X_3 < \frac{\gamma_{th}}{c_1 X_1}\right) = \int_0^\infty F_{X_3}\left(\frac{\gamma_{th}}{c_1 z}\right)f_{X_1}(z)dz = \int_0^\infty \left(1 - e^{-\frac{\gamma_{th}}{\lambda_5 c_1 z}}\right)\frac{1}{\lambda_1}e^{-\frac{z}{\lambda_1}}dz \tag{30}$$

$$= 1 - 2\sqrt{\frac{\gamma_{th}}{\lambda_1\lambda_5 c_1}}K_1\left(2\sqrt{\frac{\gamma_{th}}{\lambda_1\lambda_5 c_1}}\right).$$

$$I_5 = 1 - e^{-\frac{\gamma_{th}}{\lambda_4 b_3}}. \tag{31}$$

$$I_6 = \Pr\left(Y_3 < \frac{c_3\gamma_{th}}{c_2}\right) + \Pr\left(Y_3 < \frac{\gamma_{th}}{c_2 Y_1 - c_3\gamma_{th}}, Y_1 > \frac{c_3\gamma_{th}}{c_2}\right)$$

$$= F_{Y_3}\left(\frac{c_3\gamma_{th}}{c_2}\right) + \int_{\frac{c_3\gamma_{th}}{c_2}}^\infty F_{Y_3}\left(\frac{\gamma_{th}}{c_2 z - c_3\gamma_{th}}\right)f_{Y_1}(z)dz \tag{32}$$

$$= 1 - 2e^{-\frac{c_3\gamma_{th}}{\lambda_2 c_2}}\sqrt{\frac{\gamma_{th}}{\lambda_2\lambda_6 c_2}}K_1\left(2\sqrt{\frac{\gamma_{th}}{\lambda_2\lambda_6 c_2}}\right).$$

Notice that K_v is the modified Bessel function of the second kind and vth order [19].

Substituting Eqs. (27)–(32) into Eq. (20), we obtain the exact closed-form expression of outage probability for this RF-EH NOMA relaying system as follows:

$$P_{out} = \begin{cases} I_1 + I_2 + I_3 I_4[I_5 + (1 - I_5)I_6], & \gamma_{th} < \dfrac{a_n}{a_m} - 1. \\[3mm] I_1 + I_3 I_4[I_5 + (1 - I_5)I_6], & \dfrac{a_n}{a_m} - 1 < \gamma_{th} < \dfrac{a_n}{a_m}. \\[3mm] 1, & \gamma_{th} > \dfrac{a_n}{a_m}. \end{cases} \tag{33}$$

Throughput

At this point, we analyze throughput (Φ) at the destination node for delay- limited transmission mode. It is found out by evaluating outage probability at a fixed source transmission rate—Ω bps/Hz. We observe that the source transmit information at the rate of Ω bps/Hz and the effective communication time from the source to the destination in the block time T is $(1-\alpha)T/2$. Therefore, throughput Φ at the destination is defined as follows:

$$\Phi = (1 - P_{out})\Omega \frac{(1 - \alpha)T/2}{T} = \frac{(1 - \alpha)(1 - P_{out})\Omega}{2}. \tag{34}$$

Substituting the result of Pout as Eq. (33) in Section 3.1 into Eq. (34), we obtain the exact closed-form expression of throughput for this RF-EH NOMA relaying system.

These derivations are similar to [20]. By using these expressions for programming, we can investigate the behaviors of this considered system, and then we can adjust the inputs to achieve the optimal performance for this network.

Numerical results and discussion

In this section, we provide the numerical results according to the system key parameters (i.e., the average transmit SNR γ_o and γ_s, number of users, time switching ratio, and power allocation coefficients) to clarify the performance of proposed protocol for this considered RF-EH NOMA relaying system. Furthermore, we also provide Monte Carlo simulation results to verify our analytical results. The simulation parameters are shown in **Table 1**.

Table 1. *Simulation parameters.*

Parameters	System values
Environment	Rayleigh
Number of antennas of each node	1
Fixed rate (Ω)	1 bps/Hz
Number of users (M)	4, 6, 8
Energy conversion efficiency (η)	0.9
Distances (d)	1
Path loss exponent (θ)	2

Figure 5 depicts P_{out} of this considered system versus the average transmit power of power station with different numbers of users M. This figure shows that when we increase the transmit power of power station P, P_{out} of this system decreases. Similarly, **Figure 6** shows the variation of throughput with respect to γ_o for different values of M. This figure also shows that the throughput of this system increases when increasing transmit power of power station P. These mean that we can improve the performance by increasing transmit power to provide more energy to users or reducing the NOMA of users.

Figures 7 and 8 plot the curves of P_{out} and throughput Φ of this system versus time switching ratio for different values of average transmit power of S, respectively. From these figures, we found that when the time switching ratio α is small, α increases, then P_{out} decreases, and Φ increases. This can be explained by that there is more time for the user and relay to harvest energy as α grows. When α continues to increase, P_{out} inversely increases, and Φ decreases. The reason is that there is less time for message transmission phases when α is greater than α* value. When α is greater than $1 - 2\Omega/\log_2(a_n/a_m + 1)$, then P_{out} reaches 1. From this analysis, there exists a specific value of α* that leads P to obtain the lowest value and leads Φ to reach the highest value. Obviously, we can select the best time switching ratio α to achieve the optimal performance of this system. From these figures, we also found that the performance of this system can be improved by increasing the transmit power of source S.

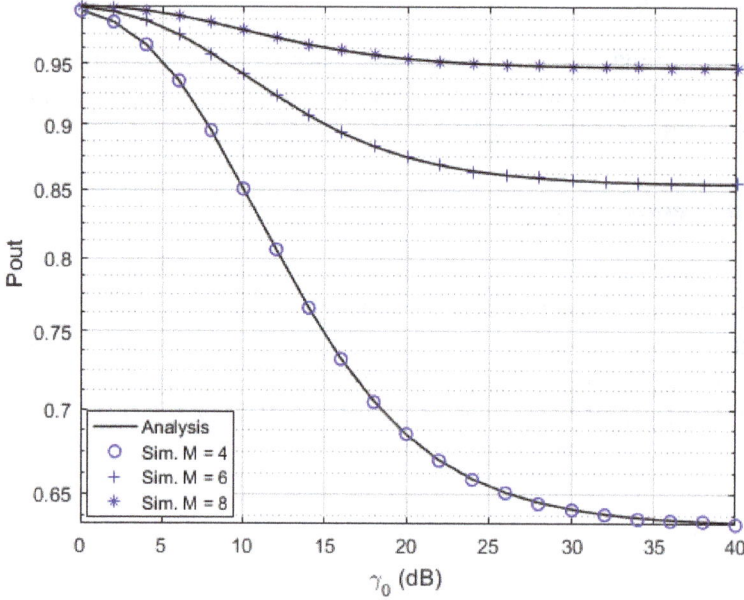

Figure 5. P_{out} vs. average transmit SNR of P with different numbers of users M with $\gamma_S = 20$ dB, $a_n = 0.9$, m = 2, n = 3, $\Omega = 1$ bps/Hz, α = 0.3, η = 0.9, $d_{PDm} = d_{PR} = d_{SDm} = d_{SR} = d_{mn} = d_{RDn} = 1$, θ = 2.

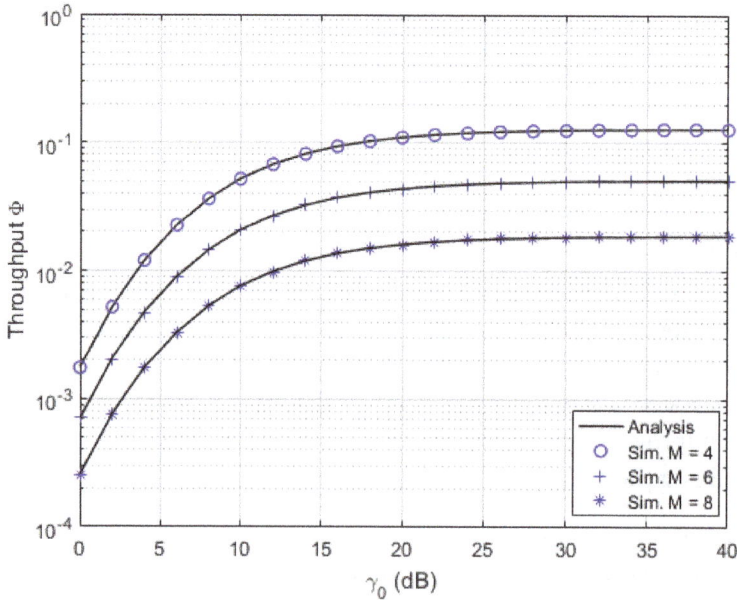

Figure 6. Throughput Φ vs. average transmit SNR of P with different numbers of users M with $\gamma_S = 20$ dB, $a_n = 0.9$, m = 2, n = 3, $\Omega = 1$ bps/Hz, α = 0.3, η = 0.9, $d_{PDm} = d_{PR} = d_{SDm} = d_{SR} = d_{mn} = d_{RDn} = 1$, θ = 2.

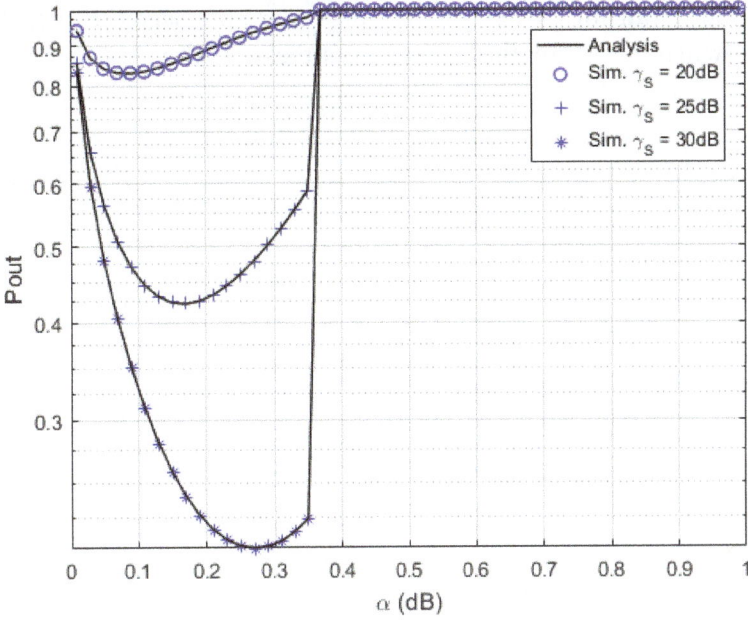

Figure 7. P_{out} with respect to time switching ratio for different values of average transmit SNR of S with $\gamma_0 = 20$ dB, M = 8, m = 2, n = 3, $\Omega = 1$ bps/Hz, $a_n = 0.9$, $\eta = 0.9$, $d_{PDm} = d_{PR} = d_{SDm} = d_{SR} = d_{mn} = d_{RDn} = 1$, $\theta = 2$.

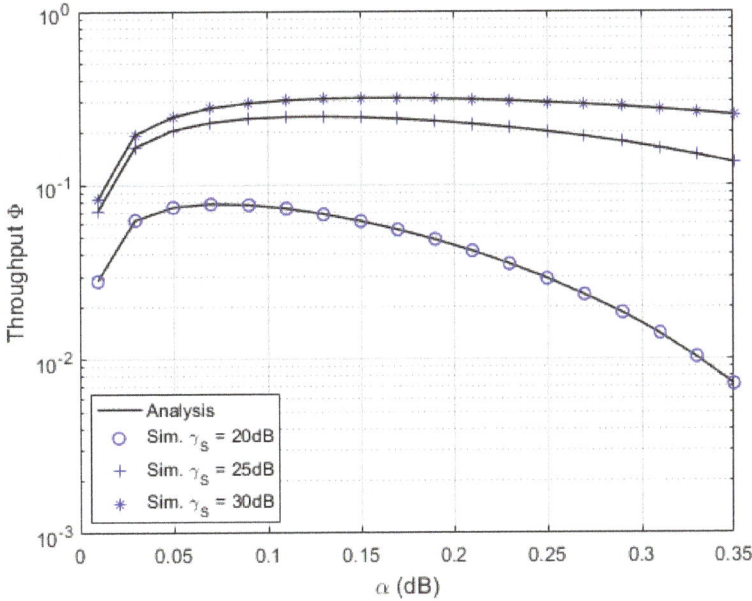

Figure 8. Throughput Φ with respect to time switching ratio for different values of average transmit SNR of S with $\gamma_0 = 20$ dB, M = 8, m = 2, n = 3, $\Omega = 1$ bps/Hz, $a_n = 0.9$, $\eta = 0.9$, $d_{PDm} = d_{PR} = d_{SDm} = d_{SR} = d_{mn} = d_{RDn} = 1$, $\theta = 2$.

Figures 9 and 10 show the variation of P_{out} and throughput Φ with respect to power allocation coefficient a_n for different values of average transmit SNR of S, respectively. From these figures, we can see that when $a_n \to 1$, the performance of this system degrades. Due to the constrain of γ_{th} (i.e., $\gamma_{th} = 2^{2\Omega/(1-\alpha)}-1 < a_n/a_m$), by the given value of R and α, the P_{out} reaches 1 when $a_n/a_m < 2^{2\Omega/(1-\alpha)}-1$. According to these figures, the performance can be improved when $a_n \to 0.89$ for $\Omega = 1$ bps/Hz, $\alpha = 0.1$. In order to improve the performance of this system, we can allocate more transmit power for the worse user's message (i.e., s_n). However, at that time the power which leaves for the better user's message (i.e., s_m) will be smaller, and it should satisfy $a_n/a_m > 2^{2\Omega/(1-\alpha)}-1$.

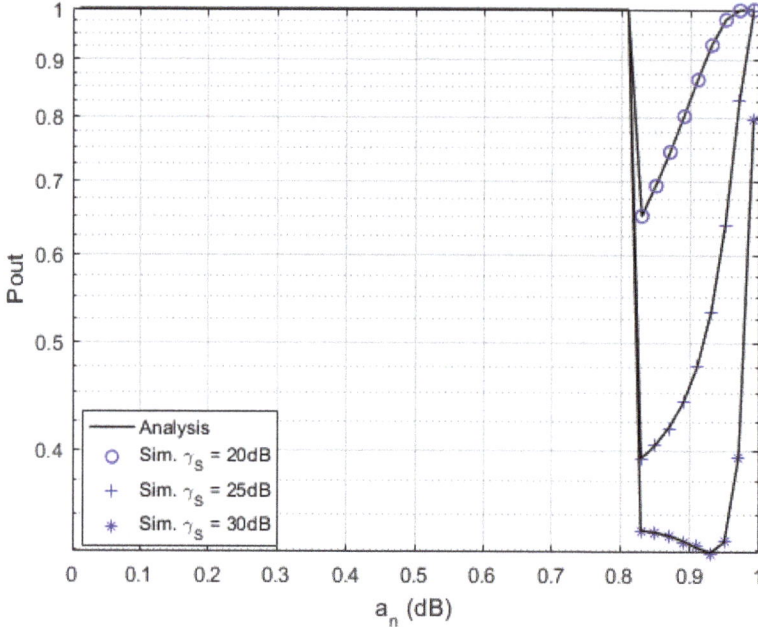

Figure 9. P_{out} with respect to power allocation coefficient a_n for different values of average transmit SNR of S with $\gamma_0 = 20$ dB, $M = 8$, $m = 2$, $n = 3$, $\Omega = 1$ bps/Hz, $\alpha = 0.1$, $\eta = 0.9$, $d_{PDm} = d_{PR} = d_{SDm} = d_{SR} = d_{mn} = d_{RDn} = 1$, $\theta = 2$.

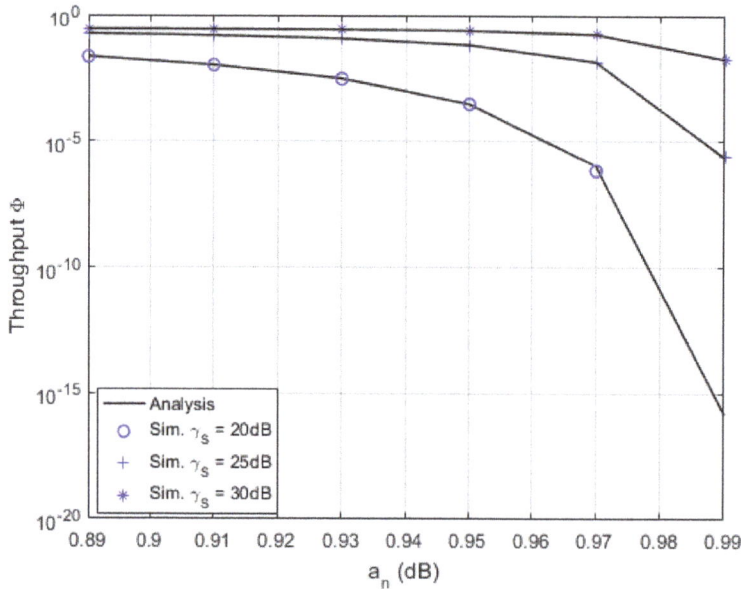

Figure 10. Throughput Φ with respect to power allocation coefficient for different values of average transmit SNR of S with $\gamma_0 = 20$ dB, $M = 8$, $m = 2$, $n = 3$, $\Omega = 1$ bps/Hz, $\alpha = 0.1$, $\eta = 0.9$, $d_{PDm} = d_{PR} = d_{SDm} = d_{SR} = d_{mn} = d_{RDn} = 1$, $\theta = 2$.

In addition, **Figure 11** plots the curves of P_{out} with and without relay versus average transmit SNR of power station P with different numbers of users M. In this figure, we can observe that P_{out}

of relaying scheme is lower than P_{out} without relaying. In other words, this result confirms that relaying method can improve the performance of this considered system.

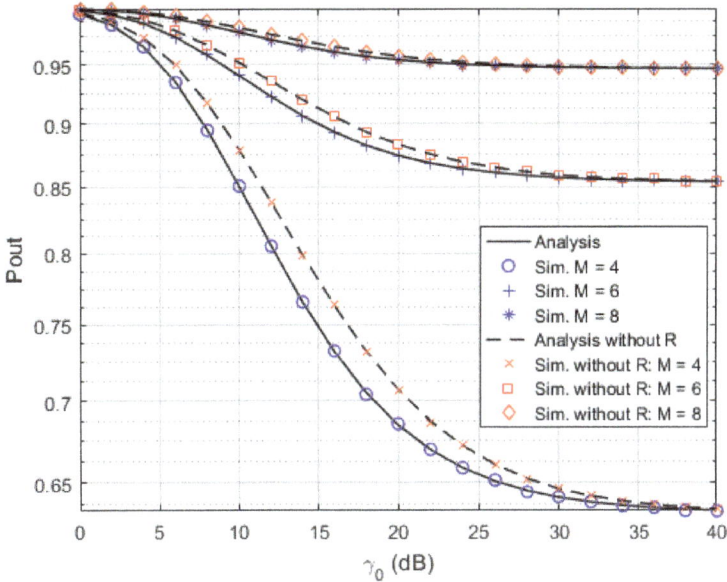

Figure 11. P_{out} with and without relay vs. average transmit SNR of P with different numbers of users M with $\gamma_S = 20$ dB, $a_n = 0.9$, $m = 2$, $n = 3$, $\Omega = 1$ bps/Hz, $\alpha = 0.3$, $\eta = 0.9$, $d_{PDm} = d_{PR} = d_{SDm} = d_{SR} = d_{mn} = d_{RDn} = 1$, $\theta = 2$.

Finally, we can observe from the above figures that the analysis and simulation results are good matching. This confirms the correctness of our analysis.

Conclusions

In this chapter, we have presented the performance analysis of downlink RF-EH NOMA relaying network with triple-phase harvest-transmit-forward transmission protocol in terms of outage probability and throughput. The exact closed-form expressions of outage probability and throughput for this proposed system have been derived. We have found that the performance of this considered system is enhanced by applying relaying technique or increasing the transmit power for energy harvesting and/or increasing the transmit power for information transmission. Moreover, the existence of best time switching ratio is proven to achieve the optimal performance of this system. We will solve the best time switching ratio searching problem in the future work.

Author details

Dac-Binh Ha[1]* and Jai P. Agrawal[2]

Duy Tan University, Da Nang, Vietnam

Purdue University Northwest, Hammond, IN, USA

*Address all correspondence to: hadacbinh@duytan.edu.vn

References

[1] Bi S, Zeng Y, Zhang R. Wireless powered communication networks: An overview. IEEE Wireless Communications. 2016;23(2):10-18. DOI: 10.1109/MWC.2016.7462480

[2] Adu-Manu KS, Adam N, Tapparello C, Ayatollahi H, Heinzelman W. Energy-harvesting wireless sensor networks (EH-WSNs): A review. ACM Transactions on Sensor Networks. 2018;14(2):1-50. DOI: 10.1145/3183338

[3] Sing K, Moh S. Energy harvesting in wireless sensor networks: Techniques and issues. International Journal of Control and Automation. 2017;10(6): 71-84. DOI: 10.14257/ijca.2017.10.6.08 [Accessed: 2019-05-15]

[4] Cheng J, Chen W, Tao F, Lin C-L. Industrial IoT in 5G environment towards smart manufacturing. Journal of Industrial Information Integration. 2018;10:10-19. DOI: 10.1016/j. jii.2018.04.001. [Accessed: 2019-05-18]

[5] Akpakwu GA, Silva BJ, Hancke GP, Abu-Mahfouz AM. A survey on 5G networks for the internet of things: Communication technologies and challenges. IEEE Access. 2018;6: 3619-3647. DOI: 10.1109/ACCESS. 2017.2779844

[6] Guanyao D, Xiong K, Qiu Z. Outage analysis of cooperative transmission with energy harvesting relay: Time switching versus power splitting. Mathematical Problems in Engineering. 2015;2015:1-9. DOI: 10.1155/2015/ 598290. [Accessed: 2019-03-18]

[7] Chakrabarti A, Sabharwal A, Aazhang B. Cooperative Communications: Fundamental Limits and Practical Implementation. Springer Publishers; 2006. Availabe from: https:// scholarship.rice.edu/handle/1911/19781. [Accessed: 2019-09-14]

[8] Nasir AA, Zhou X, Durrani S, Kennedy RA. Relaying protocols for wireless energy harvesting and information processing. IEEE Transactions on Wireless Communications. 2015;12(7): 3622-3636. DOI: 10.1109/ TWC.2013.062413.122042

[9] Ha D-B, Tran D-D, Tran-Ha V, Hong E-K. Performance of amplify-and- forward relaying with wireless power transfer over dissimilar channels. Elektronika ir Elektrotechnika Journal. 2015;21(5):90-95. DOI: 10.5755/j01. eee.21.5.13331

[10] Cvetkovic A, Blagojevic V, Ivaniš P. Performance analysis of nonlinear energy-harvesting DF relay system in interference-limited Nakagami-m fading environment. ETRI Journal. 2017; 39(6):803-812. DOI: 10.4218/etrij.2017-0096. [Accessed: 2019-03-18]

[11] Islam SMR, Avazov N, Dobre OA, Kwak K-s. Power-domain non- orthogonal multiple access (NOMA) in 5G systems: Potentials and challenges. IEEE Communication Surveys and Tutorials. 2017;19(2):721-742. DOI: 10.1109/COMST.2016.2621116

[12] Ha D-B, Nguyen SQ. Outage performance of energy harvesting DF relaying NOMA networks. Mobile Networks and Applications. 2018;23(6): 1572-1585. DOI: 10.1007/s11036-017-0922-x. [Accessed: 2019-03-18]

[13] Tran D-D, Ha D-B, Vo VN, So-In C, et al. Performance analysis of DF/AF cooperative MISO wireless sensor networks with NOMA and SWIPT over Nakagami-m fading. IEEE Access. 2018; 6:56142-56161. DOI: 10.1109/ACCESS. 2018.2872935

[14] Tran DD, Tran HV, Ha DB, Kaddoum G. Cooperation in NOMA networks under limited user-to-user communications: Solution and analysis. In: Proceedings of 2018 IEEE Wireless Communications and Networking Conference (WCNC'18); Barcelona, Spain: 15-18 April 2018. 2018

[15] Ha D-B, Nguyen SQ, Nguyen HT. Cooperative cognitive non-orthogonal multiple access under unreliable backhaul connections. Mobile Networks and Applications. 2019;24(2):596-617. DOI: 10.1007/s11036-018-1161-5. [Accessed: 2019-03-18]

[16] Tran D-D, Tran H-V, Ha D-B, Kaddoum G. Secure transmit antenna selection protocol for MIMO NOMA networks over Nakagami-m channels. IEEE Systems Journal (Early Access). 2019:1-12. DOI: 10.1109/JSYST.2019. 2900090

[17] Ding Z, Peng M, Poor HV. Cooperative non-orthogonal multiple access in 5G systems. IEEE Communications Letters. 2015;19(8): 1462-1465. DOI: 10.1109/LCOMM.2015. 2441064

[18] Men J, Ge J. Performance analysis of non-orthogonal multiple access in downlink cooperative network. IET Communications. 2015;9(18):2267-2273. DOI: 10.1049/iet-com.2015.0203

[19] Gradshteyn IS, Ryzhik IM. Table of Integrals, Series and Products. 7th ed. Academic Press; 2007. Available from: https://www.sciencedirect.com/book/9780123849335/table-of-integrals- series-and-products

[20] Ha DH, Ha DB, Voznak M. On secure cooperative non-orthogonal multiple access network with RF power transfer. In: Proceedings of 2019—5th EAI International Conference on Industrial Networks and Intelligent Systems; Hochiminh City, Vietnam; 19 August. 2019

2

Long-Distance Wireless Power Transfer Based on Time Reversal Technique

Bin-Jie Hu, Zhi-Wu Lin and Peng Liao

Abstract

Wireless power transfer (WPT) using microwave irradiation can set human free from the annoying wires. However, WPT has low energy efficiency due to electro- magnetic wave diffraction in the case of indoor non-line-of-sight (NLOS) and causes electromagnetic radiation pollution around the room in the case of indoor line-of-sight (LOS). Time reversal (TR) technique is an inverse operation of time- domain signals and makes full use of the multipath effect. TR technique can improve the efficiency and reduce the pollution due to its unique temporal–spatial focusing effect. We will detail the principles of TR with the finite TR arrays. What's more, we propose a sequential convex programming (SCP) algorithm based on diode circuit to obtain the optimal frequency point amplitude to further improve energy efficiency. The simulation result shows that the TR-SCP-WPT system model will get the significant energy gain.

Keywords: wireless power transfer, time reversal, sequential convex programming

Introduction

With the development of economy and science technology, more and more applications of wireless electronic products have been made. People have higher requirements on power transmission mode, efficiency, security, and so on. The traditional battery supply mode limits the using time of electronic products. However, reducing energy consumption or increasing battery capacity is not the most essential solutions to which. The wired cable power supply method is also not suitable for communication electronic products with strong mobility and wide distribution. Therefore, wireless power transfer (WPT) is a trend.

There are four types of WPT: magnetic induction coupling, magnetic coupling resonance, laser, and microwave. Among them, microwave power transmission has a bright prospect, since it is not limited by distance and it does not require a precise angle [1]. But WPT has low energy efficiency due to electromagnetic wave diffraction in the case of indoor non-line-of-sight (NLOS) and causes electromagnetic radiation pollution around the room in the case of indoor line-of-sight (LOS). The transmission efficiency of microwave power transmission is mainly composed of three parts: microwave and direct current (DC) conversion efficiency, antenna receiving efficiency, and electromagnetic wave spatial transmission efficiency.

Time reversal (TR) technique is an inverse operation of time-domain signals and makes full use of the multipath effect, which can improve the efficiency by the temporal effect and reduce the pollution by the spatial effect. In [2], it conducted a TR indoor experiment with a nanosecond pulse with a carrier frequency of 2.45 GHz. It is proven that TR can achieve an energy gain of 30 dB, and when continuous wave is used, the TR scheme can avoid indoor fading phenomenon. In [3], it explores the spatial profile of TR-WPT for energy reception at the vertical direction of the focus point in a metal cavity and gives a closed-form expression. In [4], it uses the tempo-ral–spatial focusing effect of TR, combined with coils containing metamaterials, to illuminate centimeter-level LED lamps, thus demon- strating the precise control of near-field electromag-netic waves.

TR can improve the transmission efficiency, and we also consider the microwave and DC con-version efficiency. The conversion efficiency depends on rectenna. It is not only a function of rectenna design but also a function of input waveform, i.e., input power and shape. This results in that the conversion efficiency is not a constant, but a nonlinear function of the input wave-form [5]. Therefore, we consider a nonlinear rectenna circuit model in our system model with TR technique. Then we propose a sequential convex programming (SCP) to get the optimal frequency amplitude for the better conversion efficiency.

The potential of TR and SCP to handle efficiency issues for WPT systems is investigated in this chapter. We introduce the background of WPT and TR in section 1. Then we detail the principles of TR in section 2. We describe the TR-SCP- WPT system model in section 3. Finally, the energy efficiency performance of the WPT system is proposed.

Principles of TR

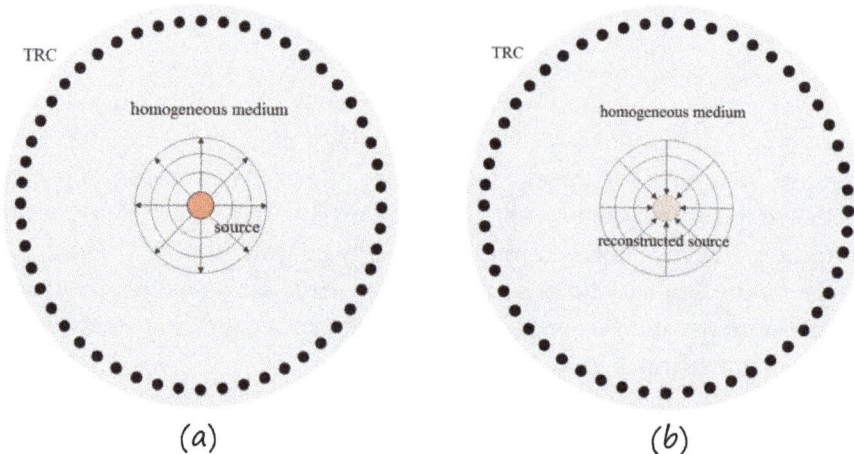

Figure 1. *The process of TR (a) detection stage (b) TR stage.*

Time reversal is a digital signal processing technique. If applied to waveform design, it can be used as a beamforming technique. According to Fink's time reversal cavity (TRC) in [6], the time reversal technique uses the detection signal to obtain all the information of the spatial channel so that the source signal can be perfectly reconstructed in the original position. Consid-ering the actual situation, Fink uses the Huygens principle to transform the three-dimensional

TRC into a two-dimensional and finite number of time reversal mirrors (TRM). The basic idea of time reversal technique is to reverse the signal in the time domain or phase- conjugate the signal in the frequency domain. The process of TR contains two stages as shown in **Figure 1**. The first stage is detection stage. The source transmits the detection signal. In the second stage, TR stage, TRC records the signal with the channel impulse response (CIR) and then reverses it. The time reversal signal will arrive at the same position of the source through the same CIR.

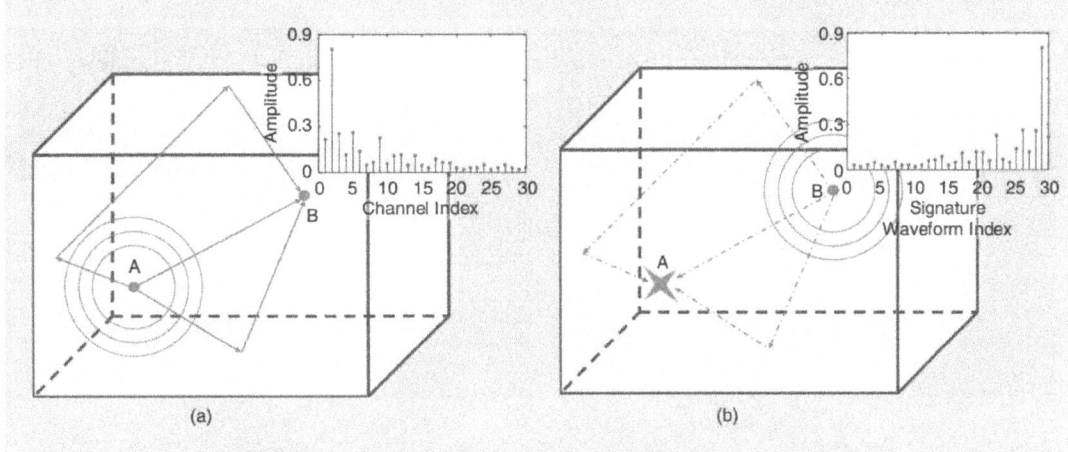

Figure 2. *An illustration of TR (a) the channel probing phase and (b) the data transmission and focusing phase [7].*

The CIR goes through by the multipaths. To more detail, let us imagine that there are two points A and B in the space of the metal box like in **Figure 2**. When A transmits radio signals, its radio waves bounce back and forth in the box. Some of them bounce back and forth through B. After a period of time, the energy level decreases and can no longer be observed. Meanwhile, B can record the multipath distribution of arrival wave as time distribution. Then, such multipath distribution is inverted (and conjugated) by B and emitted accordingly, the last one and the last one. Through channel reciprocity, all waves following the original path will arrive at A at the same specific time and add up in a perfect and constructive way. This is called focusing effect [7] (**Figure 2**).

The specific operation of TR is to reverse the time-domain signal on the time axis or to adopt phase conjugation for the complex frequency domain signal. In our research, we choose the operation in time domain.

Temporal focusing effect of TR

The temporal–spatial focusing effect of TR utilizes the principle of channel reciprocity, which means that in the two stages, the channel information is time invariant. Channel reciprocity requires a high correlation between the CIR of the forward link and the backward link, while channel stability requires that the CIR be stationary for at least one of the detection and TR stages. Experiments in the laboratory area show that the correlation of CIR between forward and backward links is as high as 0.98 in [8]. Therefore, TR can play an important role. The temporal focusing effect of TR refers to that the signal originally arriving last will be transmitted first while the signal originally arriving first will be transmitted last, and finally the signal of all paths arrive at the same moment.

We consider a long distance and wireless environment, namely, rich multipath channel can be represented as

$$h(t) = \sum_{l=0}^{L-1} \alpha_l \delta(t - \tau_l) \qquad (1)$$

where L is the whole number of multipath, a_l is the amplitude of the path, τ_l is the delay of the path, and $\delta()$ is the Dirichlet function.

According to the process of TR, the TR signal containing the reversed CIR will go through the same CIR. Therefore, there is an equivalent channel of TR, such as

$$h_{eq}(t) = h(-t) \otimes h(t)$$

$$= \sum_{l=0}^{L-1} \alpha_l \delta(-t - \tau_l) \otimes \sum_{k=0}^{L-1} \alpha_k \delta(t - \tau_k)$$

$$= \sum_{l=0}^{L-1} \sum_{k=0}^{L-1} \alpha_l \alpha_k \delta(t - (\tau_l - \tau_k)) \qquad (2)$$

$$= \sum_{l=0}^{L-1} \alpha_l^2 \delta(t) + \sum_{l=0}^{L-1} \sum_{k=0, \, k \neq l}^{L-1} \alpha_l \alpha_k \delta(t - (\tau_l - \tau_k))$$

$$= A_{focuse} + A_{sidelobe}$$

According to Eq. (2), the equivalent channel is focused to zero moment and is superposed by L paths. In addition, this will generate a lot of sidelobe information as $A_{sidelobe}$. Since the multipath delay range is large and the difference of $(\tau_l - \tau_k)$ is the same, the equivalent channel has many smaller side lobes on the time axis. We prove the temporal focusing effect of TR by the indoor channel, IEEE802.15.3a in [9] like in **Figure 3**.

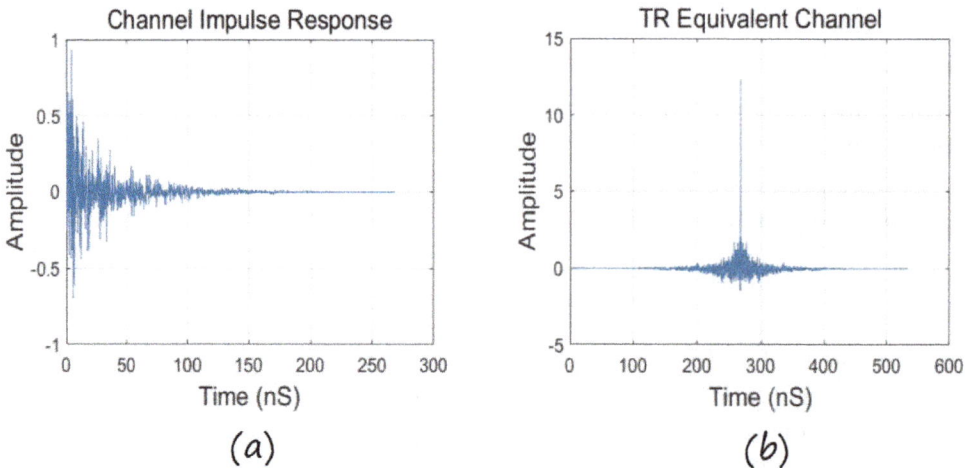

Figure 3. *Temporal focusing effect of TR (a) CIR (b) TR channel.*

According to Eq. (2), at zero-focus moment, the effect of delay is just offset, and the L multipath components add up, resulting in an increase in peak value at that time. **Figure 3a** shows the normalized indoor channel, and the TR equivalent channel of **Figure 3b** is time reversed by the channel of **Figure 3a**. It can be seen from the peaks of the two that the TR equivalent channel has an order of amplitude higher than the peak, which is consistent with Eq. (2). Multipath delays can cause phase shifts in the signal. Therefore, for the transmitted signal, it is equivalent to the fact that at the time of focusing, the signal information is superimposed in phase, and a focus peak appears. This is the temporal focusing effect of the TR.

Spatial focusing effect of TR

The spatial focusing effect is illustrated by the time reversal simulation experiment in a reverberant metal cavity through the XFDTD. As shown in **Figure 4**, there is the source node A at the center of reverberant metal cavity (0, 0, and 0 mm) and four time reversal mirror (TRM) array elements namely TRM1 (200, 0, and 0 mm), TRM2 (-200, 0, and 0 mm), TRM3 (0, 200, and 0 mm), TRM4 (0, -200, and 0 mm). The distance between the source node A and the TRM array elements is 200 mm, which is in the far field($L > \lambda$, λ is the component of the signal included the shortest wavelength).

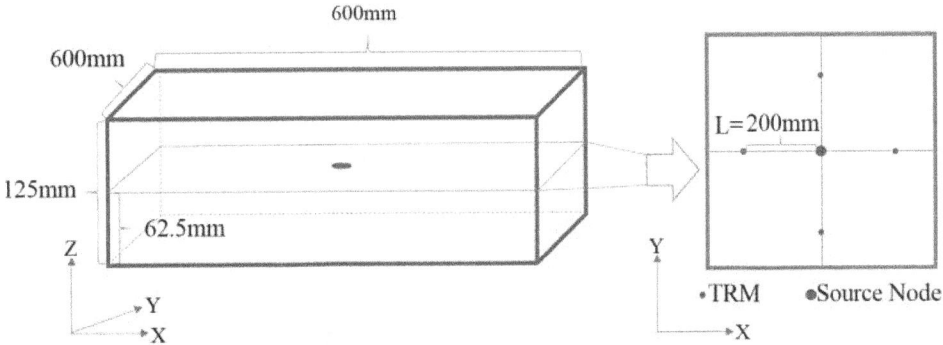

Figure 4. *A reverberant metal cavity with one source node and four TRM nodes.*

The basic TR process in this reverberant metal cavity proceeds as follows: first, the detection signal $x(t)$ is injected into the cavity at the source node A, and the received signal at point B of the TRM array element is $y(t)$ after passing through the channel $h_{AB}(t)$:

$$y(t) = x(t) \otimes h_{AB}(t) \tag{3}$$

where $\otimes$ represents the convolution. Second, the received signal $y(t)$ is time reversed at point B and injected into the cavity again. The received signal at point A of source node is as follows:

$$R_A(t) = y(-t) \otimes h_{BA}(t)$$
$$= x(-t) \otimes h_{AB}(-t) \otimes h_{BA}(t) \tag{4}$$
$$= x(-t) \otimes h_{eq}(t)$$

where $h_{eq}(t)$ is an autocorrelation function of $h_{AB}(t)$, which can be regarded as the equivalent channel impulse response of TR.

When the channel impulse response between point C (at a certain distance from point source node A) and point B of TRM array unit is recorded as $h_{BC}(t)$, the expression of received signal at non-source node can be obtained as follows:

$$R_C(t) = y(-t) \otimes h_{BC}(t)$$

$$= x(-t) \otimes h_{AB}(-t) \otimes h_{BC}(t) \tag{5}$$

Because of the maximum value range of coherent superposition of multipath signals, the spatial focusing effect of TR can make the multipath signals superpose coherently at the source node A at a certain time, thus enhancing the electric field intensity at the source node to produce a peak. As shown in **Figure 5**, it can be seen that TR processing results in a lower signal amplitude in an area other than the location of the source node and the recovery of $x(t)$ occurs at the primary source node A, and the peak value of $R_A(t)$ is higher than the peak value of $R_C(t)$. Because $h_{AB}(-t)$ and $h_{BC}(t)$ are non-autocorrelation, there will be coherent cancelation in the signal. The obtained $R_C(t)$ and $x(t)$ signals are completely different. The peak power ratio of the signal that the source node and the non-source node can receive is determined by the total multipath gain of the channel and the correlation coefficient between the channels. Even if the multipath gain of the source node and the non-source node is the same or slightly higher, the ratio of peak power will still have a large gap when the channel correlation coefficient is very small. In the scattering-rich environment, as long as the distance is enough, the correlation of impulse response of multipath channel decreases to a very low level. The spatial focusing effect of TR is adaptive and can reduce electromagnetic radiation pollution.

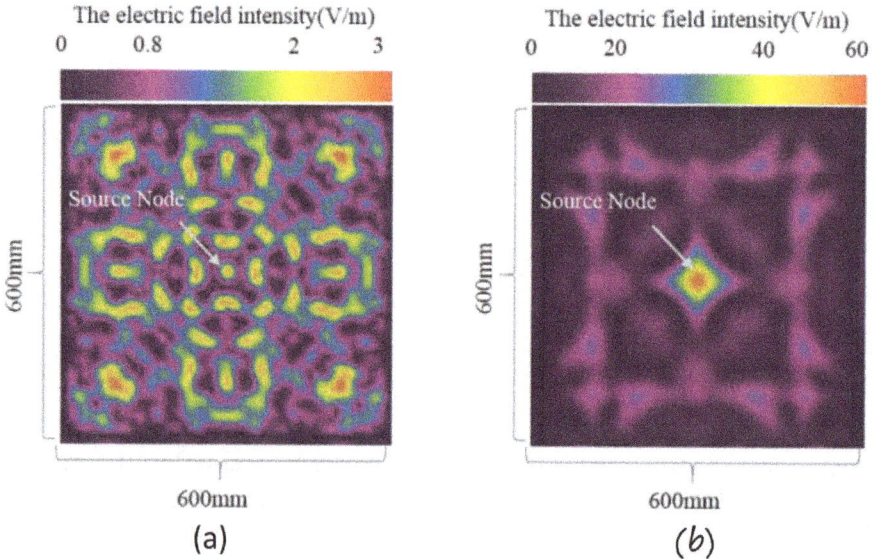

Figure 5. *Spatial focusing effect of time reversal electromagnetic wave: Electric field distribution Z = 0 plane of the source node (a) in NO-TR system (b) in TR-WPT system.*

WPT system model and analysis

Through section 2, we can get the idea that TR technique can increase power gain by the temporal focusing effect and reduce the electromagnetic radiation pollution by the spatial focusing effect. In WPT system, the important part is the rectifier antenna. According to [5, 10], the rectifier antenna consists of a diode and a low-pass filter to transform radio frequency (RF) input signal into direct current (DC). The collected DC energy is a function of input power level and RF-DC conversion efficiency. RF-DC conversion efficiency is not only a function of rectifier antenna design but also a function of input waveform, that is, input power and shape. Therefore, we consider to design the frequency point amplitude to improve energy efficiency based on the diode circuit with TR technique.

System model

Through the above introduction, we consider the system model as **Figure 6**. Because the two stages of TR, the transmitter and receiver both have two antennas. In the transmitter, the channel estimation module can extract the channel state information (CSI) between two devices. Then the time reversal module can use the CSI to obtain the time reversal signal. According to the rectenna circuit model, a non-convex problem about the received signal and the output voltage can be extracted. Usually, a non-convex problem should be converted into a convex problem [11]. We propose that the sequential convex programming algorithm can solve the problem. More detail can be found in [12] due to the space limitations. Finally, the sequential convex programming module will get the optimal frequency point amplitude according to rectenna circuit.

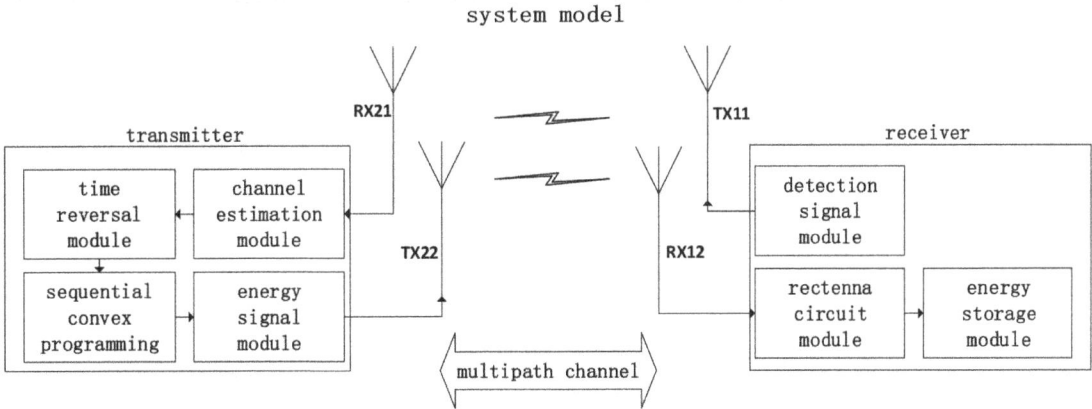

Figure 6. *TR-SCP-WPT system model.*

Simulation

Firstly, we consider a multipath channel response based on a metal cavity like in **Figure 7**, with the measurement in frequency domain by Vector Network Analyzer (VNA) [13].

In [14], it proposes that the larger the bandwidth, the more the number of multipath like in **Figure 8**. In the above analysis and Eq. (2), the performance of TR is related to the multipath. In addition, the factor of SCP is frequency point numbers. Therefore, we compare three system models, namely, direct transmission (DT), TR, and TR-SCP, in different bandwidth (20–200 MHz) and frequency point numbers (2–8).

Result and analysis

Figure 7. *The metal cavity.*

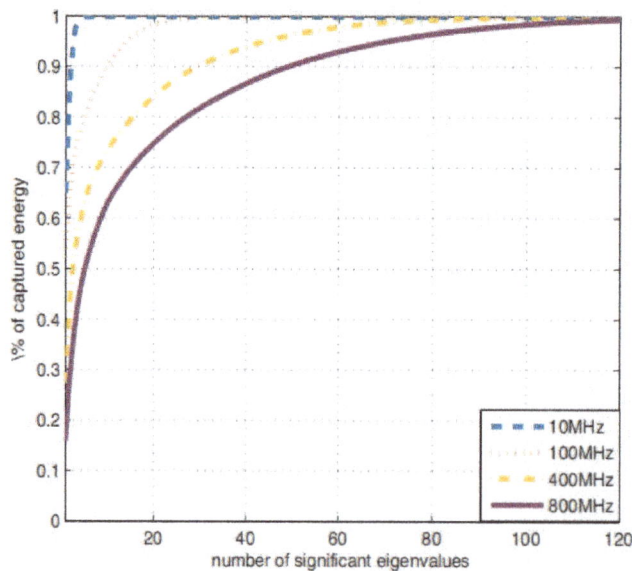

Figure 8. *Percentage of captured energy versus the number of significant eigenvalues with a single antenna [14].*

In the simulation, we set the carrier frequency as 5.8GHz and the emitting power as 36 dBm. Then we combine the measured channel with Matlab module like in **Figure 6** to perform the experiment. As in **Figure 9a**, the TR-SCP and TR systems perform much better than the DT system because the TR technique can take advantage of the multipath channel. What's more, as the bandwidth becomes bigger, their performance is better. As in **Figure 9b**, because the TR-SCP can make use of CSI based on the diode circuit, its performance will be better than TR system. And when the number of frequency point is from 2 to 8, the system can get more information about CSI. Therefore, in **Figure 9b**, the TR-SCP system will be better and better than the TR system as the number increases.

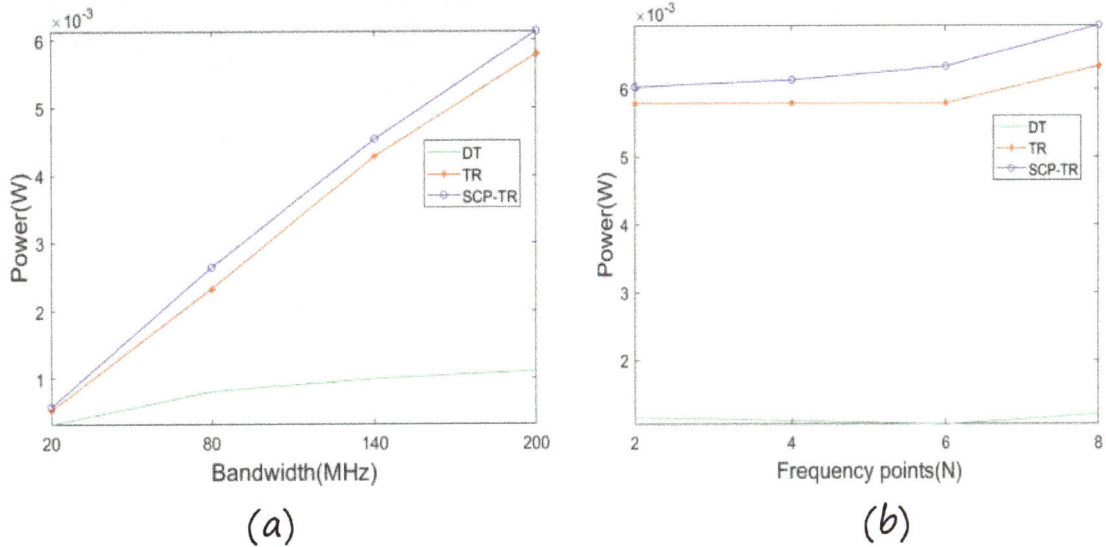

Figure 9. *Result of the simulation (a) power and bandwidth (b) power and frequency points.*

Conclusion

This chapter firstly introduces the principle of TR technique. Then according to the focusing effect of TR, we propose TR be applied to WPT. Because the conversion efficiency is associated with the waveform, we propose SCP algorithm based on diode circuit to design the amplitude of the emitting signal. Finally, we combine the SCP with the TR to propose a SCP-TR system model. Based on the results of the simulation, our design model can achieve better performance on wireless power transmission. Therefore, we believe that our model is robust and stable in a complex and rich multipath environment and can adapt to a variety of wireless power transmission scenarios.

Acknowledgements

This work was supported by the National Natural Science Foundation of China (NSFC) under Grant (61871193) and by the key project of Guangdong Natural Science Foundation under Grant (2018B030311049).

Author details

Bin-Jie Hu*, Zhi-Wu Lin and Peng Liao

School of Electronic and Information Engineering, South China University of Technology, Guangzhou, China

*Address all correspondence to: eebjiehu@scut.edu.cn

References

[1] Strassner B, Chang K. Microwave power transmission: Historical milestones and system components. Proceedings of the IEEE. 2013;101: 1379-1396. DOI: 10.1109/JPROC.2013.2246132

[2] Ibrahim R, Voyer D, Bréard A. Experiments of time-reversed pulse waves for wireless power transmission in an indoor environment. IEEE Transactions on Microwave Theory and Techniques. 2016;64:2159-2170. DOI: 10.1109/TMTT.2016.2572679

[3] Cangialosi F, Grover T, Healey P, et al. Time reversed electromagnetic wave propagation as a novel method of wireless power transfer. In: IEEE 2016 Wireless Power Transfer Conference (WPTC); 5–6 May 2016; Aveiro. Portugal: IEEE; 2016. pp. 1-4

[4] Chabalko M, Sample A. Electromagnetic time reversal focusing of near field waves in metamaterials. Applied Physics Letters. 2016;109: 263901. DOI: 10.1063/1.4973210

[5] Valenta CR, Morys MM, Durgin GD. Theoretical energy-conversion efficiency for energy-harvesting circuits under power-optimized waveform excitation. IEEE Transactions on Microwave Theory and Techniques. 2015;63:1758-1767. DOI: 10.1109/TMTT.2015.2417174

[6] Fink M. Time reversal of ultrasonic fields. I. Basic principles. IEEE Transactions on Ultrasonics, Ferroelectrics, and Frequency Control. 1992;39:555-566. DOI: 10.1109/58.156174

[7] Chen Y, Wang BB, Han Y, et al. Why time reversal for future 5G wireless? [perspectives]. IEEE Signal Processing Magazine. 2016;33:17-26. DOI: 10.1109/MSP.2015.2506347

[8] Su ZQ, Shao GF, Liu HP.Synchronization-free model with signal repeater for timing-based localization. In: IEEE 2016 84th Vehicular Technology Conference (VTC-Fall); 18–21 September 2016. Montreal, QC: IEEE; 2016. pp. 1-5

[9] Foerster J. Channel modeling sub- committee report final. IEEE. 2003. DOI: p802.15–02/368r5-SG3a

[10] Collado A, Georgiadis A. Optimal waveforms for efficient wireless power transmission. IEEE Microwave and Wireless Components Letters. 2014;24: 354-356. DOI: 10.1109/LMWC.2014. 2309074

[11] Boyd S, Vandenbergh L. Convex Optimization. New York: Cambridge University Press; 2004. pp. 1-287. DOI: 10.1017/CBO9780511804441

[12] Lin Z, Hu B, Wei Z, Liao P. An optimal time reversal waveform based on sequential convex programming for wireless power transmission. In: IEEE 18th International Conference on Communication Technology (ICCT); 8–11 October 2018. Chongqing: IEEE; 2018, 2019. pp. 916-920

[13] Chong C, Kim Y, Lee S. UWB indoor propagation channel measurements and data analysis in various types of high- rise apartments. In: 2004 IEEE 60th Vehicular Technology Conference; 26–29 September 2004. Los Angeles, CA: IEEE; 2004. pp. 150-154

[14] Han Y, Chen Y, Wang B, Liu KJR. Realizing massive MIMO effect using a single antenna: A time-reversal approach. In: 2016 IEEE Global Communications Conference (GLOBECOM). Washington, DC: IEEE; 2016. pp. 1-6

3

Transceiver Design for Wireless Power Transfer for Multiuser MIMO Communication Systems

Anming Dong and Haixia Zhang

Abstract

This chapter describes transceiver design methods for simultaneous wireless power transmission (WPT) and information transmission in two typical multiuser MIMO networks, that is, the MIMO broadcasting channel (BC) and interference channel (IC) networks. The design problems are formulated to minimize the trans- mit power consumption at the transmitter(s) while satisfying the quality of service (QoS) requirements of both the information decoding (ID) and WPT of all users. The mean-square error (MSE) and the signal-to-interference-noise ratio (SINR) criteria are adopted to characterize the ID performance of the BC network and the IC network, respectively. The designs are cast as nonconvex optimization problems due to the coupling of multiple variables with respect to transmit precoders, ID receivers, and power splitting factors, which are difficult to solve directly. The feasibility conditions of these deign problems are discussed, and effective solving algorithms are developed through alternative optimization (AO) framework and semidefinite programming relaxation (SDR) techniques. Low-complexity algo- rithms are also developed to alleviate the computation burden in solving the semidefinite programming (SDP) problems. Finally, simulation results validating those proposed algorithms are included.

Keywords: wireless power transfer (WPT), energy harvesting, multiuser MIMO, transceiver design, alternating optimization, semidefinite programming relaxation (SDR)

Introduction

Wireless power transfer (WPT) through radio frequency (RF) signals has been redeemed as one of the promising techniques to provide perpetual and cost- effective power supplies for mobile devices [1–4]. Compared with traditional energy harvesting (EH) methods depending on external sources, such as solar power and wind energy, the RF WPT is able to power the wireless devices at any time. Moreover, since RF signals carry energy as well as information, wireless devices can be charged while communicating. These merits of WPT bring great convenience and provide quality of service (QoS) guarantee for wireless devices.

On the other hand, multiple-input, multiple-output (MIMO) techniques are widely used in many wireless communication systems such as WiFi and the fifth generation mobile (5G) systems, due to their potential in providing increased link capacity and spectral efficiency combined with improved link reliability. Moreover, the evolution of MIMO techniques to the massive MIMO systems, where tens or hundreds of antennas are equipped at transmitters or/and receivers, accompanied by shrinking coverage of base stations (BSs) in the future wireless systems, makes it possible to transmit wireless power with higher efficiency. It is envisaged that the power line connected to the mobile devices would be eliminated completely in future wireless communications by combing WPT with MIMO wireless information transmission (WIT) system [5–9].

Transceiver design plays a very important role in achieving this vision. The objective of the transceiver design is to improve the energy and spectral efficiency of the transmitter by optimizing the beam patterns of the transmitting antennas and the filters at the information decoding (ID) receivers. However, it is not a trivial task to design transceivers for multiuser MIMO systems operating in simultaneous wireless information and power transfer (SWIPT) mode due to the presence of inter-user interference. The interference makes the whole design complicated since it is harmful to WIT but beneficial to WPT, and it is very challenging to balance the role of interference in ID and EH. And what makes things worse is that the interference and the PS factors are coupled together, which makes the joint transceiver design and power splitting (JTPS) problems nonconvex. These problems are NP- hard in general, so effective algorithms should be found to get feasible solutions.

In this chapter, we will discuss transceiver design methods for SWIPT in two typical multi-user MIMO scenarios, that is, the broadcasting channel (BC) network and interference channel (IC) network. We focus on the QoS-constrained problems that are formulated as minimizing the transmit power consumption subject to both the minimum ID and EH requirements. The mean-square error (MSE) and the signal-to-interference-noise ratio (SINR) criteria are adopted to characterize the ID performance of the two kinds of network, respectively. The formulated optimization problems are nonconvex with respect to the optimization variables, that is, the parameters of transmit precoders, ID receivers, and PS factors. In order to develop effective solutions, the feasibility is first investigated and found to be independent with EH constraints and PS factors. Based on this, we develop an effective initializing procedure for the design problems. Then, effective iterative solving algorithms are developed based on alternative optimization (AO) framework and semidefinite programming relaxation (SDR) techniques. Specifically, we find that the original problems can be equivalently reformulated as convex semidefinite programming (SDP) with respect to the transceivers and PS ratios when the receivers are fixed.

On the other hand, when the transmitters and PS factors are fixed, the original problems degenerate to the classical linear MSE minimization receiver design problem for the BC network and the SINR maximization receiver design problem for the IC network, respectively. Since the SDP problems can be solved exactly in polynomial time, feasible solutions can be obtained for the proposed algorithms effectively.

However, the SDP solving is not computationally efficient for large number of variables case [10], and the computational complexity of the SDP-based algorithms is prohibitively high for large number of antenna and user. This greatly restricts its application. To break this, low-complexity schemes should be developed. In this chapter, closed-form power splitting factors with given the transceivers designed from traditional transceiver design algorithms are developed.

Notations: $\mathcal{C}$ represents the complex and positive real field. Bold uppercase and lowercase letters represent matrix and column vectors, respectively. Nonbold italic letters represent scalar values. $\mathbf{I}_N$ is an $N \times N$ identity matrix. $\mathbf{A}^H$, $\mathbf{A}^T$ and $\mathbf{A}^{-1}$ represent the Hermitian transpose, transpose, and inverse of $\mathbf{A}$, respectively. $\mathrm{Tr}\,(\mathbf{A})$ and $rank(\mathbf{A})$ are the trace and rank of matrix $\mathbf{A}$, respectively. $|\mathbf{A}|$ denotes the determinant of matrix $\mathbf{A}$. $\mathbf{A} = \mathrm{diag}(a_1, \dots, a_i, a_{i+1}, \dots, a_N)$ is a $N \times N$ diagonal matrix with the i-th diagonal elements being a_i. $\mathbb{E}\,[\cdot]$ denotes the statistical expectation. $\| \cdot \|_2$ and $\| \cdot \|_F$ denote the two-norm and Frobenius norm, respectively.

Joint transceiver design and power splitting optimization based on MSE criterion for MIMO BC channel

System model

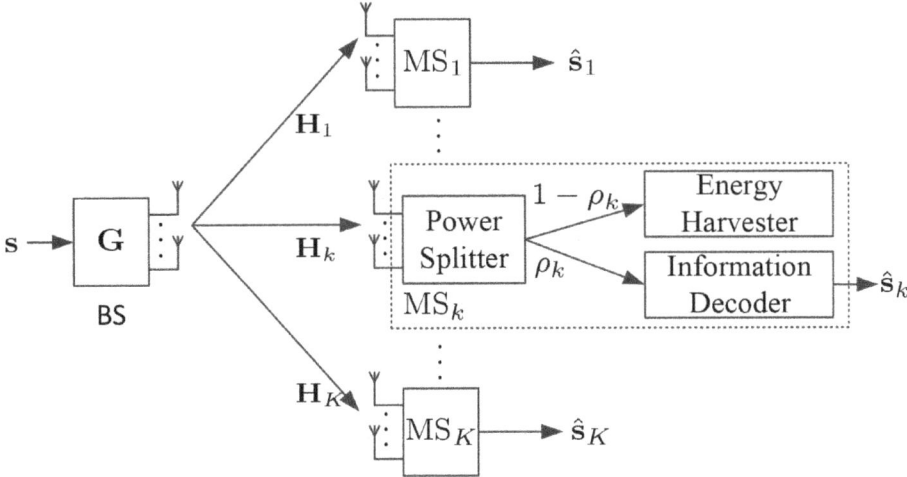

Figure 1. *Downlink MU-MIMO SWIPT system.*

A downlink MIMO BC channel network shown in **Figure 1**, where one base station (BS) serves K mobile stations (MSs) simultaneously through spatial multiplexing, is considered. The number of antennas of the BS and the kth ($k \in \{1, 2, \dots, K\}$) user are denoted as M and L_k, respectively. N_k represents the number of data stream for the kth user, and the total number of data streams served by BS is $d = \sum_{k=1}^{K} N_k \leq M$. Let $\mathbf{s}_k \in \mathcal{C}^{N_k \times 1}$ denote the data vector transmitted to the user k, the data vector transmitted by BS can be expressed as

$\mathbf{s} = \left[\mathbf{s}_1^H, \dots, \mathbf{s}_K^H\right]^H \in \mathcal{C}^{d \times 1}$. It is assumed that $\mathbb{E}\left[\mathbf{ss}^H\right] = \mathbf{I}_d$.

The BS transmits the signal s to users, and the received baseband signal at the kth receiver is

$$\mathbf{r}_k = \mathbf{H}_k \mathbf{G} \mathbf{s} + \mathbf{n}_k, \tag{1}$$

where $\mathbf{G} \in \mathcal{C}^{M \times d}$ denote the transmit precoding matrix, $\mathbf{H}_k \in \mathcal{C}^{L_k \times M}$ denotes the channel propagation matrix from the BS to the kth user and $\mathbf{n}_k \in \mathcal{C}^{L_k \times 1}$ denotes the noise vector, elements of which are assumed to be independent and identically (i.i.d.) zero mean complex Gaussian random variables with variance $\sigma^2_{\mathbf{n}_k}$. The total transmit power of the BS can be calculated as $P = \|\mathbf{G}\|^2_F$.

As shown in **Figure 1**, each user divides the received signal into two parts through power splitters, one part is for information decoder (ID) and the other is for EH. For easy analysis, solution, and implementation, we adopt the uniform PS model [11] in this chapter, that is, the power splitters of all the antennas of a user have the same PS factor. Denote $0 \leq \rho_k \leq 1$ as the PS factors for the kth user, the signal received at the ID of the kth user is

$$\mathbf{r}_k^{ID} = \sqrt{\rho_k}(\mathbf{H}_k\mathbf{G}\mathbf{s} + \mathbf{n}_k) + \mathbf{w}_k, \tag{2}$$

where $\mathbf{w}_k \in \mathcal{C}^{L_k \times 1}$ is the noise caused by power splitter, elements of which are assumed to be i.i.d. zero mean complex Gaussian random variables with variance $\sigma^2_{\mathbf{w}_k}$. The signal received by the EH receiver of user k is written as

$$\mathbf{r}_k^{EH} = \sqrt{1 - \rho_k}(\mathbf{H}_k\mathbf{G}\mathbf{s} + \mathbf{n}_k). \tag{3}$$

At the ID receiver, a filter $\mathbf{F}_k \in \mathbf{C}^{N_k \times L_k}$ is employed to process the received signal and the detected signal is written as

$$\hat{\mathbf{s}}_k = \frac{1}{\sqrt{\rho_k}}\mathbf{F}_k\mathbf{r}_k^{ID} = \mathbf{F}_k\mathbf{H}_k\mathbf{G}\mathbf{s} + \mathbf{F}_k\mathbf{n}_k + \frac{1}{\sqrt{\rho_k}}\mathbf{F}_k\mathbf{w}_k. \tag{4}$$

It is noted that a scaling factor $\frac{1}{\sqrt{\rho_k}}$ is introduced to the received signal, which makes the problem modeling and solving more convenient.

Consequently, the MSE at the kth ID receiver can be expressed as

$$\begin{aligned}
\mathrm{MSE}_k &= \mathbb{E}\left[\|\hat{\mathbf{s}}_k - \mathbf{s}_k\|^2_2\right] \\
&= \mathrm{Tr}(\mathbf{I}_{N_k}) + \mathrm{Tr}\left(\mathbf{H}_k^H\mathbf{F}_k^H\mathbf{F}_k\mathbf{H}_k\mathbf{G}\mathbf{G}^H\right) - \mathrm{Tr}\left(\boldsymbol{\Xi}_k\mathbf{G}^H\mathbf{H}_k^H\mathbf{F}_k^H\right) \\
&\quad - \mathrm{Tr}\left(\mathbf{F}_k\mathbf{H}_{kk}\mathbf{G}\boldsymbol{\Xi}_k^H\right) + \left(\sigma^2_{\mathbf{n}_k} + \frac{\sigma^2_{\mathbf{w}_k}}{\rho_k}\right)\mathrm{Tr}\left(\mathbf{F}_k^H\mathbf{F}_k\right).
\end{aligned} \tag{5}$$

where $\boldsymbol{\Xi}_k = \left[\mathbf{0}_{N_k \times \sum_{l=1}^{k-1}N_l}, \mathbf{I}_{N_k}, \mathbf{0}_{N_k \times \sum_{l=k+1}^{K}N_l}\right]$. At the same time, the energy harvested by the kth EH receiver is expressed as

$$P_k^{EH} = \xi_k(1 - \rho_k)\left(\|\mathbf{H}_k\mathbf{G}\|^2_F + L_k\sigma^2_{\mathbf{n}_k}\right), \tag{6}$$

where $0 \leq \xi_k \leq 1$ denotes the energy conversion efficiency.

Problem formulation and feasibility analysis

Problem formulation

A case that each MS has its dedicated ID and EH QoS requirements and the BS has to satisfy all the users with minimum transmit power consumed is considered. This scenario can be modeled by the following QoS constrained power minimization problem

$$\min_{\{\mathbf{G}, \mathbf{F}_k, \rho_k, \forall k\}} \quad \text{Tr}\left(\mathbf{G}\mathbf{G}^{\mathrm{H}}\right)$$

$$\text{s.t. :} \quad \text{MSE}_k \leq \varepsilon_k,$$

$$P_k^{\mathrm{EH}} \geq \psi_k,$$

$$0 < \rho_k < 1, \forall k = 1, \dots, K, \tag{7}$$

where $\varepsilon_k > 0$ and $\psi k > 0$ are the ID MSE target and the EH threshold of user k, respectively.

Obviously, (7) is nonconvex with respect to the precoders, receivers, and power splitters and thus difficult to be solved directly. Before developing an effective algorithm for it, its feasibility should be analyzed first.

Feasibility analysis

Some sufficient and necessary conditions for the feasibility of the original problem (7) can be given by the following propositions.

Proposition 1 *Problem (7) is feasible if and only if the following problem is 05 feasible:*

$$\min_{\{\mathbf{G}, \mathbf{F}_k, \rho_k, \forall k\}} \quad \text{Tr}\left(\mathbf{G}\mathbf{G}^{\mathrm{H}}\right)$$

$$\text{s.t. :} \ \left\| \mathbf{F}_k \mathbf{H}_k \mathbf{G} - \mathbf{\Xi}_k \right\|_{\mathrm{F}}^2 + \left(\sigma_{\mathbf{n}_k}^2 + \frac{\sigma_{\mathbf{w}_k}^2}{\rho_k} \right) \|\mathbf{F}_k\|_{\mathrm{F}}^2 \leq \varepsilon_k, \tag{8}$$

$$\forall k = 1, \dots, K.$$

Proposition 2 *Problem (8) is feasible if and only if the following problem is feasible:*

$$\min_{\{\mathbf{G}, \mathbf{F}_k, \forall k\}} \quad \text{Tr}\left(\mathbf{G}\mathbf{G}^{\mathrm{H}}\right)$$

$$\text{s.t. :} \ \left\| \mathbf{F}_k \mathbf{H}_k \mathbf{G} - \mathbf{\Xi}_k \right\|_{\mathrm{F}}^2 + \left(\sigma_{\mathbf{n}_k}^2 + \sigma_{\mathbf{w}_k}^2 \right) \|\mathbf{F}_k\|_{\mathrm{F}}^2 \leq \varepsilon_k, \tag{9}$$

$$\forall k = 1, \dots, K.$$

For the sake of brevity, we omit the proof to these propositions. Interested readers are suggested to refer to [12–16]. Proposition 1 reveals that the feasibility of the original problem (7) is irrelevant to the EH constraints, while Proposition 2 further shows that the EH constraints are irrelevant to the feasibility. Since the feasibility of the formulated problem depends on neither the EH constraints nor the PS factors, checking its feasibility can be simplified to checking the feasibility of (9), which is a traditional MSE-based multiuser MIMO transceiver design problem [17–19]. So, in the following, we assume that problem (9) is feasible under the given MSE QoS requirements and focus on how to solve it.

Alternative optimization solution based on semidefinite programming relaxation

Alternative optimization framework

By reviewing the MSE expression (5), we know that it is convex with respect to either G or F_k. So, we can develop an iterative algorithm for the original problem based on the optimization (AO) framework, that is, optimizing the transmit precoder together with PS factors and the receivers iteratively.

Specifically, when the receiver F_k, $\forall k$ is fixed, the optimization problem is reduced to a joint transmitter design and power splitting (JTDPS) subproblem, which is

$$\min_{\{G, \rho_k, \forall k\}} \quad \mathrm{Tr}(GG^H)$$

$$\text{s.t.} : \quad MSE_k \leq \varepsilon_k, \tag{10}$$

$$P_k^{EH} \geq \psi_k,$$

$$0 \leq \rho_k \leq 1, \forall k = 1, \dots, K.$$

It is noted that problem (10) is not convex in its current form. We will further process it based on SDR techniques in the next subsection.

When the precoders and PS factors are fixed, the transmit power at the BS and the EH power are fixed. Considering that only MSE at the ID receiver of user k is relevant to F_k, we can optimize the ID receiver by minimizing the MSE. The optimization problem can be formulated as

$$\min_{\{F_k\}} \quad MSE_k. \tag{11}$$

Problem (11) is the traditional unconstrained MSE minimization problem and its closed-form solution can be given by

$$F_k = \left(H_k G \Xi_k^H\right)^H \left[H_k GG^H H_k^H + \left(\sigma_{n_k}^2 + \frac{\sigma_{w_k}^2}{\rho_k}\right)I\right]^{-1}. \tag{12}$$

Therefore, by alternatively optimizing the transmitter together with PS factors according to (10) and the receivers according to (12), an iterative optimization framework is established and is summarized in Algorithm 1.

Algorithm 1 Alternating optimization framework for JTDPS.

1: Initialize the receivers $F_k = \tilde{F}_k$, $\forall k$ and the power splitting factors ρ_k, $\forall k$.
2: Optimize the transmitter G and the PS factors ρ_k, $\forall k$ by solving problem (10).
3: Optimize the receive filters F_k, $\forall k$ according to (12).
4: Repeat 2 and 3 until convergence or the maximum number of iterations is reached.

Convergence analysis

For the AO framework, it is vital to analyze its convergence property. The following proposition reveals this property.

Proposition 3 For the initial receivers $\tilde{\mathbf{F}}_k$, $\forall k$, if problem (10) is feasible and its optimal solution can be obtained, then the proposed Algorithm 1 is convergent.

The proof can be found in [12]. According to Proposition 3, two critical prerequisites should be satisfied in order to guarantee finding a feasible solution for problem (7) through Algorithm 1. 1) the subproblem (10) should be feasible and 2) the initialization of the receivers $\mathbf{F}_k$, $\forall k$ should be carefully chosen such that proper transmitters and the PS factors can be obtained in the first iteration of the Algorithm 1. This means that it is vital to find the optimal solution for the subproblem (10). Therefore, before proceeding, the feasibility of the subproblem (10) is investigated in the following subsection.

Feasibility of the transmitter design subproblem

By checking the MSE constraints of (9), a necessary condition for the feasibility of problem (9) is established as

$$\varepsilon_k - \left(\sigma_{\mathbf{n}_k}^2 + \sigma_{\mathbf{w}_k}^2\right)\|\mathbf{F}_k\|_{\mathrm{F}}^2 \geq 0. \tag{13}$$

This condition shows that the Frobenius norm of the receiver should be small enough to make the problem feasible. In order to satisfy this condition, we introduce a positive scaling parameter p_k to the receiver $\mathbf{F}_k$ in the transceiver design model (10), that is, $\frac{1}{p_k}\mathbf{F}_k$. The expression of MSE is then recast as

$$\mathrm{MSE}_k = \|\frac{1}{p_k}\mathbf{F}_k\mathbf{H}_k\mathbf{G} - \mathbf{\Xi}_k\|_{\mathrm{F}}^2 + \left(\sigma_{\mathbf{n}_k}^2 + \frac{\sigma_{\mathbf{w}_k}^2}{\rho_k}\right)\frac{\|\mathbf{F}_k\|_{\mathrm{F}}^2}{p_k^2} \tag{14}$$

After replacing the MSE constraints with (14), problem (10) can then be reformulated as

$$\min_{\{\mathbf{G}, p_k, \rho_k, \forall k\}} \quad \mathrm{Tr}(\mathbf{GG}^{\mathrm{H}})$$

$$\mathrm{s.t.}: \left\|\frac{1}{p_k}\mathbf{F}_k\mathbf{H}_k\mathbf{G} - \mathbf{\Xi}_k\right\|_{\mathrm{F}}^2 + \left(\sigma_{\mathbf{n}_k}^2 + \frac{\sigma_{\mathbf{w}_k}^2}{\rho_k}\right)\frac{1}{p_k^2}\|\mathbf{F}_k\|_{\mathrm{F}}^2 \leq \varepsilon_k, \tag{15}$$

$$\xi_k(1 - \rho_k)\left(\|\mathbf{H}_k\mathbf{G}\|_{\mathrm{F}}^2 + L_k\sigma_{\mathbf{n}_k}^2\right) \geq \psi_k,$$

$$\forall k = 1, \ldots, K.$$

It can be proved that a sufficient and necessary condition for the feasibility of problem (15) is given by the following proposition [12].

Proposition 4 Problem (15) is feasible if and only if the following problem is feasible:

$$\min_{\{\mathbf{G}, p_k, \forall k\}} \quad \mathrm{Tr}(\mathbf{GG}^{\mathrm{H}})$$

$$\mathrm{s.t.}: \left\|\frac{1}{p_k}\mathbf{F}_k\mathbf{H}_k\mathbf{G} - \mathbf{\Xi}_k\right\|_{\mathrm{F}}^2 + \left(\sigma_{\mathbf{n}_k}^2 + \sigma_{\mathbf{w}_k}^2\right)\frac{1}{p_k^2}\|\mathbf{F}_k\|_{\mathrm{F}}^2 \leq \varepsilon_k, \tag{16}$$

$$\forall k = 1, \ldots, K.$$

Proposition 4 reveals that the feasibility of the transmitter and PS problem (15) does not depend on the EH constraints either. Therefore, the feasibility of the joint transmitter design and PS problem (15) can be simply verified by checking whether problem (16) is feasible or not. To guarantee the feasibility of problem (16), the following proposition is proposed.

Proposition 5 Fix $\mathbf{F}_k$, $\forall k$, problem (16) can be reformulated as a convex SDP problem given by

$$\min_{\{\mathbf{G}, p_k, \forall k\}} \quad \|\mathbf{G}\|_{\mathrm{F}}^{\dot{c}}$$

$$\text{s.t. :} \quad \begin{bmatrix} p_k \sqrt{\varepsilon} & \mathbf{a}_k^{\mathrm{H}} & \beta_k \\ \mathbf{a}_k & p_k \sqrt{\varepsilon} \mathbf{I} & \mathbf{0} \\ \beta_k & \mathbf{0} & p_k \sqrt{\varepsilon} \end{bmatrix} \succeq \mathbf{0} \ , \forall k = 1, \dots, K, \quad (17)$$

where $\beta_k = \sqrt{\sigma_{\mathbf{n}_k}^2 + \sigma_{\mathbf{w}_k}^2} \|\mathbf{F}_k\|_{\mathrm{F}}$, and $\mathbf{a}_k = \mathrm{vec}(\mathbf{F}_k \mathbf{H}_k \mathbf{G} - p_k \Xi_k)$ is affine jointly in $\mathbf{G}$ and p_k.

Proof. The MSE constraints in problem (16) can be recast as

$$p_k^2 \mathrm{MSE}_k = \left\| \mathbf{F}_k \mathbf{H}_k \mathbf{G} - p_k \Xi_k \right\|_{\mathrm{F}}^2 + \left(\sigma_{\mathbf{n}_k}^2 + \sigma_{\mathbf{w}_k}^2 \right) \|\mathbf{F}_k\|_{\mathrm{F}}^2$$

$$= \left\| \mathrm{vec}\left(\mathbf{F}_k \mathbf{H}_k \mathbf{G} - p_k \Xi_k \right) \right\|_2^2 + \left(\sigma_{\mathbf{n}_k}^2 + \sigma_{\mathbf{w}_k}^2 \right) \|\mathbf{F}_k\|_{\mathrm{F}}^2 \quad (18)$$

$$= \left\| \begin{bmatrix} \mathbf{a}_k \\ \beta_k \end{bmatrix} \right\|_2^2 \leq p_k^2 \varepsilon$$

Table 1. *The initializing procedure for algorithm 1.*

1: Given the MSE requirements ε_k, $\forall k$.

2: Generate random matrix $\tilde{\mathbf{F}}_k \in \mathbf{C}^{M \times N}$, $\forall k$.

3: By fixing $\mathbf{F}_k = \tilde{\mathbf{F}}_k$, solve (17) to obtain the optimized receivers $\mathbf{G}$ and PS factors p_k, $\forall k$.

4: Return $\mathbf{F}_k = \frac{1}{p_k} \tilde{\mathbf{F}}_k$, $\forall k$.

According to the Schur complement lemma, the inequality (18) is equivalent to

$$\begin{bmatrix} p_k \sqrt{\varepsilon} & \mathbf{a}_k^{\mathrm{H}} & \beta_k \\ \mathbf{a}_k & p_k \sqrt{\varepsilon} \mathbf{I} & \mathbf{0} \\ \beta_k & \mathbf{0} & p_k \sqrt{\varepsilon} \end{bmatrix} \succeq \mathbf{0} \ . \quad (19)$$

The proposition is obtained.

Problem (17) is convex, and its optimal solution can be obtained. Therefore, problem (16) is feasible. With the solution of problem (16), an effective initialization procedure for problem (16) can be constructed. Specifically, the receiver $\mathbf{F}_k$ can be initialized by any randomly generated matrix, that is, $\mathbf{F}_k = \tilde{\mathbf{F}}_k \in \mathcal{C}^{L_k \times M}$, $\forall k$. Then, by solving (17), p_k is obtained. Finally, the receiver $\mathbf{F}_k$ is constructed to be $\frac{1}{p_k} \tilde{\mathbf{F}}_k$, such that the problem (16) is feasible. The initialization process is summarized in **Table 1**.

Through Proposition 4 and Proposition 5, it is known that the joint transmitter design and power splitting subproblem is feasible. However, Algorithm 1 cannot be carried out in its current form, since problem (10) is still nonconvex. In the following, the SDP relaxation is adopted to reform it in to convex form.

Algorithm description

By introducing two variables c_k and d_k with $\frac{\sigma_{\mathbf{w}_k}^2}{\rho_k} \leq c_k$ and $\frac{\psi_k}{\xi_k(1-\rho_k)} \leq d_k, \forall k = 1, \ldots, K$, (10) can be rewritten as

$$\min_{\{\mathbf{G}, \rho_k, c_k, d_k, \forall k\}} \quad \mathrm{Tr}(\mathbf{G}\mathbf{G}^{\mathrm{H}})$$

$$\text{s.t.} : \mathrm{Tr}(\mathbf{G}^{\mathrm{H}}\mathbf{H}_k^{\mathrm{H}}\mathbf{F}_k^{\mathrm{H}}\mathbf{F}_k\mathbf{H}_k\mathbf{G}) - \mathrm{Tr}(\Xi_k\mathbf{G}^{\mathrm{H}}\mathbf{H}_k^{\mathrm{H}}\mathbf{F}_k^{\mathrm{H}}) - \mathrm{Tr}(\mathbf{F}_k\mathbf{H}_{kk}\mathbf{G}\mathbf{X}_k^{\mathrm{H}})$$

$$+\left(\sigma_{\mathbf{n}_k}^2 + c_k\right)\mathrm{Tr}(\mathbf{F}_k^{\mathrm{H}}\mathbf{F}_k) \leq \varepsilon_k - N_k,$$

$$\mathrm{Tr}(\mathbf{G}^{\mathrm{H}}\mathbf{H}_k^{\mathrm{H}}\mathbf{H}_k\mathbf{G}) + L_k\sigma_{\mathbf{n}_k}^2 \geq d_k, \tag{20}$$

$$\frac{\sigma_{\mathbf{w}_k}^2}{\rho_k} \leq c_k,$$

$$\frac{\psi_k}{\xi_k(1-\rho_k)} \leq d_k,$$

$$0 \leq \rho_k \leq 1, \forall k = 1, \ldots, K.$$

Adopting the Schur complement lemma, the constraints $\frac{\sigma_{\mathbf{w}_k}^2}{\rho_k} \leq c_k$ and $\frac{\psi_k}{\xi_k(1-\rho_k)} \leq d_k$ can be reformulated as $\begin{bmatrix} c_k & \sigma_w \\ \sigma_w & \rho_k \end{bmatrix} \geq 0$ and $\begin{bmatrix} d_k & \sqrt{\psi_k/\xi_k} \\ \sqrt{\psi_k/\xi_k} & 1-\rho_k \end{bmatrix} \geq 0$, respectively. Then, problem (21) can be further rewritten as

$$\min_{\{\mathbf{G}, \rho_k, c_k, d_k, \forall k\}} \quad \mathrm{Tr}(\mathbf{G}\mathbf{G}^{\mathrm{H}})$$

$$\text{s.t.} : \mathrm{Tr}(\mathbf{G}^{\mathrm{H}}\mathbf{H}_k^{\mathrm{H}}\mathbf{F}_k^{\mathrm{H}}\mathbf{F}_k\mathbf{H}_k\mathbf{G}) - \mathrm{Tr}(\Xi_k\mathbf{G}^{\mathrm{H}}\mathbf{H}_k^{\mathrm{H}}\mathbf{F}_k^{\mathrm{H}}) - \mathrm{Tr}(\mathbf{F}_k\mathbf{H}_{kk}\mathbf{G}\Xi_k^{\mathrm{H}})$$

$$+\left(\sigma_{\mathbf{n}_k}^2 + c_k\right)\mathrm{Tr}(\mathbf{F}_k^{\mathrm{H}}\mathbf{F}_k) \leq \varepsilon_k - N_k,$$

$$\mathrm{Tr}(\mathbf{G}^{\mathrm{H}}\mathbf{H}_k^{\mathrm{H}}\mathbf{H}_k\mathbf{G}) + L_k\sigma_{\mathbf{n}_k}^2 \geq d_k,$$

$$\begin{bmatrix} c_k & \sigma_w \\ \sigma_w & \rho_k \end{bmatrix} \geq 0, \tag{21}$$

$$\begin{bmatrix} d_k & \sqrt{\psi_k/\xi_k} \\ \sqrt{\psi_k/\xi_k} & 1-\rho_k \end{bmatrix} \geq 0,$$

$$0 \leq \rho_k \leq 1, \forall k = 1, \ldots, K.$$

Problem (21) is a nonconvex inhomogeneous quadratically constrained quadratic program (QCQP) [20, 21], which are NP-hard [10, 22–25]. In order to solve it effectively, SDR is utilized. Specifically, a new variable $\mathbf{X} = \mathbf{G}\mathbf{G}^{\mathrm{H}}$ is defined and relaxed as $\mathbf{X} \geq \mathbf{G}\mathbf{G}^{\mathrm{H}}$, which is equivalent to $\begin{bmatrix} \mathbf{X} & \mathbf{G} \\ \mathbf{G}^H & \mathbf{I}_d \end{bmatrix} \geq 0$, problem (21) can be relaxed as

$$\min_{\{\mathbf{X}\succeq 0,\,\mathbf{G},\,\rho_k,\,c_k,\,d_k,\,\forall k\}} \quad \mathrm{Tr}(\mathbf{X})$$

$$\text{s.t.}:\ \mathrm{Tr}\!\left(\mathbf{H}_k^{\mathrm{H}}\mathbf{F}_k^{\mathrm{H}}\mathbf{F}_k\mathbf{H}_k\mathbf{X}\right) - \mathrm{Tr}\!\left(\mathbf{G}^{\mathrm{H}}\mathbf{H}_k^{\mathrm{H}}\mathbf{F}_k^{\mathrm{H}}\mathbf{\Xi}_k\right) - \mathrm{Tr}\!\left(\mathbf{\Xi}_k^{\mathrm{H}}\mathbf{F}_k\mathbf{H}_{kk}\mathbf{G}\right) + \left(\sigma_{\mathbf{n}_k}^2 + c_k\right)\mathrm{Tr}\!\left(\mathbf{F}_k^{\mathrm{H}}\mathbf{F}_k\right) \le \varepsilon_k - N_k,$$

$$\mathrm{Tr}\!\left(\mathbf{H}_k^{\mathrm{H}}\mathbf{H}_k\mathbf{X}\right) + L_k\sigma_{\mathbf{n}_k}^2 \ge d_k,$$

$$\begin{bmatrix} c_k & \sigma_w \\ \sigma_w & \rho_k \end{bmatrix} \succeq 0,$$

$$\begin{bmatrix} d_k & \sqrt{\psi_k/\xi_k} \\ \sqrt{\psi_k/\xi_k} & 1-\rho_k \end{bmatrix} \succeq 0,$$

$$0 \le \rho_k \le 1, \forall k = 1, \ldots, K,$$

$$\begin{bmatrix} \mathbf{X} & \mathbf{G} \\ \mathbf{G}^H & \mathbf{I}_d \end{bmatrix} \succeq 0. \tag{22}$$

Problem (22) is a convex SDP with respect to $\mathbf{G} \in \mathcal{C}^{M \times d}$, positive semidefinite Hermitian symmetric variable $\mathbf{X} \in \mathcal{C}^{M \times M}$ and nonnegative variables ρk, c_k, d_k, $\forall k$ and thus can be solved efficiently by using traditional convex optimization techniques [23, 26].

It is noted that the optimal objective of (22) is a lower bound of that of the nonconvex QCQP problem (21), since the same objective function is minimized over a larger set [27]. Let $\mathbf{X}_{\mathrm{SDR}}$ and $\mathbf{G}_{\mathrm{SDR}}$ denote the optimal solution of the SDR problem (22), if $\mathbf{X}_{SDR} = \mathbf{G}_{\mathrm{SDR}}\mathbf{G}_{SDR}^H$, then $\mathbf{G}_{\mathrm{SDR}}$ must be optimal for (21). Although not yet proven, simulation results show that the relaxation is always tight, that is, the equality in the relaxation is always satisfied. Replacing $\mathbf{G}$ in step 2 of Algorithm 1 with $\mathbf{G}_{\mathrm{SDR}}$ results in an SDP-based JTDPS (SDP-JTDPS) algorithm, a practical algorithm solving problem (10) is finally obtained.

The complexity of Algorithm 1 is mainly introduced by the SDP (22). Given a solution accuracy $\epsilon > 0$, the computational complexity solving SDP is about $\mathcal{O}(N_{\mathrm{Iter}}M^{4.5}\log(1/\epsilon))$ [10], where N_{Iter} is the iteration number. Therefore, the computational complexity of the algorithm is prohibitively high when the system becomes large in number of antennas and users. So, it is necessary to develop low-complexity algorithms.

Low-complexity design scheme

In this section, a low-complexity algorithm is derived by first designing ID transceivers to satisfy the MSE constraints and then optimizing the transmit power together with PS factors with the designed transceivers. The scheme is of quite low computational complexity.

It is noted that the MSE-constrained transceiver design problem (9) can be solved efficiently by existing methods proposed in [17]. So, let $\left\{\hat{\mathbf{G}}, \hat{\mathbf{F}}_k, \forall k\right\}$ denote its solution, amplify the precoder $\hat{\mathbf{G}}$ by a positive scaling factor $\sqrt{\alpha} > 1$, and decrease the receiver $\hat{\mathbf{F}}_k$ by the factor $\sqrt{\alpha} > 1$, the problem (9) can be rewritten as

$$\min_{\{\alpha,\,\rho_k,\,\forall k\}} \quad \alpha\mathrm{Tr}\!\left(\hat{\mathbf{G}}\hat{\mathbf{G}}^{\mathrm{H}}\right)$$

$$\text{s.t.} : \left\|\hat{\mathbf{F}}_k \mathbf{H}_k \hat{\mathbf{G}} - \Xi_k\right\|_F^2 + \left(\sigma_{\mathbf{n}_k}^2 + \frac{\sigma_{\mathbf{w}_k}^2}{\rho_k}\right)\frac{1}{\alpha}\left\|\hat{\mathbf{F}}_k\right\|_F^2 \leq \varepsilon_k,$$

$$\xi_k(1 - \rho_k)\left(\alpha\|\hat{\mathbf{H}}_k\hat{\mathbf{G}}\|_F^2 + L_k\sigma_{\mathbf{n}_k}^2\right) \geq \psi_k, \tag{23}$$

$$\alpha > 1,$$

$$0 \leq \rho_k \leq 1,$$

$$\forall k = 1, \dots, K,$$

where the scaling factor α and the PS factors in (7) are jointly optimized to satisfy both the MSE and EH constraints. Problem (23) can be solved in closed form, which is shown by the following proposition.

Proposition 6 The optimal solution of problem (23) is given in closed form by

$$\alpha^* = \max_{\forall k} \alpha_k^*, \tag{24}$$

$$\rho_k^* = \frac{\sigma_{\mathbf{w}_k}^2}{\alpha^* c_k - \sigma_{\mathbf{n}_k}^2}, \tag{25}$$

where $\alpha_k^* = \frac{B_k + \sqrt{B_k^2 - 4A_k C_k}}{2A_k}$ with $c_k = \frac{\varepsilon_k - \|\hat{\mathbf{F}}_k\mathbf{H}_k\hat{\mathbf{G}} - \Xi_k\|_F^2}{\|\hat{\mathbf{F}}_k\|_F^2}$, $d_k = \|\hat{\mathbf{H}}_k\hat{\mathbf{G}}\|_F^2$, $A_k = c_k\xi_k d_k$,

$B_k = \xi_k d_k\left(\sigma_{\mathbf{n}_k}^2 + \sigma_{\mathbf{w}_k}^2\right) + c_k\psi_k - c_k\xi_k L_k\sigma_{\mathbf{n}_k}^2$, and $C_k = \psi_k\sigma_{\mathbf{n}_k}^2 - \xi_k L_k\sigma_{\mathbf{n}_k}^2\left(\sigma_{\mathbf{n}_k}^2 + \sigma_{\mathbf{w}_k}^2\right)$.

Proof. Problem(23) can be transformed to

$$\min_{\{\alpha, \rho_k, \forall k\}} \alpha \tag{26}$$

$$\text{s.t.} : \rho_k \geq \frac{\sigma_{\mathbf{w}_k}^2}{\alpha c_k - \sigma_{\mathbf{n}_k}^2}, \tag{27}$$

$$1 - \rho_k \geq \frac{\psi_k}{\xi_k\left(\alpha d_k + L_k\sigma_{\mathbf{n}_k}^2\right)}, \tag{28}$$

$$\alpha > 1, \tag{29}$$

$$0 < \rho_k < 1, \forall k = 1, \dots, K. \tag{30}$$

By summing (27) and (28) for user k, problem (26–30) is equivalent to

$$\min_{\{\alpha\}} \alpha$$

$$\text{s.t.} : F_k(\alpha) \leq 1, \forall k = 1, \dots, K, \tag{31}$$

$$\alpha > 1,$$

where $F_k(\alpha) = \frac{\sigma_{\mathbf{w}_k}^2}{\alpha c_k - \sigma_{\mathbf{n}_k}^2} + \frac{\psi_k}{\xi_k\left(\alpha d_k + L_k\sigma_{\mathbf{n}_k}^2\right)}$. Since the MSE constraints are satisfied with equality when

$\alpha = 1$, there exists $\frac{\sigma_{\mathbf{w}_k}^2}{c_k - \sigma_{\mathbf{n}_k}^2} = 1$ or $\sigma_{\mathbf{w}_k}^2 + \sigma_{\mathbf{n}_k}^2 = c_k$, and $F_k(1) = 1 + \frac{\psi_k}{\xi_k\left(d_k + L_k\sigma_{\mathbf{n}_k}^2\right)} > 1$. It is known that

$F_k(\alpha)$ will monotonically decrease when $\alpha > \frac{\sigma_{\mathbf{n}_k}^2}{c_k} = \frac{\sigma_{\mathbf{n}_k}^2}{\sigma_{\mathbf{n}_k}^2 + \sigma_{\mathbf{w}_k}^2}$. Thus, $F_k(\alpha)$ decreases monotonically when $\alpha > 1$.

When $\alpha > 1$, the function $F_k(\alpha) = 1$ has a unique solution $\alpha_k^* = \frac{B_k + \sqrt{B_k^2 - 4A_k C_k}}{2A_k}$. Problem (31) is then equivalent to

$$\min_{\{\alpha\}} \quad \alpha$$
$$\text{s.t.} : \alpha \geq \alpha_k^*, \forall k. \tag{32}$$

The optimal solution of problem (32) is given by $\alpha* = \max_{\forall k} \alpha_k^*$.

For the optimal α^*, $F_k(\alpha^*) = \frac{\sigma_{w_k}^{\prime}}{\alpha^* c_k - \sigma_{n_k}^2} + \frac{\psi_k}{\xi_k\left(\alpha^* d_k + L_k \sigma_{n_k}^2\right)} \leq 1$. Let $\rho_k^* = \frac{\sigma_{w_k}^{\prime}}{\alpha^* c_k - \sigma_{n_k}^2}$, we have

$\frac{\psi_k}{\xi_k\left(\alpha^* d_k + L_k \sigma_{n_k}^2\right)} \leq 1 - \rho_k^*$. Thus, $\{\alpha^*, \rho_k^*, \forall k\}$ is an optimal solution for (26–30).

Given $\alpha*$ and $\left\{\hat{\mathbf{G}}, \hat{\mathbf{F}}_k, \forall k\right\}$, the transceiver which is feasible to problem (7) can be determined by $\left\{\mathbf{G} = \sqrt{\alpha^*}\hat{\mathbf{G}}, \mathbf{F}_k = \frac{1}{\sqrt{\alpha^*}}\hat{\mathbf{F}}_k, \forall k\right\}$. The design process is summarized in Algorithm 2.

Algorithm 2 Low-complexity closed-form PS (CF-PS) algorithm.

1: Solve problem (9) to obtain $\left\{\hat{\mathbf{G}}, \hat{\mathbf{F}}_k, \forall k\right\}$ based on the traditional MSE QoS constraint power minimization algorithm proposed in [17].

2: Optimize the optimal scaling factor α^* and the PS factors ρ_k^* according to (24) and (25).

3: Return the feasible solution $\left\{\mathbf{G} = \sqrt{\alpha^*}\hat{\mathbf{G}}, \mathbf{F}_k = \frac{1}{\sqrt{\alpha^*}}\hat{\mathbf{F}}_k, \rho_k^*, \forall k\right\}$ to problem (7).

The main computational complexity of Algorithm 2 comes from solving problem (9), which is of $\mathcal{O}(N_{\text{Iter}}d^3)$ [17]. Thus, the complexity of Algorithm 2 is quite lower than that of Algorithm 1.

Simulation results and analysis

The performance of the described algorithms is validated through simulations. The user number is set to be $K = 3$. The channel matrices $\mathbf{H}_k, \forall k$ are set as i.i.d. zero mean complex Gaussian random variables with variance $r_k^{-\beta}$, where r_k is the distance in meter between the jth transmitter and the kth receiver, and β is the path loss factor. The following parameters are set unless otherwise noted, M = 8, $L_k = L = 4$, $N_k = N = L/2$, $r_k = r = 5$, $\beta = 2.7$, $\sigma_{n_k}^2 = \sigma_n^2 = 10{-}3$, and $\sigma_{w_k}^2 = \sigma_w^2 = 10^{-2}$. The MSE and EH thresholds are set nonuniformly as $\varepsilon_1 = 0.01$, $\varepsilon_2 = 0.02$, $\varepsilon_3 = 0.2$, $\psi 1 = 20$, $\psi 2 = 25$, and $\psi 3 = 30$dBm. CVX toolbox [26] is adopted to solve SDP problems. The traditional MSE QoS constraint (MSE-QoS- TRAD) power minimization algorithm [17] is adopted as a performance benchmark.

The convergence property of the described scheme is shown in **Figure 2**. It can be seen that the optimized transmit power and the PS factors decrease monotonically along with the increase of the number of iterations. It needs special explanation that $\mathbf{X}_{\text{SDR}} = \mathbf{G}_{\text{SDR}}\mathbf{G}_{\text{SDR}}^H$ is always satisfied during the simulations. This verifies that the SDP solution of the relaxed problem (22) is also optimal for the joint transmitter and PS subproblem (10).

Figure 3 shows the per-user MSE and harvested energy of the proposed scheme along with iterations. Similar to the MSE-QoS-TRAD scheme, the SDP-JTDPS scheme satisfies the MSE QoS requirements in each iteration. Moreover, the EH requirements can also be satisfied.

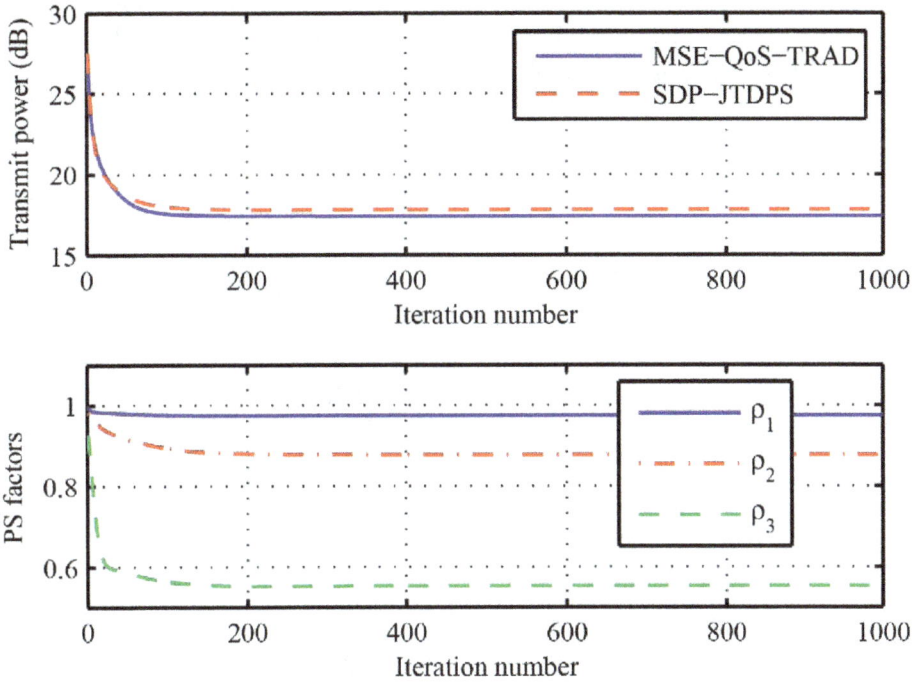

Figure 2. *Transmit power and PS factors versus iterations with the SDP-JTDPS algorithm.*

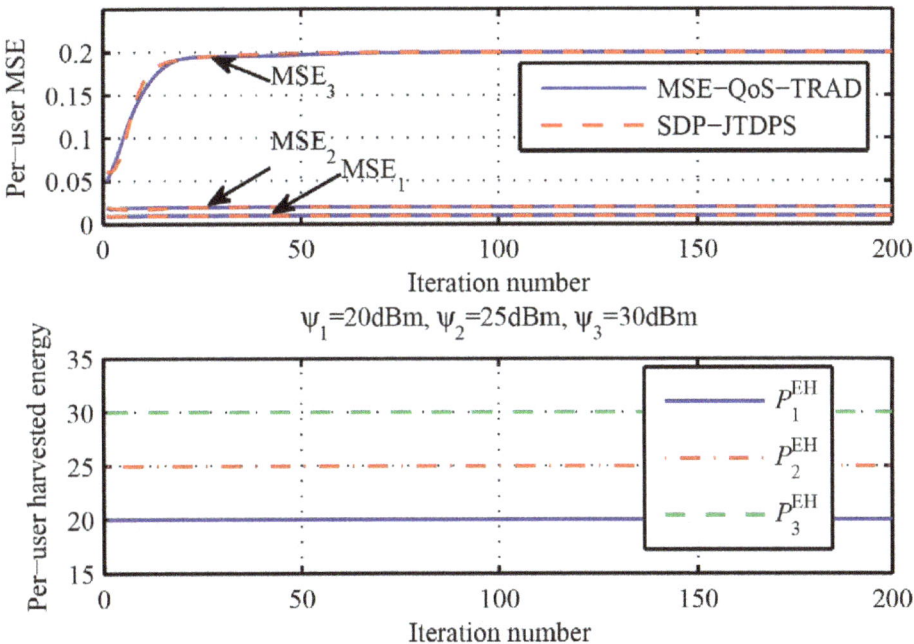

Figure 3. *Achieved per-user MSE and harvested energy versus iterations by the proposed SDP-TDPS scheme. The titles show the QoS targets.*

The performance of SDP-JTDPS and CF-PS algorithms are compared in **Figure 4**. It can be observed that all of them achieve the MSE QoS requirements exactly. This is consistent with the analysis that the MSE constraints can be satisfied with equality for both schemes. It is also shown that the SDP-JTDPS can exactly reach the EH targets of all users, which implies that the EH constraints are satisfied with equality. Different from the SDP-JTDPS, the CF-PS harvests more energy than the predefined threshold, but at the expense of more transmit power, which is shown in **Figure 5**.

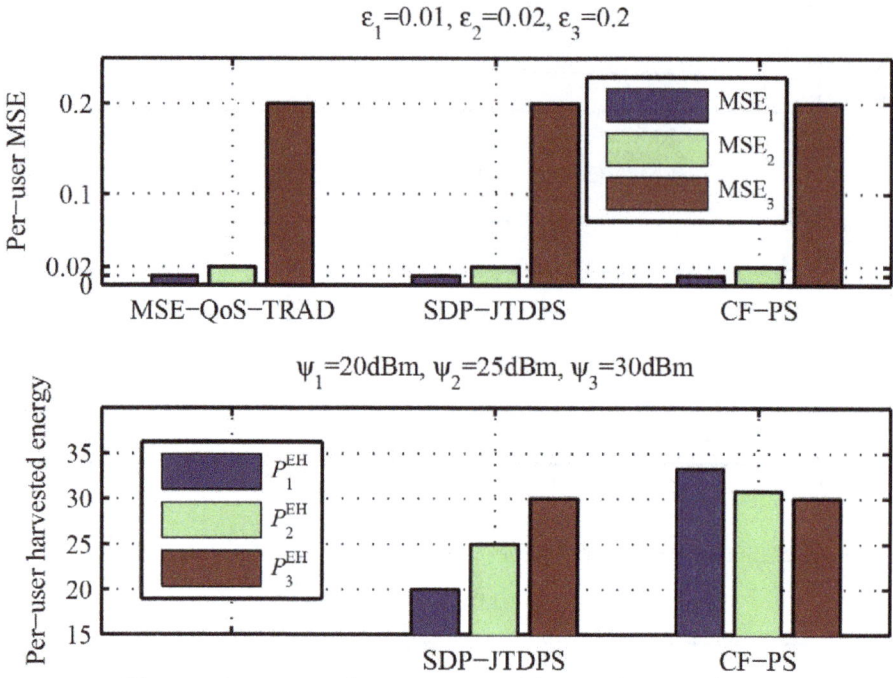

Figure 4. *Comparison of per-user MSE and harvested energy achieved by the proposed algorithms. The titles show the QoS targets.*

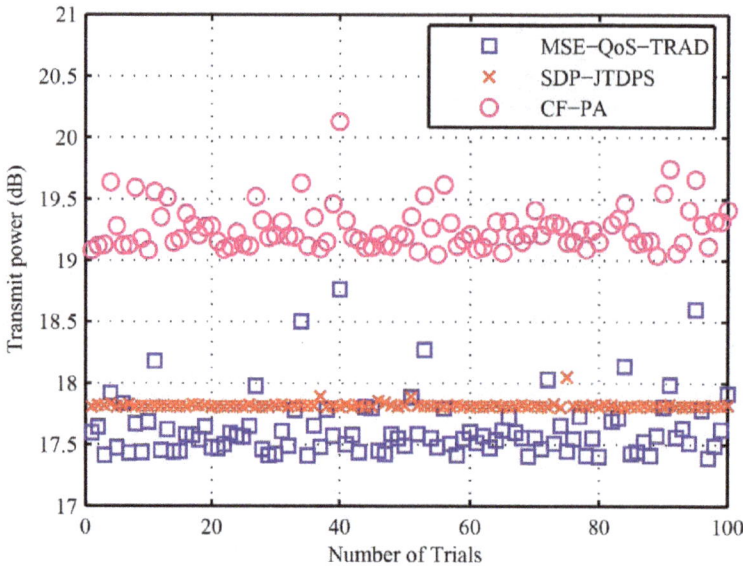

Figure 5. *Transmit power versus number of trials.*

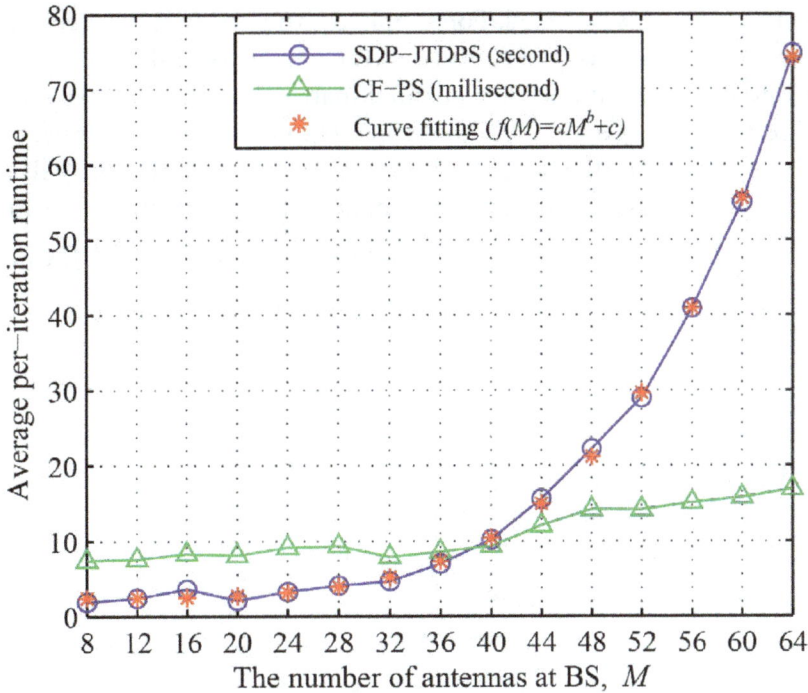

Figure 6. *Comparison of computational complexity.*

The optimized transmit power achieved by the algorithms are compared in **Figure 5**. During the simulations, all algorithms run at the same independently generated initial receivers in each trail. It can be observed that SDP-JTDPS and CF-PS consume higher transmit power than the traditional MSE-QoS-TRAD at most of the trials, and CF-PS consumes more power than SDP-JTDPS. This is obvious because more power is needed to satisfy the EH requirements. CF-PS consumes more transmit power than SDP-JTDPS does, which has been mentioned in the previous section that the low complexity is achieved at the cost of high transit power.

As shown in **Figure 6**, CF-PS performs the best with respect to the computational complexity. For the SDP-JTDPS scheme, its execution time is well fitted as a power function on the number of transmit antennas M with $a = 2.662 \times 10\text{-}7$, $b = 4.608$ and $c = 2.408$. This is exactly coherent with the complexity analysis in Section 2.3.4.

Joint transceiver design and power splitting optimization based on SINR criterion for MIMO IC channel

In this section, we further consider the joint transceiver design and wireless power transfer for MIMO IC networks.

System model

A K-user MIMO IC network as shown in **Figure 7** is considered. Without loss of generality, a symmetric configuration, that is, each user consists of a pair of transmitters with n_t transmit antennas and nr receive antennas and each user transmits d data streams from its transmitter to its receiver, is assumed. For simplicity, the MIMO IC network is henceforth denoted by $(nt, nr, d)^K$. It is assumed that the transceiver pairs share the same frequency band and operate in SWIPT mode. The channel matrix from transmitter j to receiver k is denoted by $\mathbf{H}_{kj} \in \mathcal{C}^{nr \times nt}$, $\forall k, j \in \{1, \dots, K\}$, in which $\mathbf{H}_{kk}$ describes the channel coefficients of the desired direct link of the kth user pair, and $\mathbf{H}_{kj}$, $\forall j \neq k$ constitutes channel coefficients of all interference links. It is assumed that $\mathbf{s}_j \in \mathcal{C}^{d \times 1}$ denotes the data vector of the jth transmitter and assumes $\mathbb{E}\left[\mathbf{s}_j \mathbf{s}_j^H\right] = \mathbf{I}_d$. After being precoded by a matrix $\mathbf{V}_k \in \mathcal{C}^{n_t \times d}$, it will be launched over the wireless channel. The transmit power of the kth transmitter can be calculated as $\mathbb{E}\left[\mathrm{Tr}\left(\mathbf{V}_j \mathbf{V}_j^H\right)\right] = P_j$.

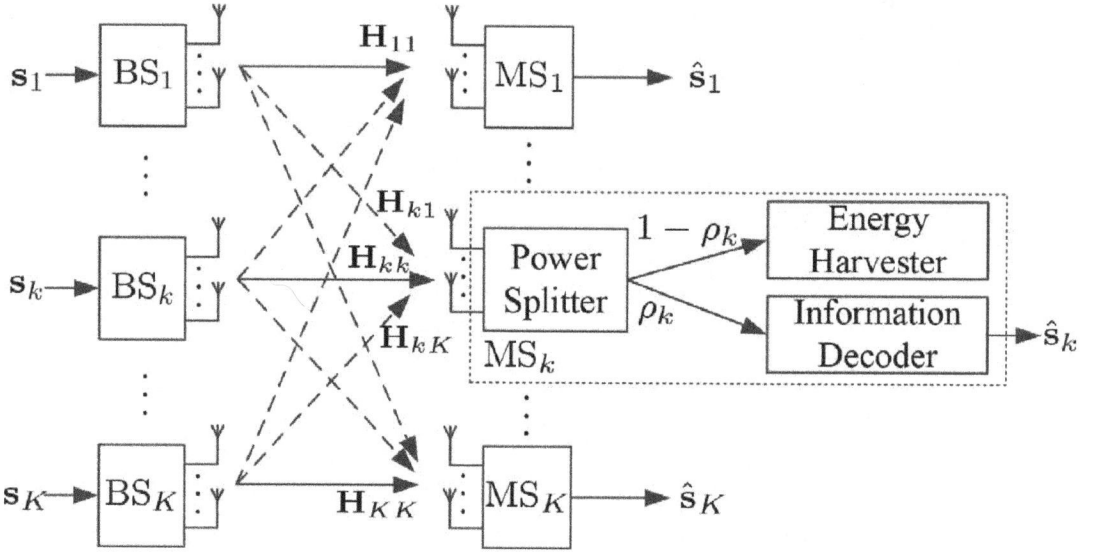

Figure 7. *MIMO interference channel SWIPT system.*

The received baseband signal at the kth receiver is written as

$$\mathbf{r}_k = \mathbf{H}_{kk}\mathbf{V}_k\mathbf{s}_k + \sum_{j=1,\, j \neq k}^{K} \mathbf{H}_{kj}\mathbf{V}_j\mathbf{s}_j + \mathbf{n}_k, \tag{33}$$

where $\mathbf{n}_k \in \mathcal{C}^{nr \times 1}$ is the noise vector at the kth receiver, whose elements are assumed to be i.i.d. complex Gaussian random variables with variance $\sigma_{\mathbf{n}_k}^2$.

Similar to Section 2.1, the received signal at each antenna is then divided into two parts via a power splitter; one part is for information decoding, and the other is transformed to stored energy. The signal split into the ID receiver at the kth user is expressed as

$$\mathbf{r}_k^{ID} = \sqrt{\rho_k}\left(\mathbf{H}_{kk}\mathbf{V}_k\mathbf{s}_k + \sum_{j=1,\, j \neq k}^{K} \mathbf{H}_{kj}\mathbf{V}_j\mathbf{s}_j + \mathbf{n}_k\right) + \mathbf{w}_k, \tag{34}$$

where $\mathbf{w}_k \sim \mathcal{CN}\left(0, \sigma_{\mathbf{w}_k}^2 \mathbf{I}_{nr}\right)$ is the additive complex Gaussian noise introduced by the power splitter.

Define U_k to be the receive filter for information decoding at the kth receiver, the detected signal is written as

$$\hat{s}_k = \sqrt{\rho_k} U_k^H r_k^{ID}$$

$$= \underbrace{\sqrt{\rho_k} U_k^H H_{kk} V_k s_k}_{\text{Desired Signal}} + \underbrace{\sum_{j=1,\, j\neq k}^{K} \sqrt{\rho_k} U_k^H H_{kj} V_j s_j}_{\text{Interference}} + \underbrace{\sqrt{\rho_k} U_k^H n_k + U_k^H w_k}_{\text{Noise}}. \tag{35}$$

The SINR of the lth data stream of the kth user is defined as

$$\text{SINR}_{kl} = \frac{\rho_k \left| u_{kl}^H H_{kk} v_{kl} \right|^2}{\sum_{(j,m)\neq(k,l)} \rho_k \left| u_{kl}^H H_{kj} v_{jm} \right|^2 + \left(\rho_k \sigma_n^2 + \sigma_w^2 \right) u_{kl}^H u_{kl}}, \tag{36}$$

where u_{kl} and v_{kl} denote the lth column vector in U_k and V_k, respectively.

Let $\xi_k \in (0, 1]$ be the energy conversion efficiency, the harvested energy at the kth receiver writes

$$P_k^{EH} = \xi_k (1 - \rho_k) \left[\sum_{j=1}^{K} \text{Tr}\left(H_{kj} V_j V_j^H H_{kj}^H \right) + n_r \sigma_{n_k}^2 \right]. \tag{37}$$

Problem formulation

To minimize the transmit power under the given QoS constraints, the joint transceiver design and power splitting problem is formulated as

$$\min_{\{U_k, V_k, \rho_k\}} \sum_{k=1}^{K} \sum_{l=1}^{d} \|v_{kl}\|_2^2 \tag{38}$$

$$s.t.: \quad \text{SINR}_{kl} \geq \gamma_{kl}, \tag{39}$$

$$P_k^{EH} \geq \psi_k, \tag{40}$$

$$0 \leq \rho_k \leq 1, \forall (k, l). \tag{41}$$

Here, SINR is adopted to measure the QoS of ID. Eqs. (38–41) are nonconvex, and thus, it is very difficult to obtain its optimal solution. Similar to Section 2.1, the AO framework can be adopted to develop an iterative algorithm. In order to achieve this, the concept of interference alignment can be utilized. Therefore, some preliminaries on IA are introduced in the following section.

Interference alignment

IA is a ground-breaking interference management method for IC networks. The idea of IA is to coordinate the transmitters so that the interference received at each receiver can be aligned into a subspace with a small dimension and thus leaves the interference-free subspace for signal [28]. IA has the ability to achieve the maximum degrees of freedom (DoF) of the K-user IC networks.

As shown in **Figure 1**, when the EH receivers are removed, the system degenerates into traditional symmetric MIMO IC networks. The DoF of such MIMO IC network is min $(n_t, n_r)K/2$ [28,

29]. To achieve IA, the following feasibility condition should be satisfied [30].

$$\mathbf{U}_k^H \mathbf{H}_{kj} \mathbf{V}_j = \mathbf{0} \ , \forall j, k \in \{1, \dots, K\}, j \neq k \tag{42}$$

$$rank\left(\mathbf{U}_k^H \mathbf{H}_{kk} \mathbf{V}_k\right) = d, \tag{43}$$

When the condition (42) is satisfied, the inter-user interference can be completely suppressed. While the condition (43) guarantees that sufficient dimensions are left for signal subspace. For the considered system, in order to achieve IA, the maximum number of streams for each user should no more than $d \leq (n_t + n_r)/(K + 1)$ [31]. Since the IA condition is over constrained, it is not trivial to develop closed-form solution for it. In literature, a lot of iterative algorithms have been developed, such as the MIL algorithm [30], max-SINR algorithm [30], and MMSE algorithm [32].

If IA conditions are perfectly satisfied, and the received signal of user k reduces to

$$\hat{\mathbf{s}}_k = \mathbf{U}_k^H \mathbf{r}_k = \overline{\mathbf{H}}_k \mathbf{s}_k + \overline{\mathbf{n}}_k, \tag{44}$$

where $\overline{\mathbf{H}}_k = \mathbf{U}_k^H \mathbf{H}_{kk} \mathbf{V}_k$ and $\overline{\mathbf{n}}_k = \mathbf{U}_k^H \mathbf{n}_k$ denote the effective channel matrix and the effective noise vector at receiver k, respectively.

Eq. (44) means that the system is equivalent to a traditional point-to-point MIMO system after IA and the ergodic achievable rate of the kth user is

$$\mathcal{R}_k = \mathbb{E}\left[\log_2 \left| \mathbf{I}_d + \frac{\overline{\mathbf{H}}_k \overline{\mathbf{H}}_k^H}{\sigma_{\mathbf{n}_k}^2} \right| \right]. \tag{45}$$

In the following sections, the feasibility of the formulated problem (38) will be discussed, and suboptimal schemes solving the problem will be developed.

Feasibility analysis

The feasibility of the formulated problems (38–41) can be given by the following propositions [14].

Proposition 7 Problem (38) is feasible if and only if the following problem is feasible.

$$\begin{aligned} \text{Find}: \quad & \{\mathbf{U}_k, \mathbf{V}_k, \rho_k\} \\ \text{Such that}: \quad & \text{SINR}_{kl} \geq \gamma_{kl}, \\ & 0 \leq \rho_k \leq 1, \forall(k, l). \end{aligned} \tag{46}$$

Proposition 8 Problem (46) is feasible if and only if the following problem is feasible.

$$\begin{aligned} \text{Find}: \quad & \{\mathbf{U}_k, \mathbf{V}_k\} \\ \text{Such that}: \quad & \text{SINR}'_{kl} \geq \gamma_{kl}, \forall(k, l), \end{aligned} \tag{47}$$

where

$$\text{SINR}'_{kl} = \frac{\left|\mathbf{u}_{kl}^H \mathbf{H}_{kk} \mathbf{v}_{kl}\right|^2}{\sum_{(j,m) \neq (k,l)} \left|\mathbf{u}_{kl}^H \mathbf{H}_{kj} \mathbf{v}_{jm}\right|^2 + \left(\sigma_n^2 + \sigma_w^2\right) \mathbf{u}_{kl}^H \mathbf{u}_{kl}}. \tag{48}$$

Proposition 7 and Proposition 8 show that the feasibility of (38–41) is independent of the EH constraints and the PS factors. Proposition 8 is a sufficient and necessary condition for the feasibility of problem (38–41). However, it is not hard to solve (47) [33, 34] directly since SINRs are overconstrained. As an alternative, a sufficient condition for the feasibility of problem (47) is derived based on IA by the following proposition.

Proposition 9 (47) is feasible for any given SINR constraints if the system is interference unlimited, that is, the interference can be completely eliminated by the linear transceivers.

Proof. If interference is completely eliminated, given the transceivers U_k, V_k, $\forall k$, that is,

$$\mathbf{u}_{kl}^H \mathbf{H}_{kj} \mathbf{v}_{jm} = \mathbf{0}, \forall (j,m) \neq (k,l), \tag{49}$$

$$\mathbf{u}_{kl}^H \mathbf{H}_{kk} \mathbf{v}_{kl} \neq 0, \forall (k,l), \tag{50}$$

SINR (48) becomes

$$\text{SINR}_{kl}' = \frac{\left| \mathbf{u}_{kl}^H \mathbf{H}_{kk} \mathbf{v}_{kl} \right|^2}{(\sigma_n^2 + \sigma_w^2) \mathbf{u}_{kl}^H \mathbf{u}_{kl}} = \frac{p_{kl} \left| \mathbf{u}_{kl}^H \mathbf{H}_{kk} \bar{\mathbf{v}}_{kl} \right|^2}{(\sigma_n^2 + \sigma_w^2) \|\mathbf{u}_{kl}\|_2^2}, \tag{51}$$

where $\bar{\mathbf{v}}_{kl} = \frac{\mathbf{v}_{kl}}{\|\mathbf{v}_{kl}\|_2}$ is the normalized precoding vector, p_{kl} is the transmit power along the beamforming direction $\bar{\mathbf{v}}_{kl}$ and $\mathbf{v}_{kl} = p_{kl} \bar{\mathbf{v}}_{kl}$. According to (51), the SINR constraints in problem (47) can always be satisfied by increasing the transmit power p_{kl}, if the interference is completely suppressed.

Based on Proposition 9 and the IA feasibility condition (35), problems (38–41) must be feasible if the system is IA feasible. In the following, it is assumed that the considered MIMO IC network is IA feasible.

Alternative optimization solution based on semidefinite programming relaxation

An iterative algorithm for (38–41) can be developed based on AO framework, that is, alternatively optimizing the transmitters V_k, $\forall k$ together with the PS factors ρk, $\forall k$ and the receivers U_k, $\forall k$.

Transmitter and power splitting optimization

When the receivers are fixed, problems (38–41) are reduced to the following joint transmit precoders and PS factors optimization problem

$$\min_{\{\mathbf{v}_{kl}, \rho_k, \forall (k,l)\}} \sum_{k=1}^{K} \sum_{l=1}^{d} \|\mathbf{v}_{kl}\|_2^2$$

$$\text{s.t.}: \quad \frac{\left| \mathbf{u}_{kl}^H \mathbf{H}_{kk} \mathbf{v}_{kl} \right|^2}{\mathbf{u}_{kl} \mathbf{B}_{kl} \mathbf{u}_{kl}^H} \geq \gamma_{kl},$$

$$\sum_{j=1}^{K} \sum_{m=1}^{d} \|\mathbf{H}_{kj} \mathbf{v}_{jm}\|_2^2 \geq \frac{\psi_k}{\xi_k(1-\rho_k)} - n_r \sigma_n^2,$$

$$0 \leq \rho_k \leq 1, \forall (k,l),$$

$$\tag{52}$$

where $\mathbf{B}_{kl} = \sum_{j=1}^{K}\sum_{m=1}^{d}\mathbf{H}_{kj}\mathbf{v}_{jm}\mathbf{v}_{jm}^{H}\mathbf{H}_{kj}^{H} - \mathbf{H}_{kk}\mathbf{v}_{kl}\mathbf{v}_{kl}^{H}\mathbf{H}_{kk}^{H} + \left(\sigma_n^2 + \frac{\sigma_w^2}{\rho_k}\right)\mathbf{I}_{n_r}$.

According to Lemma 8, Proposition 9, and [35], Proposition 9, problem (52) is feasible if the original problem is feasible. By defining $\bar{\mathbf{X}}_{kl} = \mathbf{v}_{kl}\mathbf{v}_{kl}^{H}, \mathbf{X}_{kl} \succeq \mathbf{0}$, problem (52) can be relaxed as the following convex SDP program

$$\min_{\{\mathbf{X}_{kl}, \rho_k, \forall(k,l)\}} \sum_{k=1}^{K}\sum_{l=1}^{d}\mathrm{Tr}(\mathbf{X}_{kl})$$

$$\text{s.t.:}\quad (1 + \gamma_{kl})\mathrm{Tr}\left(\mathbf{u}_{kl}^{H}\mathbf{H}_{kk}\mathbf{X}_{kl}\mathbf{H}_{kk}^{H}\mathbf{u}_{kl}\right)$$

$$-\gamma_{kl}\sum_{j=1}^{K}\sum_{m=1}^{d}\mathrm{Tr}\left(\tilde{\mathbf{u}}_{kl}^{H}\mathbf{H}_{kj}\mathbf{X}_{jm}\mathbf{H}_{kj}^{H}\mathbf{u}_{kl}\right) \geq \gamma_{kl}\left(\sigma_n^2 + \frac{\sigma_w^2}{\rho_k}\right)\|\mathbf{u}_{kl}\|_2^2, \tag{53}$$

$$\sum_{j=1}^{K}\sum_{m=1}^{d}\mathrm{Tr}\left[\mathbf{H}_{kj}\mathbf{X}_{jm}\mathbf{H}_{kj}^{H}\right] \geq \frac{\psi_k}{\xi_k(1-\rho_k)} - n_r\sigma_n^2,$$

$$0 \leq \rho_k \leq 1, \forall(k,l).$$

The convex SDP (53) can be solved efficiently. Moreover, it can be proven that there is $rank(\mathbf{X}_{kl}) = 1$ and the SDP relaxation is tight [34], meaning that the optimal solution of the relaxed problem is also optimal to the original problem. After obtaining the rank-one solution $\{\mathbf{X}_{kl}, \forall(k, l)\}$, the optimal solution $\mathbf{v}_{kl}$ to problem (52) can be further recovered from the eigen decomposition of $\mathbf{X}_{kl}$.

Receiver optimization

When the transmit precoders and PS factors are all fixed, (38–41) become separable with respect to variables $\{\mathbf{u}_{kl}, \forall(k, l)\}$. Recall that the SINR constraints have been satisfied by the solution of (52), we can further maximize the receive per-stream SINR by

$$\max_{\mathbf{u}_{kl}} \frac{\mathbf{u}_{kl}^{H}\mathbf{H}_{kk}\mathbf{v}_{kl}\mathbf{v}_{kl}^{H}\mathbf{H}_{kk}^{H}\mathbf{u}_{kl}}{\mathbf{u}_{kl}^{H}\mathbf{B}_{kl}\mathbf{u}_{kl}} \tag{54}$$

where $\mathbf{B}_{kl} = \sum_{j=1}^{K}\sum_{m=1}^{d}\mathbf{H}_{kj}\mathbf{v}_{jm}\mathbf{v}_{jm}^{H}\mathbf{H}_{kj}^{H} - \mathbf{H}_{kk}\mathbf{v}_{kl}\mathbf{v}_{kl}^{H}\mathbf{H}_{kk}^{H} + \left(\sigma_n^2 + \frac{\sigma_w^2}{\rho_k}\right)\mathbf{I}_{n_r}$. Eq. (54). Eq. (54) is a generalized Rayleigh quotient and its closed-form solution is given by

$$\mathbf{u}_{kl} = \frac{\mathbf{B}_{kl}^{-1}\mathbf{H}_{kk}\mathbf{v}_{kl}}{\|\mathbf{B}_{kl}^{-1}\mathbf{H}_{kk}\mathbf{v}_{kl}\|_2}. \tag{55}$$

Algorithm description

By alternatively optimizing the transmitters together with PS factors and the receivers, an SDP-based joint transceiver design and PS optimization scheme can be obtained, which is summarized in Algorithm 3.

Algorithm 3 Joint transceiver design and power splitting based on SDP
(SDP-JTDPS).

1: Initialize the receivers $\mathbf{U}_k, \forall k$.
2: Solve the convex problem (53) to obtain $\mathbf{X}_{kl}$ and power splitting factors
$\rho_k, \forall (k, l)$.
3: Recover $\mathbf{v}_{kl}, \forall (k, l)$ from $\mathbf{X}_{kl}$ through eigenvalue decomposition.
4: Update the receiver $\mathbf{u}_{kl}, \forall (k, l)$ by (55).
5: Repeat 2 to 4 until convergence or the maximum iteration number reached.

The convergence of Algorithm 3 is given by the following proposition [14].

Proposition 10 If (53) is feasible for the initial receivers $\mathbf{U}_k$, $\forall k$, the convergence to a locally optimal solution can be guaranteed by Algorithm 3.

According to Proposition 10, it is important to initialize the receivers $\mathbf{U}_k$, $\forall$k for the success of the algorithm. To guarantee finding a feasible solution to (53), we initialize the receivers by the IA receivers $\mathbf{U}_k^{IA}$, $\forall$k in this chapter. Similar to Algorithm 1, the complexity of Algorithm 3 is dominated by the SDP solving process, which is about $\mathcal{O}\left(n_r K d\right)^{4.5} \log \left(1/\epsilon\right)$ for one instance [10]. This complexity becomes prohibitive as the number of antennas or users increases. In the following section, a low-complexity design schemes solving this problem is developed.

Low-complexity design schemes

Two kinds of low complexity schemes are derived to solve problems (38–41) by separately designing the transceivers and power splitting factors. The transceivers are firstly designed by eigen-decomposing the effective channel matrices generated by interference alignment. Then, the transmit power and receive PS factors are optimized with the precoders and receivers fixed.

As analyzed in the previous section, to ensure that (38–41) are feasible, perfect IA should be realized. To simplify the system design, we assume that the precoders and receive filters are orthogonalized such that $\left(\mathbf{V}_k^{IA}\right)^H \mathbf{V}_k^{IA} = \mathbf{I}_d$ and $\left(\mathbf{U}_k^{IA}\right)^H \mathbf{U}_k^{IA} = \mathbf{I}_d$. Given interference alignment transceivers, the effective channel matrix for user k can be decomposed as $\overline{\mathbf{H}}_{kk} = \left(\mathbf{U}_k^{IA}\right)^H \mathbf{H}_{kk} \mathbf{V}_k^{IA} = \overline{\mathbf{U}}_k \mathbf{\Lambda}_k \overline{\mathbf{V}}_k^H$ through singular value decomposition (SVD), where $\mathbf{\Lambda}_k = \text{diag}\left(\sqrt{\lambda_{k1}}, \sqrt{\lambda_{k2}}, ..., \sqrt{\lambda_{kd}}\right)$ is a diagonal matrix. The transmit precoder matrix can be constructed as $\mathbf{V}_k' = \mathbf{V}_k^{IA} \overline{\mathbf{V}}_k$. By further multi-plying the power matrix, the transmit precoding vector is then finally constructed as $\mathbf{V}_k = \mathbf{V}_k' \mathbf{P}_k$, where $\mathbf{P}_k = \text{diag}\left(\sqrt{p_{k1}}, ..., \sqrt{p_{kd}}\right)$ is a diagonal matrix with nonnegative diagonalized elements. At the receiver side, the receive filter is constructed similarly, $\mathbf{U}_k = \mathbf{U}_k^{IA} \overline{\mathbf{U}}_k$.. Then, the following equations are established

$$\mathbf{U}_k^H \mathbf{H}_{kj} \mathbf{V}_j = \mathbf{0} \ , \forall j, k \in \{1, ..., K\}, j \neq k, \tag{56}$$

$$\mathbf{U}_k^H \mathbf{H}_{kk} \mathbf{V}_k = \mathbf{\Lambda}_k \mathbf{P}_k, \forall k. \tag{57}$$

Eq. (56) shows that interference is completely suppressed at the ID receiver by the transceiver design scheme. According to Lemma 7, Lemma 8, and Proposition 9, we know that problems (38–41) are feasible and can be reduced to the following transmit power allocation and power splitting problem

$$\min_{\{p_{kl},\rho_k,\forall k,l\}} \sum_{k=1}^{K}\sum_{l=1}^{d} p_{kl}$$

$$\text{s.t.} : \quad \frac{\rho_k \lambda_{kl} p_{kl}}{(\rho_k \sigma_n^2 + \sigma_w^2)\|\mathbf{u}_{kl}\|_2^2} \geq \gamma_{kl},$$

$$\xi_k(1-\rho_k)\left(\sum_{j=1}^{K}\sum_{m=1}^{d} p_{jm}\|\mathbf{H}_{kj}\mathbf{v}'_{jm}\|_2^2 + n_r\sigma_n^2\right) \geq \psi_k, \tag{58}$$

$$0 \leq \rho_k \leq 1, \forall(k,l).$$

In the following, two different schemes solving (58) is developed by either reformulating it as a convex problem or solved in closed form.

Optimal power allocation and power splitting scheme

After some algebraic manipulations, problem (58) can be reformed as

$$\min_{\{p_{kl},\rho_k,\forall k,l\}} \sum_{k=1}^{K}\sum_{l=1}^{d} p_{kl}$$

$$\text{s.t.} : \quad p_{kl} - \frac{\sigma_n^2 \gamma_{kl}\|\mathbf{u}_{kl}\|_2^2}{\lambda_{kl}} \geq \frac{\sigma_w^2 \|\mathbf{u}_{kl}\|_2^2 \gamma_{kl}}{\lambda_{kl}\rho_k},$$

$$\sum_{j=1}^{K}\sum_{m=1}^{d} p_{jm}\|\mathbf{H}_{kj}\mathbf{v}'_{jm}\|_2^2 \geq \frac{\psi_k}{\xi_k(1-\rho_k)} - n_r\sigma_n^2, \tag{59}$$

$$0 \leq \rho_k \leq 1, \forall(k,l).$$

Problem (59) is convex and thus can be solved optimally. Denote p_{kl}^* and ρ_k^*, $\forall k$ as its optimal solution, the transmit precoders are $\mathbf{v}_{kl} = \sqrt{p_{kl}^*}\mathbf{v}_{kl}^{\text{IA}}$.

The proposed IA-based SWIPT scheme with optimal power allocation and power splitting is summarized in Algorithm 4. The computational complexity of Algorithm 4 is mainly from solving (59) in Step 4. When the interior methods are employed, the computational complexity of Algorithm 4 is in the order of $\mathcal{O}((Kd)^3)$ [27], which is significantly lower than that of Algorithm 3.

Algorithm 4 SWIPT design with optimal transmit power allocation and receive power splitting over effective IA channel decomposing (O-PAPS).

1: Obtain IA transceivers $\{\mathbf{U}_k^{\text{IA}}, \mathbf{V}_k^{\text{IA}}, \forall k\}$ that satisfy the IA conditions.

2: Let $\bar{\mathbf{H}}_{kk} = (\mathbf{U}_k^{\text{IA}})^{\text{H}}\mathbf{H}_{kk}\mathbf{V}_k^{\text{IA}}, \forall k$, and decompose $\bar{\mathbf{H}}_{kk}$ as $\bar{\mathbf{H}}_{kk} = \bar{\mathbf{U}}_k\mathbf{L}_k\bar{\mathbf{V}}_k^{\text{H}}$ through SVD, where $\mathbf{\Lambda}_k = \text{diag}(\sqrt{\lambda_{k1}}, \sqrt{\lambda_{k2}}, ..., \sqrt{\lambda_{kd}})$.

3: Let $\mathbf{U}_k = \mathbf{U}_k^{\text{IA}}\bar{\mathbf{U}}_k$ and $\mathbf{V}'_k = \mathbf{V}_k^{\text{IA}}\bar{\mathbf{V}}_k, \forall k$.

4: Obtain the optimal transmit power p_{kl}^* and power splitting factors ρ_k^*, $\forall k,l$ by solving the convex problem (59).

5: Set $\mathbf{v}_{kl} = \sqrt{p_{kl}^*}\mathbf{v}_{kl}^{\text{IA}}, \forall k,l$.

Closed-form power allocation and power splitting scheme

Given the IA solution $\{\mathbf{U}_k^{IA}, \mathbf{V}_k^{IA}, \forall k\}$,, by discarding the EH constraints of (58), we consider the following SINR constrained power optimization problem

$$\min_{\{p_{kl}, \forall k, l\}} \quad \sum_{k=1}^{K} \sum_{l=1}^{d} p_{kl} \tag{60}$$

$$\text{s.t. :} \quad \frac{\lambda_{kl} p_{kl}}{(\sigma_n^2 + \sigma_w^2) \|\mathbf{u}_{kl}\|_2^2} \geq \gamma_{kl}, \forall(k,l).$$

According to Proposition 9, (60) is feasible. By further applying Lemma 8, (58) is feasible. Moreover, (60) can be decomposed into $\sum_{k=1}^{K} d$ parallel subproblems. For the l-th data stream of the kth user, the subproblem is expressed as

$$\min_{p_{kl}} \quad p_{kl}$$

$$\text{s.t. :} \quad \frac{\lambda_{kl} p_{kl}}{(\sigma_n^2 + \sigma_w^2) \|\mathbf{u}_{kl}\|_2^2} \geq \gamma_{kl}. \tag{61}$$

The solution of (61) is then given by

$$\hat{p}_{kl} = \frac{(\sigma_n^2 + \sigma_w^2) \|\mathbf{u}_{kl}\|_2^2 \gamma_{kl}}{\lambda_{kl}}. \tag{62}$$

Following Lemma 7, (58) can be optimized by substituting pkl with $\alpha \hat{p}_{kl}$, where $\alpha \geq 1$ is a scaling factor to be optimized. Then, (58) is reduced to a problem jointly optimizing α and PS factors ρk under SINR and EH constraints, which is

$$\min_{\alpha, \{\rho_k, \forall k, l\}} \quad \sum_{k=1}^{K} \sum_{l=1}^{d} \alpha \hat{p}_{kl}$$

$$\text{s.t. :} \quad \rho_k \geq \frac{\sigma_w^2 \|\mathbf{u}_{kl}\|_2^2 \gamma_{kl}}{\alpha \lambda_{kl} \hat{p}_{kl} - \sigma_n^2 \gamma_{kl} \|\mathbf{u}_{kl}\|_2^2}, \tag{63}$$

$$1 - \rho_k \geq \frac{\psi_k}{\xi_k \left(\sum_{j=1}^{K} \sum_{m=1}^{d} \alpha \hat{p}_{jm} \|\mathbf{H}_{kj} \mathbf{v}'_{jm}\|_2^2 + n_r \sigma_n^2\right)},$$

$$0 \leq \rho_k \leq 1, \forall(k,l),$$

$$\alpha > 1.$$

The closed-form solution of (63) can be derived and given by the following proposition [14].

Proposition 11 Given the IA transceivers $\{\mathbf{U}_k^{IA}, \mathbf{V}_k^{IA}, \forall k\}$, define $a_{kl} = \sigma_w^2 \|\mathbf{u}_{kl}\|_2^2 \gamma_{kl}$, $b_{kl} = \sigma_n^2 \gamma_{kl} \|\mathbf{u}_{kl}\|_2^2$, $c_{kl} = \lambda_{kl} \hat{p}_{kl}$, $f_k = \xi_k \sum_{j=1}^{K} \sum_{m=1}^{d} \hat{p}_{jm} \xi_k \|\mathbf{H}_{kj} \mathbf{v}'_{jm}\|_2^2$, and gk = $\xi_k n_r \sigma_n^2$. The optimal solution to (63) is given by

$$\alpha^* = \max_{\forall(k,l)} \alpha_{kl}^*, \tag{64}$$

$$\rho_k^* = \max_{\forall l} \rho_{kl}, \tag{65}$$

where α_{kl}^* is the solution to the equation $\frac{a_{kl}}{\alpha c_{kl} - b_{kl}} + \frac{\psi_k}{\alpha f_k + g_k} = 1 \ (\alpha > 1)$, $\rho_{kl} = \frac{a_{kl}}{\alpha^* c_{kl} - b_{kl}}$..

Given α^* and $\hat{p}_{kl}$, the transmit precoders are then determined by $\mathbf{v}_{kl} = \sqrt{\alpha^* \hat{p}_{kl}} \mathbf{v}'_{kl}$. The proposed IA-based SWIPT scheme with the closed-form transmit power allocation and receive power splitting is summarized in Algorithm 5.

Algorithm 5 SWIPT design with closed-form transmits power allocation and receive power splitting solutions over the effective IA channel decomposing (CF-PAPS).

1: Obtain IA transceivers $\{\mathbf{U}_k^{\mathrm{IA}}, \mathbf{V}_k^{\mathrm{IA}}, \ \forall k\}$ that satisfy the IA conditions.
2: Let $\overline{\mathbf{H}}_{kk} = (\mathbf{U}_k^{\mathrm{IA}})^{\mathrm{H}} \mathbf{H}_{kk} \mathbf{V}_k^{\mathrm{IA}}, \forall k$, and decompose $\overline{\mathbf{H}}_{kk}$ as $\overline{\mathbf{H}}_{kk} = \overline{\mathbf{U}}_k \Lambda_k \overline{\mathbf{V}}_k^{\mathrm{H}}$ through
   SVD, where $\Lambda_k = \mathrm{diag}(\sqrt{\lambda_{k1}}, \sqrt{\lambda_{k2}}, \ldots, \sqrt{\lambda_{kd}})$.
3: Let $\mathbf{U}_k = \mathbf{U}_k^{\mathrm{IA}} \overline{\mathbf{U}}_k$ and $\mathbf{V}'_k = \mathbf{V}_k^{\mathrm{IA}} \overline{\mathbf{V}}_k, \forall k$.
4: Calculate $\hat{p}_{kl} = \frac{(\sigma_n^2 + \sigma_w^2) \|\mathbf{u}_{kl}\|_2^2 \gamma_{kl}}{\lambda_{kl}}, \forall(k, l)$.
5: Obtain α^* and $\rho_k^*, \forall k$ according to (64).
6: Set $\mathbf{v}_{kl} = \sqrt{\alpha^* \hat{p}_{kl}} \mathbf{v}_{kl}^{\mathrm{IA}}, \forall(k, l)$.

The computational complexity of Algorithm 5 is determined by the IA trans- ceiver design process in Step 1. For the famous closed-from IA algorithm, the complexity is $\mathcal{O}(n^3)$. When the max-SINR or MIL algorithm [30] is applied, the complexity is about $\mathcal{O}(n_{iter} \max(n_p, n_r)^3)$. No matter which method is adopted, the complexity of Algorithm 5 is much lower than that of Algorithm 3 and Algorithm 4.

3.7 Simulation results and analysis

Simulations are done over the wireless system as described in Section 3.1, by setting the number of users $K = 3$ or $K = 4$. The entries of $\mathbf{H}_{kj}$ are assumed to be i.i. d. zero mean complex Gaussian random variables with variance $r_{jk}^{-\beta}$, where r_{kj} is the distance in meters between the jth transmitter and the kth receiver, and β is the path loss factor. The parameters of the symmetric network are set to be $n_t = n_r, r_{kl} = r = 5, \beta = 2.7, \sigma_n^2 = -70$ dBm, $\sigma_w^2 = -50$ dBm, $\xi_{kl} = \xi = 0.8, \gamma_{kl} = \gamma$ and $\psi k = \psi$. For the three-user network, the closed-form MIMO linear IA algorithm given in [28] is adopted to design IA transceivers. For the four-user network, the MIL algorithm [30] is used. The simulation results are obtained by taking the average of the simulation results of all 100 channel realizations.

Figure 8 shows the convergence performance of SDP-JTDPS with $\gamma = \{0, 20\}$ dB and $\psi = \{0, 10\}$ dBm for one channel realization of the $(4, 4, 2)^3$ network. It can be observed that the algorithm converges monotonically, which verifies the convergence analysis.

The empirical cumulative distribution function (CDF) of the output per-stream SINR for the $(4, 4, 2)^3$ network is shown in **Figure 9**. The SINR target γ is 20 dB, and the EH target is 0 and 30 dBm, respectively. The results show that the achieved SINR values exceed the given 20 dB, meaning that the proposed schemes can satisfy the SINR constraints. The difference is that the achieved SINR value can be greater than the target SINR value for SDP-JTDPS and O-PAPS, while for CF-PAPS, the achieved SINR values are always equal to the SINR target, which implies that the EH constraints are satisfied with equality in CF-PAPS.

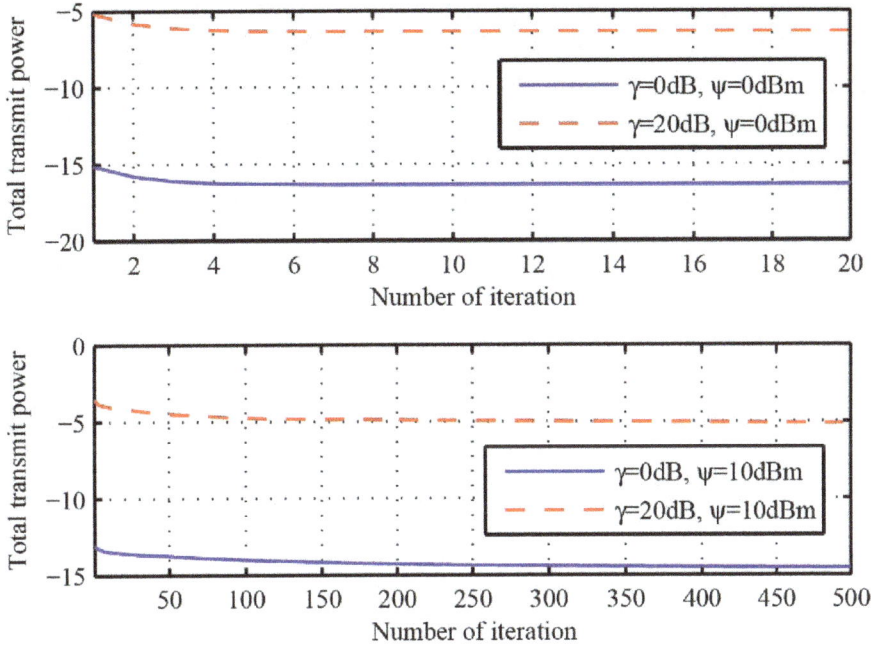

Figure 8. *Convergence property of the semidefinite programming (SDP)-joint transceiver design and power splitting (JTDPS) scheme for the $(4, 4, 2)^3$ network.*

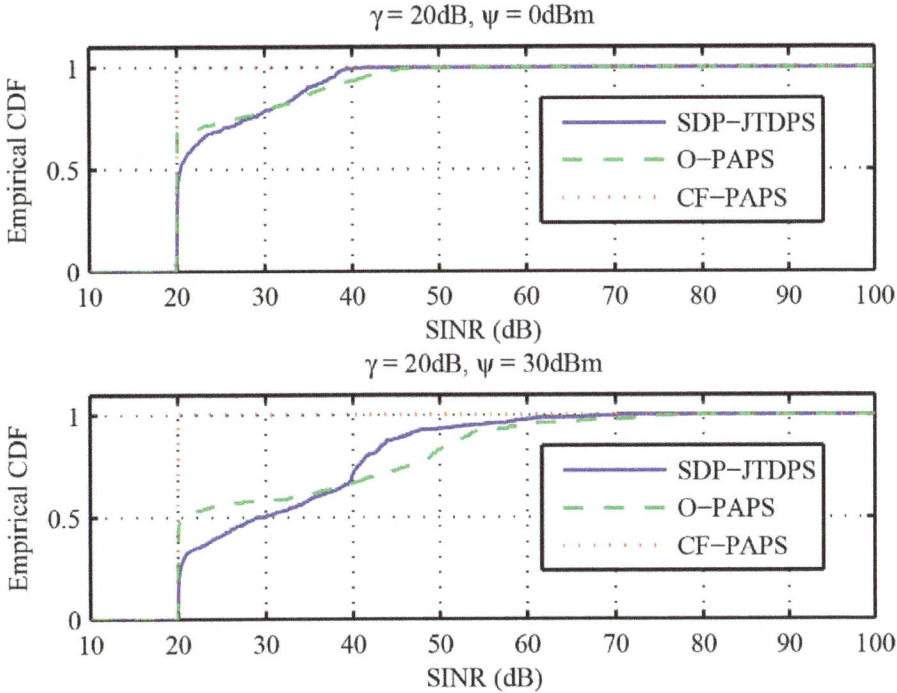

Figure 9.Empirical distribution of achieved SINR at ID receivers with different SINR and energy harvesting (EH) thresholds for the $(4, 4, 2)^3$.

Figure 10 shows the total transmit power versus SINR thresholds at different EH thresholds. It is observed that the transmit power will increase along with the increasing of the EH threshold from -10 to 30 dBm when the SINR threshold is fixed. This is because more transmit power is needed to support higher EH requirements. When the SINR threshold is low, the SDP-JTDPS performs the best, and the O-PAPS schemes achieve almost the same performance, and both of them outperform the CF-PAPS scheme. However, when the SINR threshold is high, the SDP-JTDPS scheme performs worse than the O-PAPS scheme and even worse than the CF-PAPS scheme. From the derivation of the algorithms, we expect that the SDP-JTDPS achieves the best performance, but it does not at high SINR regime. The reason is that when SINR is high, the convergence becomes slow, but we set the fixed iteration number in our simulations. Moreover, the difference between the CF-PAPS and O-PAPS will tend to zero as the SINR threshold becomes high. This implies that the performance of the CF-PAPS scheme is asymptotically the same as that of the O-PAPS scheme. The reason can be explained. Specifically, high SINR means high transmit power, and it is well known that the margin reward of the power allocation will tend to zero when the transmit power becomes high.

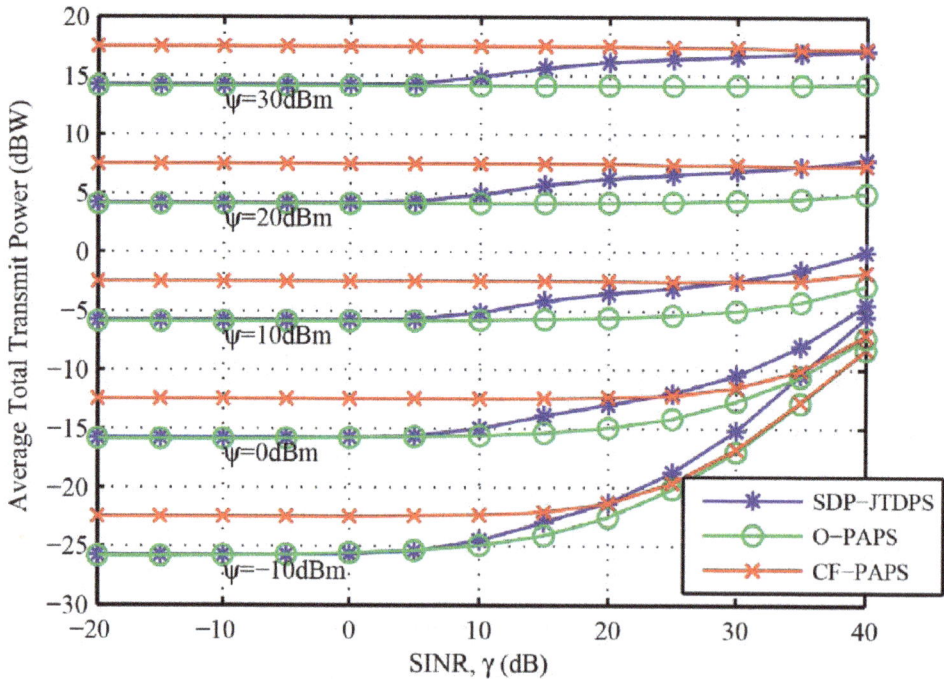

Figure 10. *Average total transmit power versus SINR thresholds for the* $(4, 4, 2)^3$.

Figure 11 shows the relationships between the average transmit power and EH thresholds given different SINR targets. It is seen that the average transmit powers asymptotically tend to be the same as the EH threshold increases for any given SINR value for both O-PAPS and CF-PAPS. For any of the three schemes, higher EH requirement means higher transmit power needed. It is also shown that SDP-JTDPS and O-PAPS achieve the same performance when the SINR threshold is relatively low (e.g., $\gamma = -20$ dB). But when the SINR threshold becomes high, SDP-JTDPS performs inferior to the other two schemes significantly at a low EH threshold. The reason is that the convergence speed of SDP-JTDPS tends to slow at that regime. In addition, the performance curves of SDP-JTDPS and CF-PAPS tend to almost the same when an extremely high EH threshold and a high SINR threshold (e.g., $EH = 30$ dBm and $\gamma = 40$ dB) are given.

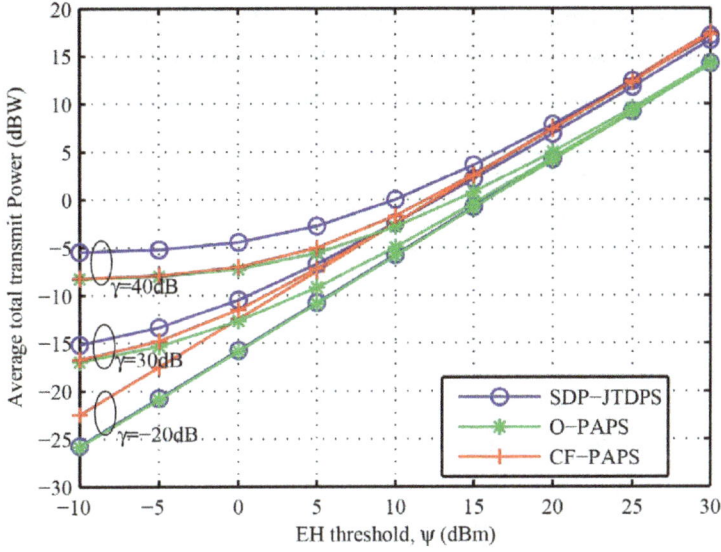

Figure 11. *Average total transmit power versus EH thresholds for the* $(4, 4, 2)^3$ *network.*

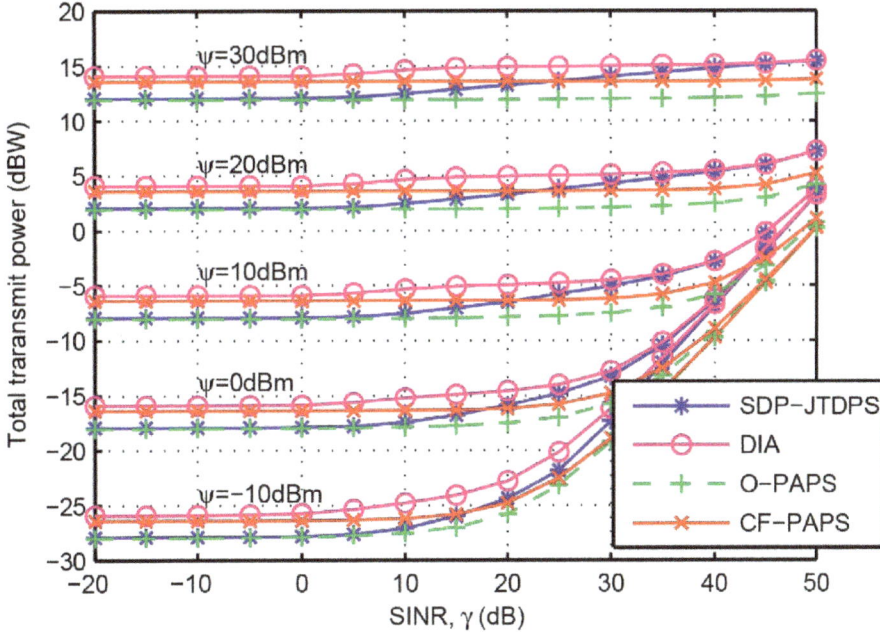

Figure 12. *Total transmit power versus SINR thresholds for the* $(6, 6, 2)^4$
network over one channel realization.

Figure 12 compares SDP-JTDPS with DIA proposed [34] in a $(6, 6, 2)^4$ network over different EH thresholds at fixed SINR values. Note that restricted by its design mechanism, only one data stream is transmitted in the DIA algorithm in the simulation. It can be seen that SDP-JTDPS performs the best, and O-PAPS performs almost the same for low SINR threshold. But when the SINR threshold becomes high, SDP-JTDPS performs worse. This phenomenon is again because SDP-JTDPS has a slower convergence speed when the SINR is high, while the maximum iteration number is set to be 10 in the simulation for saving calculation time. The DIA scheme consumes more transmit power at any given SINR and EH threshold. This is because only one beamforming vector is utilized at each transmitter in the DIA scheme. Multiplexing gain of SDP-JTDPS helps achieve better performance in transmit power than DIA.

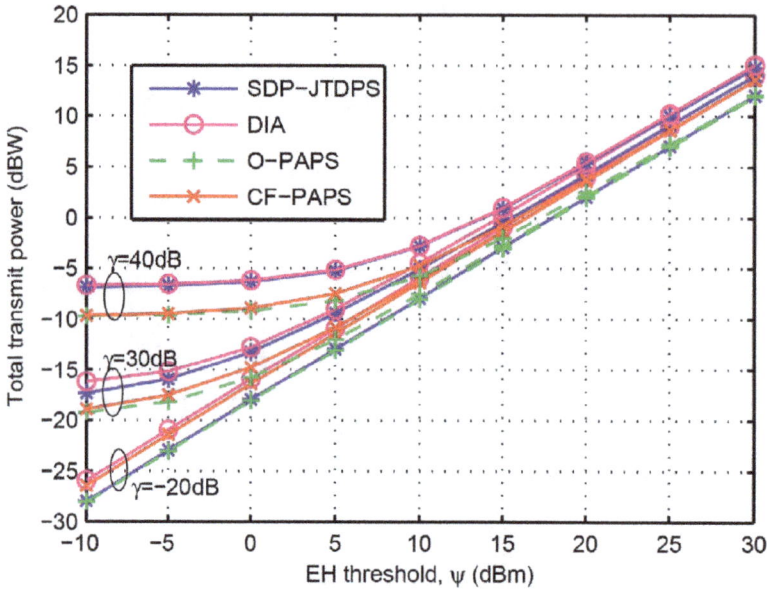

Figure 13. *Total transmit power versus EH thresholds for the* $(6, 6, 2)^4$ *network over one channel realizations.*

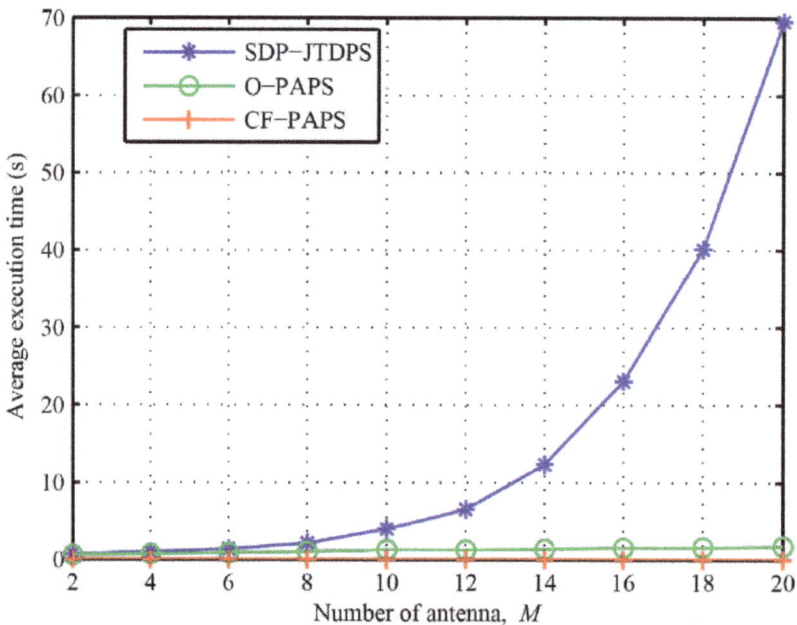

Figure 14. *Average execution time versus M at* $\gamma = 10$ *dB and* $\psi = 10$ *dBm.*

The performance of the proposed schemes is further tested in a $(6, 6, 2)^4$ network. The average transmit powers versus the EH thresholds over different SINR requirements are shown in **Figure 13**. Similar to **Figure 11**, the curves of O-PAPS and CF-PAPS asymptotically tend to be the same for a high EH threshold over all the given SINR thresholds. The curves of DIA also asymptotically tend to be the same over all the given SINR thresholds. The phenomenon reflects that CF-PAPS achieves similar performance with O-PAPS. This is somewhat like the water-filling for the power allocation in traditional MIMO systems where the average power allocation is near optimal at high SNR regimes [36]. Nevertheless, the curves of SDP-JTPDS do not tend to be the same over the observed EH scope. This is because again the convergence speed of SDP-JT-PDS will slow down when SINR becomes high. Considering its high computational complexity, the iteration number is set to be 10 in all the simulations.

Finally, **Figure 14** compares the computational complexity of the proposed schemes for different antenna numbers by assuming there are $K = 3$ users. It can be observed that the computational complexity of SDP-JTDPS increases nonlinearly with M, while those of O-PAPS and CF-PAPS increase linearly. Therefore, CF-PAPS and O-PAPS are of much lower complexity than the SDP-based scheme and thus are more attractive for practical applications.

Conclusions

The joint transceiver design and power splitting optimization for the simultaneous wireless information and power transfer of the MIMO BC network and IC network are analyzed in this chapter. For the MIMO BC network, a transmit power minimization problem subject to both the EH and MSE constraints is formulated.

While for the MIMO IC network, similar transmit power minimization problem is formulated but with the SINR QoS requirements for the ID receivers. Sufficient condition to guarantee the feasibility of nonconvex problems is derived, which reveal that the feasibility of the design problems is not dependent on the PS factors and the EH constraints. Based on the SDP relaxation, alternative solving algorithms are introduced by iteratively optimizing the transmitter together with the PS factors and the receiver. To avoid the high computational complexity of SDP-based schemes, low-complexity algorithms are developed and analyzed. Simulation results have shown the effectiveness of the proposed designs in achieving simultaneous wireless information and power transfer.

Acknowledgements

This work is supported in part by the National Natural Science Foundation of China (NSFC) under grants 61701269 and 61671278, the National Science Fund of China for Excellent Young Scholars under grant 61622111, the Natural Science Foundation of Shandong Province under grant ZR2017BF012, and the Joint Research Foundation for Young Scholars in the Qilu University of Technology (Shandong Academy of Sciences) under grant 2017BSHZ005.

Author details

Anming Dong[1,2,3] and Haixia Zhang[1]*

1 Shandong Provincial Key Laboratory of Wireless Communication Technologies and School of Control Science of Engineering, Shandong University, Jinan, China

2 School of Computer Science and Technology, Qilu University of Technology (Shandong Academy of Sciences), Jinan, China

3 Computer Science Center (National Supercomputer Center in Jinan), Shandong Provincial Key Laboratory of Computer Networks, Qilu University of Technology (Shandong Academy of Sciences), Jinan, China

*Address all correspondence to: haixia.zhang@sdu.edu.cn

References

[1] Bi S, Ho C, Zhang R. Wireless powered communication: Opportunities and challenges. IEEE Communications Magazine. 2015;53(4):117-125

[2] Xiao L, Wang P, Niyato D, In Kim D, Han Z. Wireless networks with RF energy harvesting: A contemporary survey. IEEE Communications Surveys and Tutorials. 2014;17(2):757-789

[3] Yang H, Clerckx B. Waveform design for wireless power transfer with limited feedback. IEEE Transactions on Wireless Communications. 2017;17(1): 415-429

[4] Rezaei O, Naghsh MM, Rezaei Z, Zhang R. Throughput optimization for wireless powered interference channels. IEEE Transactions on Wireless Communications. 2019;18(5):2464-2476

[5] Huang K, Zhou X. Cutting the last wires for mobile communications by microwave power transfer. IEEE Communications Magazine. 2015;53(6): 86-93

[6] Ho CK, Zhang R. Optimal energy allocation for wireless communications with energy harvesting constraints. IEEE Transactions on Signal Processing. 2012;60(9):4808-4818

[7] Gong J, Zhou S, Niu Z. Optimal power allocation for energy harvesting and power grid coexisting wireless communication systems. IEEE Transactions on Communications. 2013; 61(7):3040-3049

[8] Kang X, Chia Y-K, Ho CK, Sun S. Cost minimization for fading channels with energy harvesting and conventional energy. IEEE Transactions on Wireless Communications. 2014; 13(8):4586-4598

[9] Zhou S, Chen T, Chen W, Niu Z. Outage minimization for a fading wireless link with energy harvesting transmitter and receiver. IEEE Journal on Selected Areas in Communications. 2015;33(3):496-511

[10] Luo Z-Q, Ma W-K, So AM-C, Ye Y, Zhang S. Semidefinite relaxation of quadratic optimization problems. IEEE Signal Processing Magazine. 2010;27(3): 20-34

[11] Zhang R, Ho CK. MIMO broadcasting for simultaneous wireless information and power transfer. IEEE Transactions on Wireless Communications. 2013;12(5):1989-2001

[12] Zhang H, Dong A, Jin S, Yuan D. Joint transceiver and power splitting optimization for multiuser MIMO SWIPT under MSE QoS constraints. IEEE Transactions on Vehicular Technology. 2017;66(8):7123-7135

[13] Dong A, Zhang H, Wu D, Yuan D. QoS-constrained transceiver design and power splitting for downlink multiuser MIMO SWIPT systems. In: IEEE International Conference on Communications (ICC). 2016. pp. 1-6

[14] Dong A, Zhang H, Shu M, Yuan D. Simultaneous wireless information and power transfer for MIMO interference channel networks based on interference alignment. Entropy. 2017; 19(9):484

[15] Shi Q, Liu L, Xu W, Zhang R. Joint transmit beamforming and receive power splitting for MISO SWIPT systems. IEEE Transactions on Wireless Communications. 2014;13(6):3269-3280

[16] Shi Q, Xu W, Chang T-H, Wang Y, Song E. Joint beamforming and power splitting for MISO interference channel with SWIPT: An SOCP relaxation and decentralized algorithm. IEEE Transactions on Signal Processing. 2014; 62(23):6194-6208

17] Shi S, Schubert M, Boche H. Downlink MMSE transceiver optimization for multiuser MIMO systems: MMSE balancing. IEEE Transactions on Signal Processing. 2008;56(8):3702-3712

[18] Mezghani A, Joham M, Hunger R, Utschick W. Transceiver design for multi-user MIMO systems. Proceedings 10th International ITG/IEEE Workshop on Smart Antennas (WSA). 2006. pp. 1-8

[19] Vucic N, Boche H, Shi S. Robust transceiver optimization in downlink multiuser MIMO systems. IEEE Transactions on Signal Processing. 2009;57(9):3576-3587

[20] Beck A. Quadratic matrix programming. SIAM Journal on Optimization. 2007;17(4):1224-1238

[21] Ding Y, Ge D, Wolkowicz H. On equivalence of semidefinite relaxations for quadratic matrix programming. Mathematics of Operations Research. 2011;36(1):88-104

[22] Huang Y, Palomar DP. Randomized algorithms for optimal solutions of double-sided QCQP with applications in signal processing. IEEE Transactions on Signal Processing. 2014;62(5): 1093-1108

[23] Zhu Y, Zhou J, Shen X, Song E, Luo Y. Networked Multisensor Decision and Estimation Fusion: Based on Advanced Mathematical Methods. Boca Raton, Florida: CRC Press; 2012

[24] Wang R, Tao M, Huang Y. Linear precoding designs for amplify-and- forward multiuser two-way relay systems. IEEE Transactions on Wireless Communications. 2012;11(12): 4457-4469

[25] Vandenberghe L, Boyd S. Semidefinite programming. SIAM Review. 1996;38(1):49-95

[26] Michael Grant and Stephen Boyd. 2014CVX: Matlab Software for Disciplined Convex Programming, Version 2.1. Available from: http://cvxr. com/cvx

[27] Boyd S, Vandenberghe L. *Convex Optimization*. Cambridge: Cambridge University Press; 2004

[28] Cadambe VR, Jafar SA. Interference alignment and degrees of freedom of the K-user interference channel. IEEE Transactions on Information Theory. 2008;54(8):3425-3441

[29] Gomadam KS, Cadambe VR, Jafar SA. Reflections on interference alignment and the degrees of freedom of the K user interference channel. IEEE Information Theory Society Newsletter. 2009;59(4):5-9

[30] Gomadam KS, Cadambe VR, Jafar SA. A distributed numerical approach to interference alignment and applications to wireless interference networks. IEEE Transactions on Information Theory. 2011;57(6): 3309-3322

[31] Yetis CM, Gou T, Jafar SA, Kayran AH. On feasibility of interference alignment in MIMO interference networks. IEEE Transactions on Signal Processing. 2010; 58(9):4771-4782

[32] Shen H, Li B, Tao M, Wang X. MSE- based transceiver designs for the MIMO interference channel. IEEE Transactions on Wireless Communications. 2010; 9(11):3480-3489

[33] Chiu E, Lau VKN, Huang H, Tao W, Liu S. Robust transceiver design for K- pairs quasi-static mimo interference channels via semi-definite relaxation. IEEE Transactions on Wireless Communications. 2010;9(12):3762-3769

[34] Zong Z, Feng H, Richard Yu F, Zhao N, Yang T, Hu B. Optimal transceiver design for SWIPT in k-user MIMO interference channels. IEEE Transactions on Wireless Communications. 2016;15(1):430-445

[35] Wiesel A, Eldar YC, Shamai S. Linear precoding via conic optimization for fixed mimo receivers. IEEE Transactions on Signal Processing. January 2006;54(1):161-176

[36] Tse D, Viswanath P. Fundamentals of Wireless Communication. Cambridge: Cambridge University Press; 2005

4

Rectenna Systems for RF Energy Harvesting and Wireless Power Transfer

Mohamed Aboualalaa and Hala Elsadek

Abstract

With the rapid development of the wireless systems and demands of low-power integrated electronic circuits, various research trends have tended to study the feasibility of powering these circuits by harvesting free energy from ambient electromagnetic space or by using dedicated RF source. Wireless power transmission (WPT) technology was first pursued by Tesla over a century ago. However, it faced several challenges for deployment in real applications. Recently, energy harvesting and WPT technologies have received much attention as a clean and renewable power source. Rectenna (rectifying antenna) system can be used for remotely charging batteries in several sensor networks at internet of things (IoT) applications as commonly used in smart buildings, implanted medical devices and automotive applications. Rectenna, which is used to convert from RF energy to usable DC electrical energy, is mainly a combination between a receiving antenna and a rectifier circuit. This chapter will present several designs for single and multiband rectennas with different characteristics for energy harvesting applications. Single and multiband antennas as well as rectifier circuits with matching networks are introduced for complete successful rectenna circuit models. At the end of the chapter, a dual-band rectenna example is introduced with a detailed description for each section of the rectenna.

Keywords: dedicated RF source, directive radiation pattern, high gain antenna, rectenna (rectifying antenna), RF energy harvesting, wireless power transmission (WPT)

Introduction

Wireless power transmission (WPT) can be categorized into three different categories as depicted in **Figure 1**: near-field inductive or resonant coupling, far-field directive powering, and far-field ambient wireless energy harvesting. For the first category, it usually takes place between two coils, one is the primary and the other is the secondary. The main goal is to transfer the power from the primary coil to the secondary coil for several of centimeter as a separation distance between them [1–3]. Many defected ground structure (DGS)-based designs are proposed for this type of the wireless power transfer [4–6] to give a high efficiency coupled system.

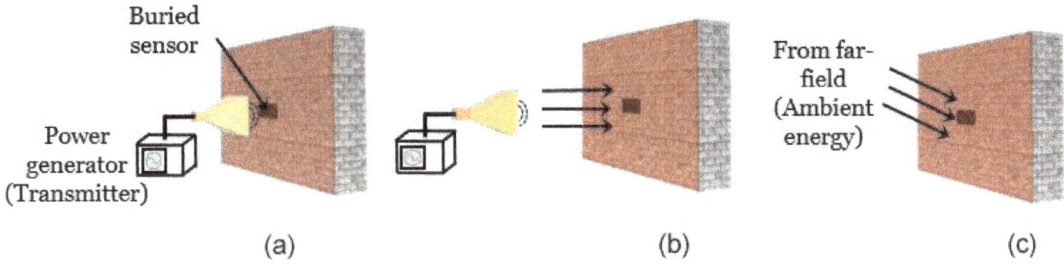

Figure 1. *WPT categories (a) near-field inductive or resonant coupling, (b) far-field directive powering and (c) far-field ambient wireless energy harvesting.*

The second category of WPT is far-field directive powering that is used with directive power transmission which means the transmission occurs in the far-field zone but with well-defined direction of the source. This sort of WPT is useful for solar power satellites (SPS) applications [7–9] or with intentional powering such as using a dedicating source with well-known direction to power a network of wireless sensors, each sensor has built-in rectenna which is used as a renewable power source to power the connected sensor. The third type is far-field energy harvesting. The receiver does not know the direction of the received power. So, one of the main goals in this type is how to increase the probability of reception by designing antennas with wide beam-width and multiple or wideband resonance frequencies.

Near-field WPT offers a solution to short range powering for electronic devices, it becomes widely commercialized for several wireless applications [10–12]. Near-field transmission can also be useful with wireless implantable devices [13–15]. Nevertheless, near-field WPT suffers from severe issue with regard to the transmitting distance, it covers only very short range distances (few centimeters); therefore this limits its applications. On the other hand, the powering scheme of far-field dedicated source or free ambient powering technique can overcome this problem because of the long-distance charging capability. Several studies are introduced in wireless energy harvesting [16–25]. Although, the great focusing on the wireless energy harvesting, there are many obstacles in the way of free source energy harvesting. One of the main issues is that low input power levels of the ambient energy. Consequently, there are many research papers introduced for rectennas at low input power levels. However, single band rectennas have a simple structures, many research studies [26–31] have investigated the multi-band rectennas as a trial for increasing the scavenging received power with the same rectenna device; various single and multi-band rectennas are presented.

Also, there are big challenges with respect to working the rectenna with fixed conversion efficiency values over a wide range of the received signal. Thus, Section 2 introduces a literature survey about single and multiband frequency operation of different rectennas; also, various rectennas' designs working at low input power and over wide input power range are discussed. Finally, in Section 3, dual-band rectenna using voltage doubler rectifier and four-section matching network is discussed as an example for a dual-band operation to illustrate the different stages of the whole rectenna system elaborately. The dual-band antenna, firstly, is designed, fabricated and measured separately to check the antenna performance. Then, the rectifier and the matching network between the antenna and the rectifying circuit are also designed and tested independently. After that the integration between the antenna and rectifier is done on the same PCB substrate.

Literature review on rectennas

Low input received power rectennas

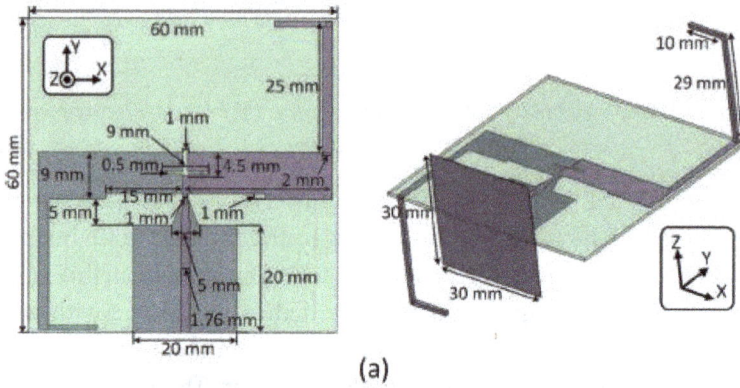

(a)

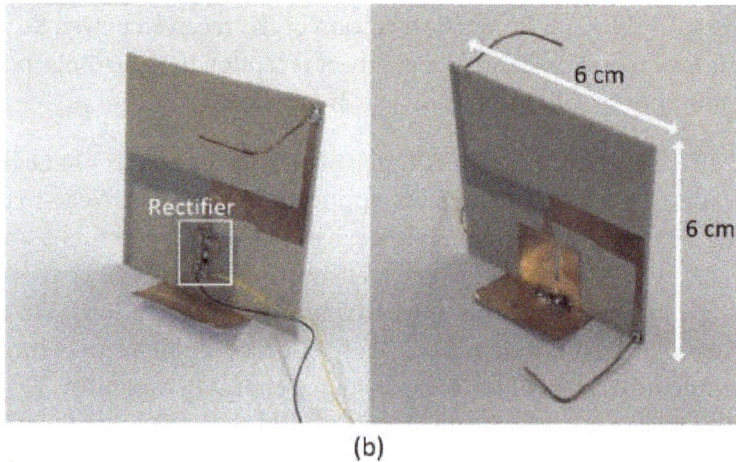

(b)

Figure 2. *Rectenna design: (a) design of the top and side view and (b) fabricated rectenna prototype [32].*

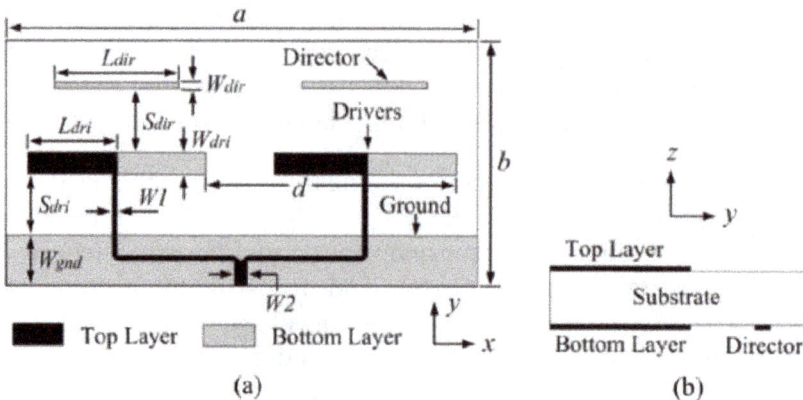

(a)   (b)

Figure 3. *Layout of the quasi-Yagi subarray. (a) Top view. (b) Side view [33].*

In [32], a compact dual-band rectenna is proposed as depicted in **Figure 2**. The rectenna has a conversion efficiency of 37 and 30% at 915 MHz and at 2.45 GHz, respectively, at input power of -9 dBm with resistive load of 2.2 kΩ. A dual-band rectenna using Yagi antenna for low input power applications shown in **Figure 3** is introduced in [33]. The rectenna offers an acceptable values for the conversion efficiencies, it reaches up to 34% at 1.84 GHz and 30% at 2.14 GHz for input power level of -20 dBm. A combination between the solar energy and RF energy harvesting is discussed in [34].

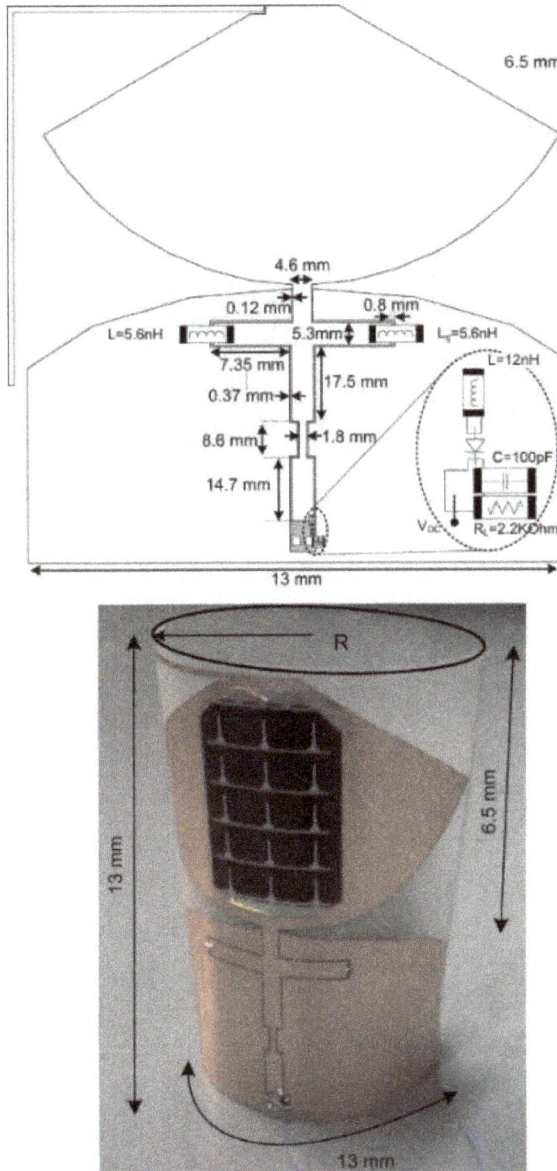

Figure 4. *Hybrid solar/EM rectenna [34].*

This solar rectenna, displayed in **Figure 4**, achieves RF-DC conversion efficiency of 15% with input power of -20 dBm at 850 MHz and 2.45 GHz. In [35], a 130 nm CMOS rectifier is proposed for ultra-low input power. **Figure 5** shows the rectenna structure. It consists of 10 stages to give the maximum efficiency of 42.8% at -16 dBm input power and output DC voltage of 2.32 V at resistive load of 0.5 MΩ. A compact co-planar waveguide-fed rectenna using single stage Cockcroft Walton rectifier and L-shaped impedance matching network, shown in **Figure 6**, is presented in [16]. The RF-DC conversion efficiency is 68% with a received input signal power of 5 dBm at 2.45 GHz. This rectenna also gives conversion efficiencies around 48 and 19% at -10 and -20 dBm, respectively.

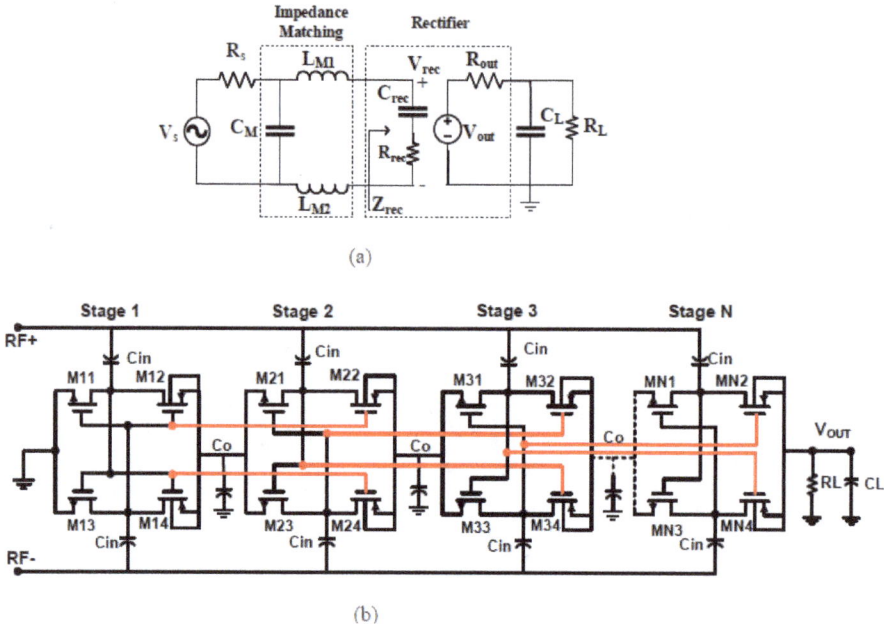

(a)

(b)

Figure 5. *(a) Proposed RF rectenna equivalent circuit, (b) self-compensated rectifier [35].*

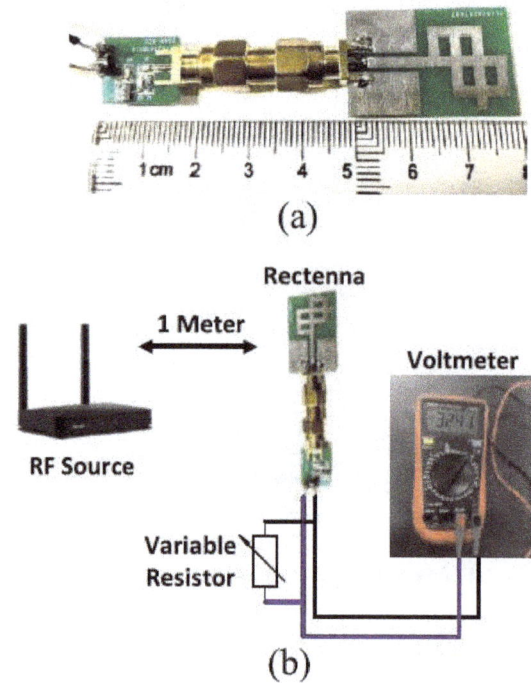

(a)

(b)

Figure 6. *(a) Complete prototype of the rectenna, (b) measurement set-up for rectenna system [16].*

2.2 Single and multi-band rectennas

The simplest way in energy harvesting is to harvest from single frequency band; this in turn makes the design of matching circuit, which is used for maximum power transmission between the receiving antenna part and the rectifying circuit, is a little bit easier. In [36], a pentagonal antenna is used with series connection single diode to produce a single band rectenna at 5 GHz. The rectenna has maximum conversion efficiency of 46% at resistive load of 2 kΩ. In [37], a 3 x 2 rectangular patch array with a gain of 10.3 dBi is used with three-stage Dickson charge pump circuit for energy harvesting. The rectenna works at 915 MHz. **Figure 7** shows the antenna array as well as the rectifying circuit. The maximum rectifier efficiency is 41% at input power of 10 dBm. A semicircular slot antenna was presented for X- band planar rectenna (at 9.3 GHz) as depicted in **Figure 8** [38]. The rectenna gives RF-to-DC conversion efficiency of about 21% at an input power density of 245 μW/cm². 35 GHz rectenna using 4 x 4 patch antenna array, displayed in **Figure 9**, is proposed in [39]. The maximum RF-to-dc conversion efficiency is 67% with input RF received power of 7 mW.

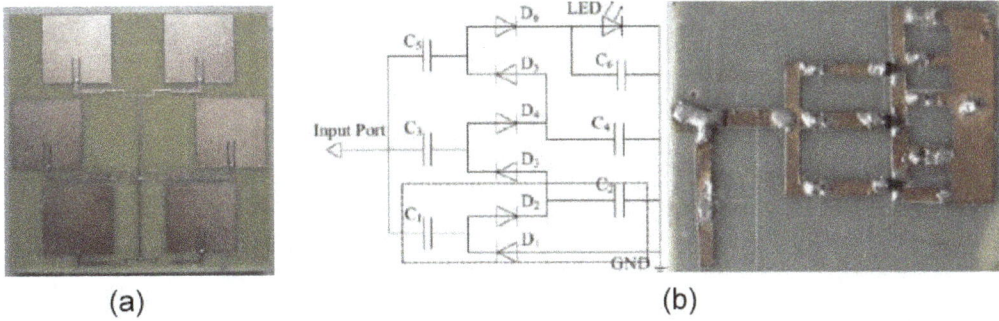

(a)   (b)

Figure 7. *Six elements antenna array (a) fabricated patch antenna array, (b) fabricated rectifier [37].*

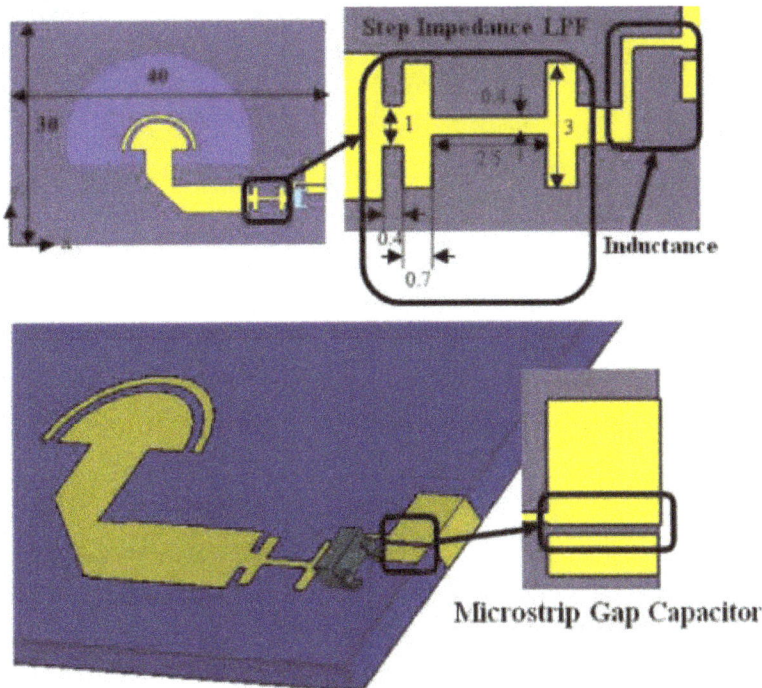

Figure 8. *Geometry of the X-band rectenna [38].*

Due to the variety in transmission bands for different wireless systems, there is a large ambient wasted energy at different frequencies. Consequently, the demand for harvesting from different bands increases. In [40], triple-band implanted rectenna is discussed. It works at 402 MHz, 433 MHz and 2.45 GHz with antenna has a stacked and spiral structure. **Figure 10** shows the antenna structure in addition to the rectifier design. It gives a conversion efficiency of 86% at input power of 11 dBm with 5 kΩ load resistor. A compact reconfigurable rectifying antenna has been presented in [41] for dual-band rectification at 5.2 and 5.8 GHz. The measured maximum conversion efficiencies of the proposed rectenna are 65.2 and 64.8% at 4.9 and 5.9 GHz, respectively, with 15 dBm input power. The rectenna fabricated prototype is shown in **Figure 11**. A dual frequency band rectenna has been developed in [42]. A planar inverted F-antenna is used with a voltage doubler circuit to configure a dual band rectenna.

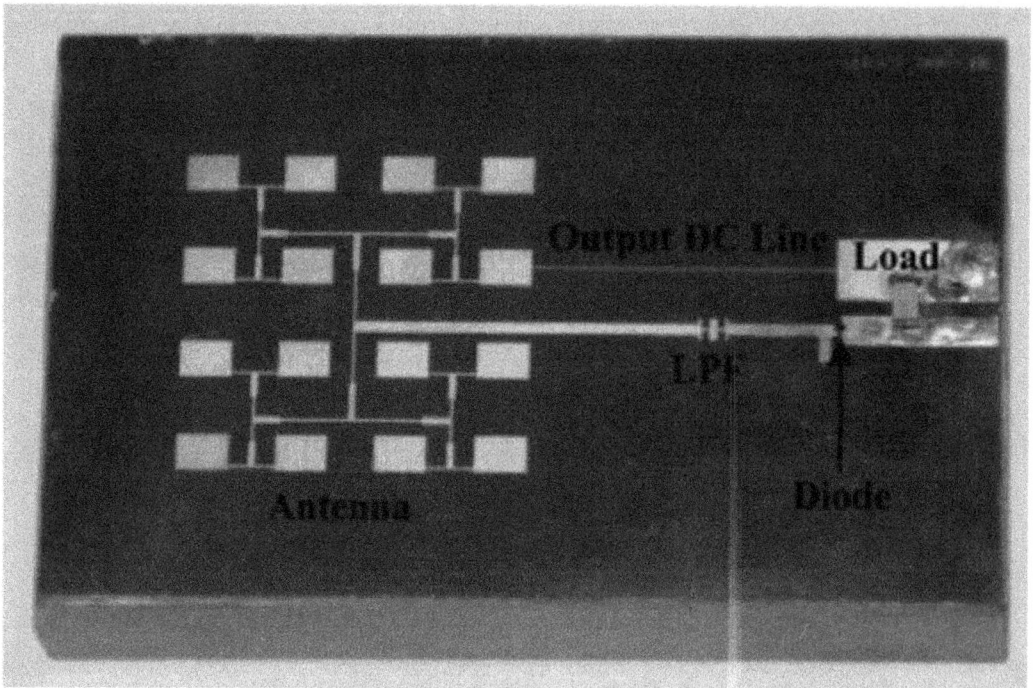

Figure 9. *Fabricated rectenna [39]*.

With increasing the number of frequencies at which rectenna can harvest, the complexity of the matching circuit and the size of the rectenna increase. Therefore, dual-band is the best choice in the designing of rectenna systems because it combines between the simplicity and the scavenging from more than one frequency band.

2.3 Wide input received power rectennas

There are several studies that are proposed to guarantee stable fixed RF-DC conversion efficiency over a wide band of the input power. In [43], dual-band rectifier with extended input power range is proposed. The rectifier schematic circuit and the fabricated design is displayed in **Figure 12**. The rectifier offers above 30% conversion efficiency with input power range from -15 to 20 dBm and the maximum value is 60% from 5 to 15 dBm. Impedance compression network (ICN) techniques is discussed in [44] to fix RF-dc conversion efficiency over a wide band of input power by maintaining the value of input impedance for the rectifier fixed regardless the value of the input power. **Figure 13** shows the rectifier configuration. The rectifier has a maximum conversion efficiency of 56% at 31.8 dBm and the input power range for efficiency over than 50% is 6.7 dBm.

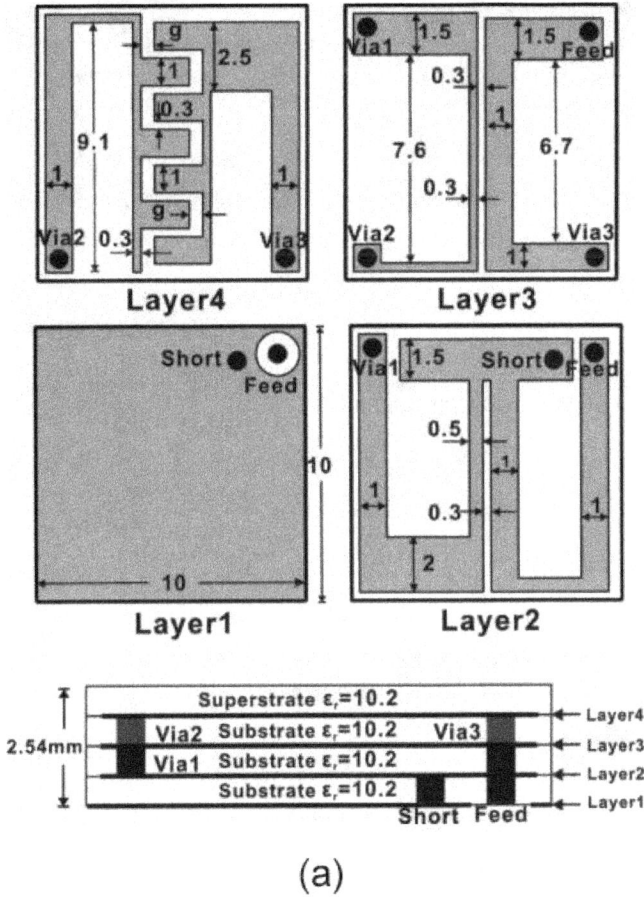

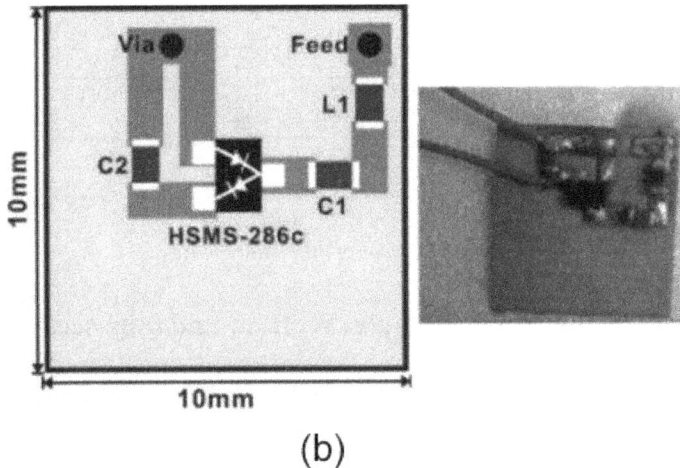

Figure 10. (a) Triple band antenna, (b) rectifier design and a photo of a fabricated rectifier [40].

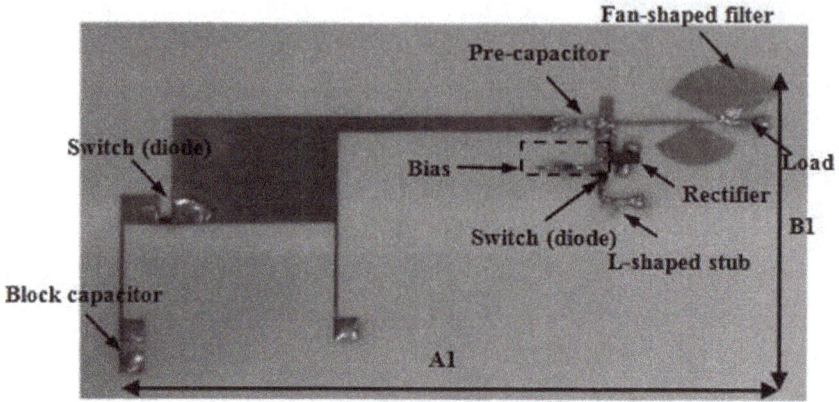

Figure 11. *Fabricated reconfigurable rectenna [41].*

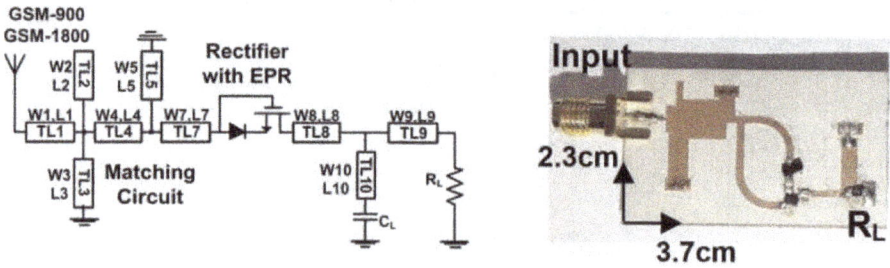

Figure 12. *Schematic diagram and fabricated circuit [43].*

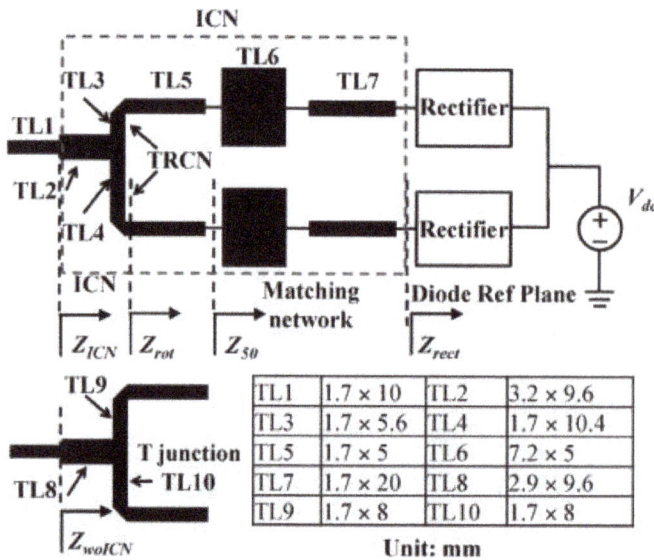

TL1	1.7×10	TL2	3.2×9.6
TL3	1.7×5.6	TL4	1.7×10.4
TL5	1.7×5	TL6	7.2×5
TL7	1.7×20	TL8	2.9×9.6
TL9	1.7×8	TL10	1.7×8

Unit: mm

Figure 13. *Layout of the rectifier [44].*

Dual-band Rectenna using voltage doubler rectifier and four-section matching network

This section introduces a dual-band rectenna with maximum measured conversion efficiency of 63 and 69% at $f_1 = 1.95$ and $f_2 = 2.5$ GHz, respectively, over wide band of the input power, 14 and 15.5 dBm for conversion efficiency above 50% at f_1 and f_2, respectively. The section arrangements are as follows: in Section 3.1, the antenna design is introduced. Then, the equivalent circuit of the antenna is discussed in Section 3.2. Antenna results (reflection coefficient as well as radiation characteristics) is discussed in Section 3.3. The rectifier-antenna matching network for the dual band is described in Section 3.4. The rectifier structure with the geometrical parameters is illustrated in Section 3.5. The rectenna experiment setup is revealed in Section 3.6. While, the rectenna performance including RF-DC conversion efficiency in addition to the DC output voltage at the two frequency bands is discussed in Section 3.7.

Antenna design

In this section, the enhanced-gain antenna design [45] is introduced to be used to configure the rectenna system. **Figure 14** shows the layout of the proposed antenna. As shown in the figure, the antenna includes two substrate layers (sub- strates 1 and 2). The two layers have the same substrate material with relative dielectric constant $\varepsilon_{r1} = \varepsilon_{r2} = 3:55$, thickness $h_1 = h_2 = 0.813$ mm and a loss tangent of 0.0027. The antenna design consists of disc antenna printed on the top layer of substrate 1. The resonance frequency of this disc is inversely proportional to the disc radius as shown in Eq. (1) [46] which can be determined from Eq. (2) [46].

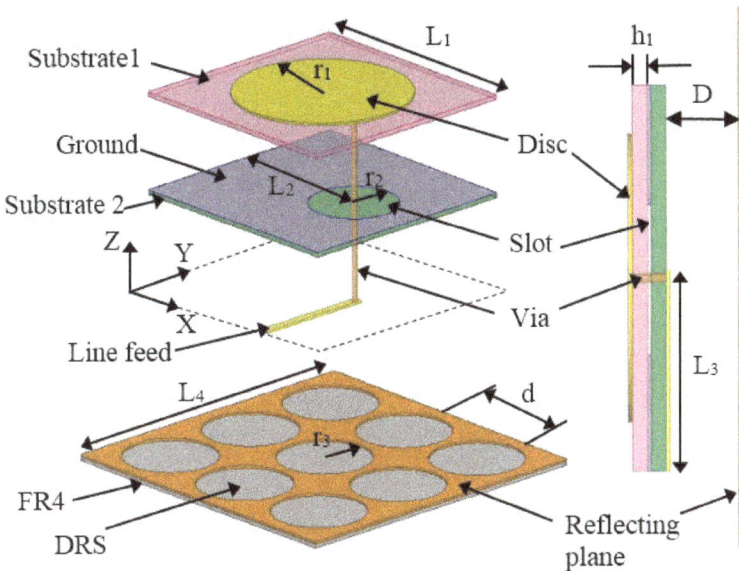

Figure 14. *3D geometry, perspective view and side view of the proposed disc antenna [45].*

This disc is directly fed by a microstrip line with 50Ω through a via with radius of 0.6 mm. Also, this disc feeds (by coupling) a circular slot on the ground plane between the feed line and the radiating patch, a square reflecting plane with defected reflector structure (DRS) placed behind the antenna at a distance of $\lambda_o/8$ to improve the antenna gain as well as enhance the front to back ratio. The reflector is built on 0.8 mm thick FR4 substrate with dielectric constant of 4.4 and a loss tangent of 0.02. The reflector is supported by a 15 mm thick layer of lightweight foam with a low dielectric constant ($\varepsilon_r = 1.06$). Substrate dimensions are 50 mm × 50 mm. The designated antenna resembles good candidate for RF energy harvesting from mobile radio waves ($f_1 = 1.95$ GHz) and from WLAN wireless communication systems ($f_2 = 2.45$ GHz).

$$f_r = \frac{1.8412v_0}{2\pi a_e \sqrt{\varepsilon_r}} \tag{1}$$

$$a_e = a\left[1 + \frac{2h}{\pi a \varepsilon_r}\left(\ln\left(\frac{\pi a}{2h}\right) + 1.7726\right)\right]^{0.5} \tag{2}$$

Equivalent circuit of the proposed antenna (modeling of patch antenna)

The first challenge of designing the equivalent circuit was to find an accurate model of the proposed antenna at f_1 and f_2. **Figure 15** shows the equivalent circuit used to model the electrical behavior of the antenna in response to an incoming RF input signal. It is useful to implement this model using basic components R1, L1, and C1, which represent the influence of the first resonant frequency (f_1), whereas R2, L2, and C2 represent the second resonant frequency (f_2). Elements L3 and C3 are included in the equivalent circuit model to represent the electrical length of the feed line and slot coupling, respectively. The resistance R1 and R2 correspond to radiating losses.

Each radiator (the disc and the slot) is represented by a resonator. Each resonator consists of parallel RLC circuit, the resonance frequency of each one can be determined from Eq. (3):

$$f_r = \frac{1}{2\pi\sqrt{LC}} \tag{3}$$

Firstly, each resonator is studied separately. S-parameters are calculated from Agilent ADS simulator. Then the resonant and cutoff frequencies (f_0 and f_c in GHz, respectively) are determined. The initial values of L and C for each one can be calculated from Eqs. (4) and (5) [47, 48].

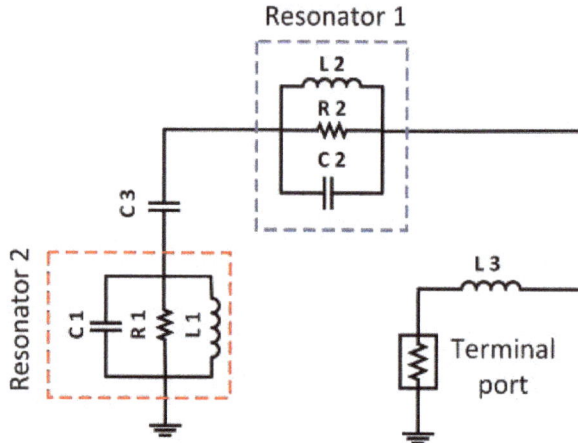

Figure 15. *Equivalent lumped-elements circuit for antenna in ADS.*

$$C_p = \frac{5f_c}{\pi\left[f_0^2 - f_c^2\right]} \, pF \tag{4}$$

$$L_p = \frac{250}{C_p\left[\pi f_0\right]^2} \, nH \tag{5}$$

where C_p is the capacitance in picofarads and L_p is the inductance in nanohenrys. **Table 1** summarizes the initial values of R and L at the two operating frequencies.

The losses resistance can be determined from the quality factor (Q)-frequency bandwidth relationship (BW) as:

Table 1. *Initial values of L and C for the two resonators.*

f_0 and f_c	$C \rightarrow (C_p = \frac{5f_c}{\pi[f_0{}^2 - f_c{}^2]} \, pF)$	$L \rightarrow (L_p = \frac{250}{C_p[\pi f_0]^2} \, nH)$
Resonator 1 (f_0 = 2.45 GHz & f_c = 2.25 GHz)	At f_c = 2.25 GHz and f_0 = 2.45 GHz, then: Cp = 3.8 pF	At f_0 = 2.45 GHz and Cp = 3.8 pF, then: Lp = 1.11 nH
Resonator 2 (f_0 = 1.95 GHz & f_c = 1.65 GHz)	At f_c = 1.65 GHz and f_0 = 1.95 GHz, then: Cp = 2.4 pF	At f_0 = 1.95 GHz and Cp = 2.4 pF, then: Lp = 2.8 nH

Table 2. *Elements values of the equivalent circuit model for the dual band antenna.*

Parameter	R1(Ω)	L1(nH)	C1(pF)	R2(Ω)	L2(nH)	C2(pF)	L3(nH)	C3(pF)
Value	750	10	1.7	500	1.39	4.15	10	0.2

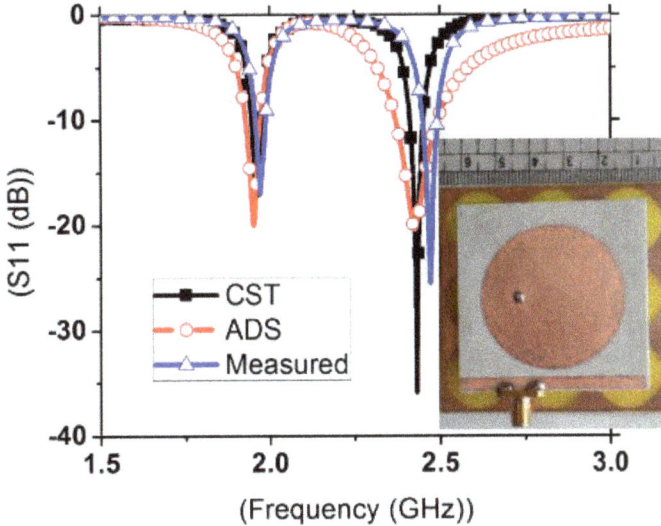

Figure 16. *Reflection coefficient of the proposed antenna.*

where the half power frequency bandwidth is evaluated from Eq. (7).

$$BW = \frac{1}{RC} \tag{7}$$

Then, the loss resistance (R) for each resonator can be determined as:

$$R = \frac{1}{BW \times C} \tag{8}$$

After combining the two resonators, taking into account the effect of losses resistances (R1 and R2) in addition to making optimization, the final equivalent circuit can be obtained. The corresponding values of the equivalent circuit elements are depicted in **Table 2**.

(a)

(b)

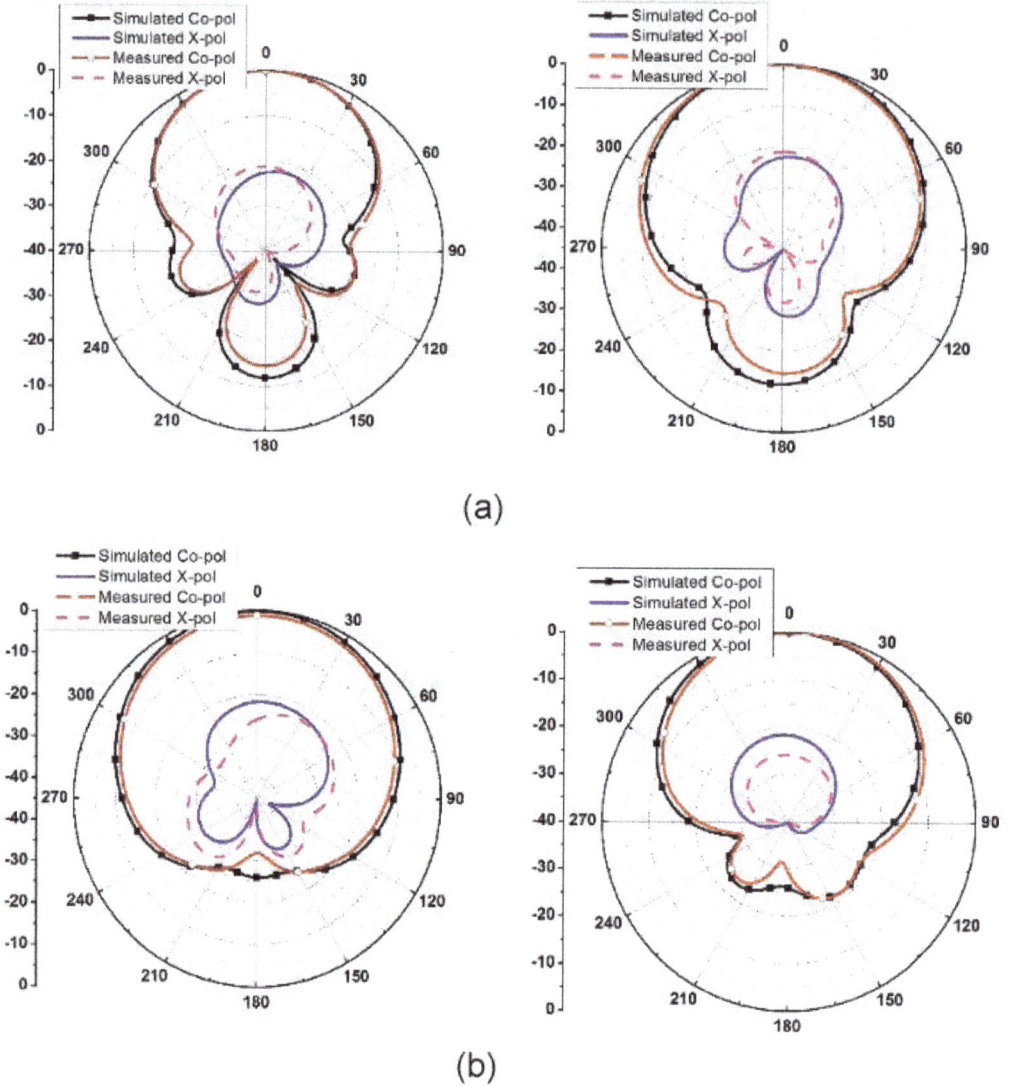

Figure 17. *2D measured and simulated results of radiation pattern for the antenna: (a) at 1.95 GHz and (b) at 2.45 GHz.*

Antenna measurement results

Designed antenna S-parameters

Figure 16 shows the reflection coefficient response of the antenna obtained from CST simulation compared with the calculated response of the equivalent circuit model by using Agilent ADS software in addition to the measured reflection coefficient. Good agreement was between the results of simulated, measured and ADS model. The antenna resonates at two bands 1.95 GHz (f_1) and 2.45 GHz (f_2). The circular patch is designed to radiate at 2.45 GHz by the direct feed with the transmission line placed behind substrate 2. Whereas, 1.95 GHz resonance frequency is designed to radiate due to the capacitive coupling between the circular patch on the top of substrate 1 and the circular slot located on the ground plane, where in 1.95 GHz case the disc antenna is considered as a feeder for the circular slot. The performance of the proposed antenna was simulated and optimized by commercial EM software CST Microwave Studio. A prototype of the proposed antenna was fabricated and tested. The reflection coefficient of the antenna was measured by R&S ZVA 67 Network Analyzer. It is noted that the simulated and measured results of the input impedances of the antenna are in good agreement. Only, a small shift in the measured S-parameters was observed due to the connector soldering, fabrication tolerance, the adhesive between the two layers of the antenna and the layers alignment in fabrication process.

Radiation characteristics

The simulated and measured results of E-plane and H-plane for the high gain antenna at f_1 and f_2 are shown in **Figure 17**(a) and (b), respectively. The measured values of gain, radiation efficiency, F/B ratio, cross polarization level, 3 dB angular beamwidth at the first resonance frequency (f_1) are 8.3 dBi, 90%, 12, –22.3 dB, 73.5°, respectively. While at the second resonance frequency, these values can be summarized as 7.8 dBi, 91.6%, 26, –21.6 dB, 79.5°, respectively. The gain and radiation pattern of the antenna were measured by the Anechoic Chamber shown in **Figure 18**. There is a good agreement between the simulated and measured results of the radiation characteristics.

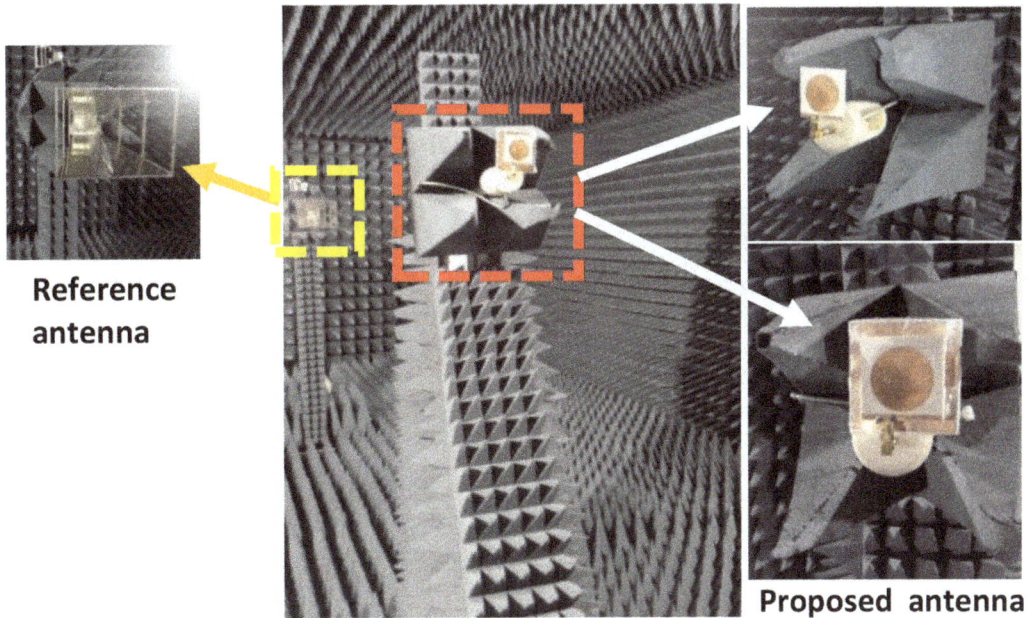

Reference antenna

Proposed antenna

Figure 18. *Antenna radiation pattern measurement set-up.*

Rectifier-antenna matching

In this design, a scheme used in [49] is employed to achieve a dual-band impedance transformation at the two frequency bands (f_1 and f_2). This scheme is used to match between a complex and frequency-dependent rectifier input impedance (Z_{Rec}) and a real impedance of the antenna (Z_{Ant}) by using four different sections (Section 1–4) as shown in **Figure 19**. The matching technique can be summarized in the following steps.

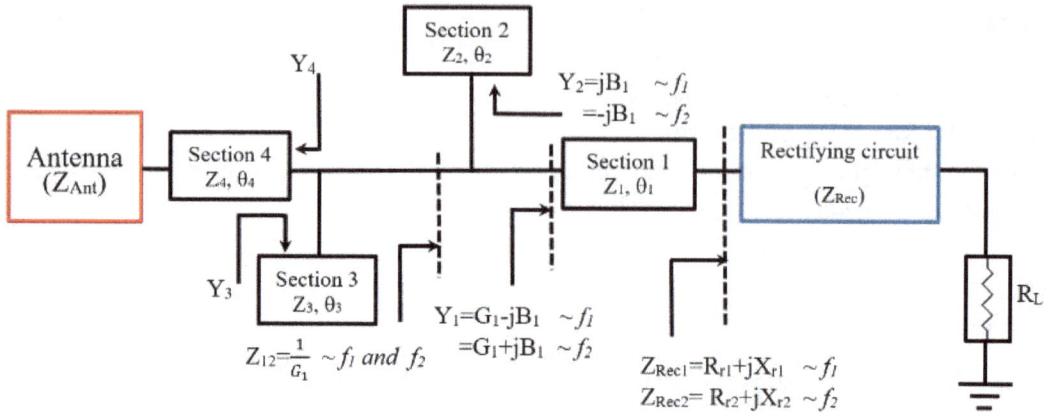

Figure 19. *Dual-band matching circuit.*

Step 1: Achieve the conjugate matching between the load values at both resonant frequencies, that is, moving the two impedance values of the load (rectifier) on the Smith chart to be located on the same real circle with the imaginary parts are equal on both sides of the Smith chart as shown in **Figure 20**.

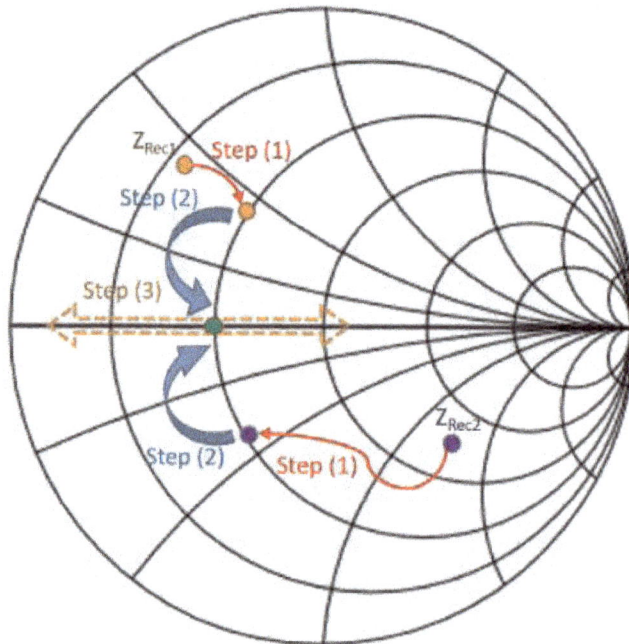

Figure 20. *Matching steps indicated on smith chart.*

Step 2: Cancel the imaginary part of the impedances at f_1 and f_2.

Step 3: Real to real impedance transformation.

Each section is characterized by two values Z and θ, where Z is the section characteristic impedance and θ is the section electrical length. The function of the first section (Section 1) is to make the real value of the rectifier input impedance at f_1 is equal to that of the rectifier input impedance at f_2 and the imaginary parts are also equal but with opposite signs (one is positive (inductive reactance) and the other one is negative (capacitive reactance)); Section 1 parameters (Z_1 and θ_1) can be calculated from [50] as:

$$Z_1 = \sqrt{\left(R_{r1}R_{r2} + X_{r1}X_{r2} + \frac{X_{r1} + X_{r2}}{R_{r2} - R_{r1}} (R_{r1}X_{r2} - R_{r2}X_{r1}) \right)} \tag{9}$$

$$\theta_1 = \frac{n\pi + arctan \frac{Z_1(R_{r1} - R_{r2})}{R_{r1}X_{r2} - R_{r2}X_{r1}}}{(m+1)} \tag{10}$$

where n is an arbitrary integer and m = f2/f1. Section 2 is used to cancel the imaginary parts of the admittance Y_1 at the two frequencies f_1 and f_2. Section 2 parameters can be determined as [49]:

$$Z_2 = \frac{\tan \theta_2}{B_1} \tag{11}$$

$$\theta_2 = \frac{(1+p)\pi}{(1+m)} \tag{12}$$

where p is an integer. Sections 3 and 4 are used for real to real impedance transformation, and their parameters can be calculated from Eqs. (13)–(17) [49, 51]

$$Z_4 = \sqrt{\left(\frac{Z_{Ant}(1 + \tan^2\theta_4)}{G_1} - Z_{Ant}^2 \right) \frac{1}{\tan \theta_4}} \tag{13}$$

$$\theta_4 = \frac{(1+s)\pi}{(1+m)} \tag{14}$$

$$Z_3 = \frac{\tan \theta_3}{B_4} \tag{15}$$

$$\theta_3 = \frac{(1+q)\pi}{(1+m)} \tag{16}$$

$$B_4 = \frac{(Z_{Ant}^2 - Z_4^2) \tan \theta_4}{Z_{Ant}^2 \times Z_4 + Z_4^3 \times \tan^2\theta_4} \tag{17}$$

where q and s are integers.

Rectifier design

Several rectifiers' topologies are used for energy harvesting, for instance, diodes in parallel connection, diodes in series connection, voltage doubler circuits, multistage voltage multiplier and so forth. Voltage multipliers generate high voltages from a low voltage power source. However, in this design a half-wave voltage doubler circuit, which is a special case of voltage multipliers, is used for the rectification to get high voltage with conservation of the design simplicity. The rectifier design as well as the matching network is depicted in **Figure 21** [52]. The voltage doubler circuit comprises two Avago HSMS2850 Schottky diodes and two SMD capacitors ($C_s = C_p$ = 100 pF). The Schottky diode has a built-in voltage (V_b) of 0.150 V and a breakdown voltage (V_{br}) of 3.8 V.

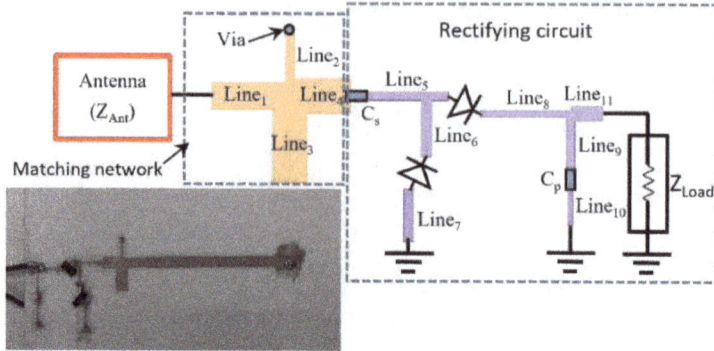

Figure 21. Rectifier layout; $L_1 = 4.5$ mm, $W_1 = 1.54$ mm, $L_2 = 4$ mm, $W_2 = 0.33$ mm, $L_3 = 5$ mm, $W_3 = 1.96$ mm, $L_4 = 3.52$ mm, $W_4 = 2$ mm, $L_5 = 4$ mm, $W_5 = 0.36$ mm, $L_6 = 4.98$ mm, $W_6 = 0.31$ mm, $L_7 = 5$ mm, $W_7 = 0.29$ mm, $L_8 = 5.1$ mm, $W_8 = 0.43$ mm, $L_9 = 4.8$ mm, $W_9 = 0.3$ mm, $L_{10} = 4.88$ mm, $W_{10} = 0.33$ mm, $L_{11} = 1.81$ mm, $W_{11} = 0.84$ mm.

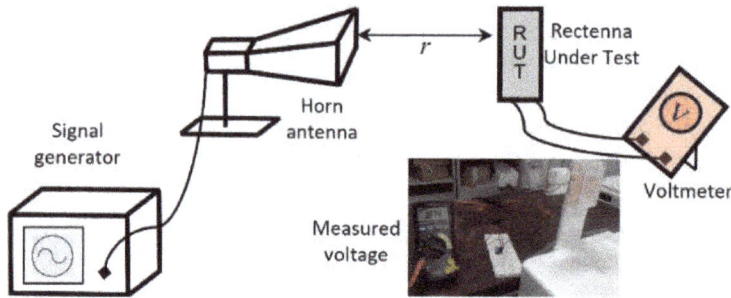

Figure 22. Rectenna measurement setup.

Due to the small values of the series resistance and the barrier capacitance (R_s of 25Ω and C_b of 0.18 pF) for the above-mentioned diodes category, so these diodes have a high cutoff frequency and high conversion efficiency. The capacitor C_s is used to store the energy in one half cycle to double the charging voltage for C_p at the other half cycle, C_s also acts as bandpass filter to block the DC voltage generated from the nonlinear diodes. C_p has two functions, it is used for bypassing the higher order modes, generated from the nonlinear diode to the ground and getting a smooth DC output voltage as well. Also the shunt connection between C_p and the load impedance R_L acts as a low pass filter.

The rectifier is designed on Rogers Duroid RO3003 with a relative permittivity (ε_r) of 3, substrate thickness (h) of 0.76 mm, a dielectric loss tangent (tanδ) of 0.0013 and a copper thickness (t) of 0.017 mm. The simulated complex rectifier input impedance (Z_{Rec}) at two frequencies are (8 -j x 28.2) and (26.1 -j x 39.7) at f_1 and f_2, respectively, with a resistive load of 1.5 kΩ. The circuit parameters have been optimized to achieve the maximum conversion efficiency at the two frequency bands for the received input power levels. The parameters of the rectifier as well as the matching circuit are illustrated in **Figure 21**, where each line is defined by the length (L) and width (W). Also, the prototype of the fabricated rectifier is shown in **Figure 21**.

Rectenna measurements

The rectifying circuit including the matching network is simulated using Keysight advanced design system (ADS), while the antenna was designed using ANSYS high-frequency structure simulator

(HFSS). The enhanced-gain antenna described in [45] is used as a receiving antenna in the proposed rectenna to increase the rectifier sensitivity. Hence, increasing rectenna capability to harvest from low input power levels. The receiving antenna and the rectifier are integrated on the same substrate, fabricated and measured in the measurement setup shown in **Figure 22**. An Agilent Technologies E8257D Analog signal generator is used to send a microwave signal which is connected to a horn antenna with 9 dBi gain at the two frequencies. On the other hand, the rectenna under test (RUT) is connected with a voltmeter to measure the DC output voltage. To take the antenna radiation characteristics into account, the antenna effective area is considered. Hence, the RF-DC conversion efficiency of the proposed rectenna (η) is calculated as the following:

$$\eta = \frac{V_{DC}^2}{P_{in} \times R_L} \times 100 \tag{18}$$

where V_{DC} is the measured DC output voltage, Pin is the received RF input power and R_L is the resistive load. P_{in} is defined in Eq. (19)

$$P_{in} = P_D \times A_{eff} \tag{19}$$

where P_D is the RF power density and A_{eff} is the antenna effective area. P_D and A_{eff} are calculated using Eqs. (20) and (21), respectively.

$$P_D = \frac{P_t G_t}{4\pi r^2} \tag{20}$$

$$A_{eff} = G_r \frac{\lambda^2}{4\pi} \tag{21}$$

P_t is the transmitting power, G_t is the horn antenna gain and r is the distance between the transmitter and the rectenna all are known. Therefore, the RF-to-DC conversion efficiency can be measured. For far-field measurements, r is chosen of 40 cm. **Figure 23** shows the photo of the rectenna measurement setup.

Rectenna results and discussion

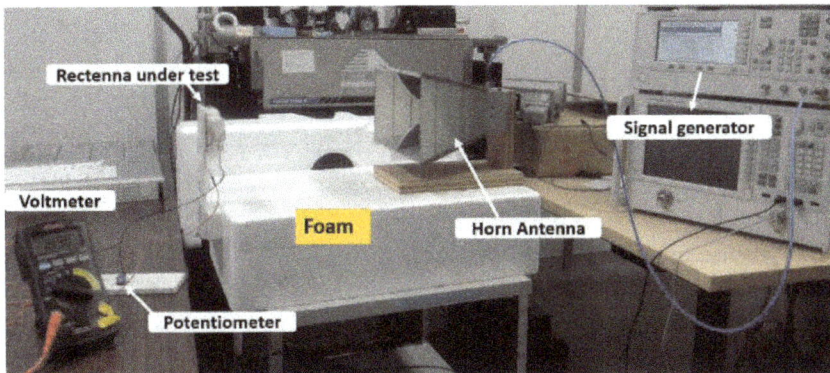

Figure 23. *Photo of the measurement setup.*

The entire system (antenna, matching circuit and rectifier) is tested over different input power levels with different resistive load values at two frequencies (f_1 and f_2); **Figure 24**(a) and

(b) show the comparison between the measured and simulated results of RF-to-DC conversion efficiency and the DC output voltage versus the input power at f_1 and f_2, respectively. The maximum measured conversion efficiency is 63% with input power range of 14 dBm (from -3.5 to 10.5 dBm) at f_1, while the measured efficiency at f_2 is 69% with input power from -4.5 to 11 dBm (15.5 dBm). There is a slight shift between the simulation and measurement results, where the maximum simulated RF-DC conversion efficiency are 66 and 73% at the same two frequencies, respectively. Due to the limitations in the experiments, the received input power is limited only up to 11 dBm.

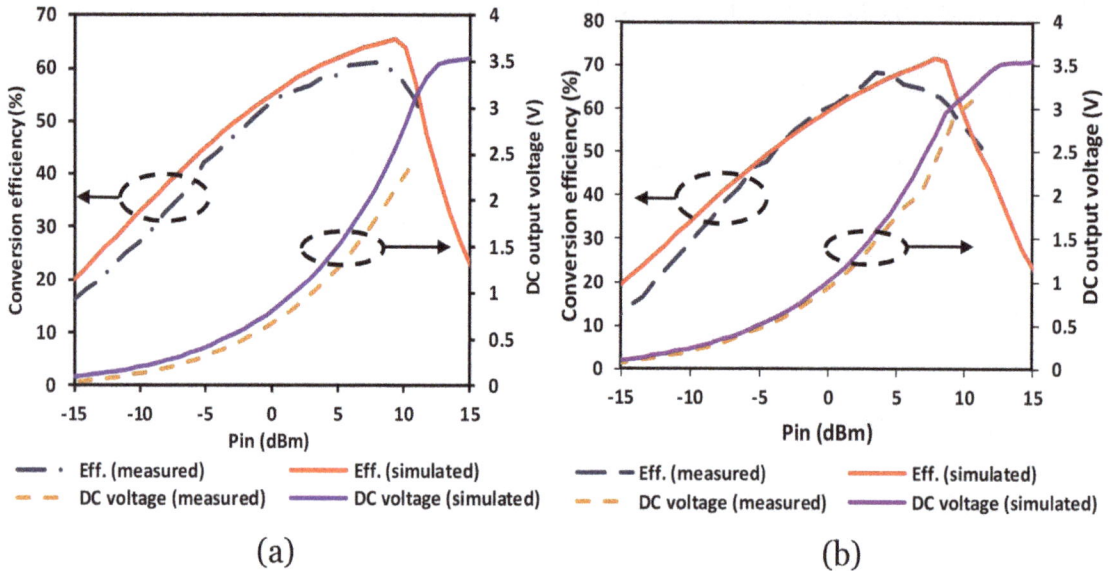

Figure 24. Simulated and measured conversion efficiency in addition to the DC output voltage versus input power (a) at f_1 (b) at f_2.

Conclusions

This chapter presents a study of rectenna systems for RF energy harvesting and wireless power transfer. A survey about employing rectennas in WPT, low input received power rectennas, single and multi-band rectennas, wide input received power rectennas are introduced. Finally, dual-band rectenna using voltage doubler rectifier and four-section matching network is discussed. The first part of the rectenna design is the dual-band disc antenna with enhanced gain in order to collect a highest amount of RF energy. It radiates at 1.95 and 2.45 GHz. The measured results showed the gain of 8.3 and 7.8 dBi at 1.95 and 2.45 GHz, respectively. The disc antenna is integrated with a dual-band rectifier with four-section matching network to introduce a dual-frequency rectenna with higher conversion efficiency over wide band of the input power for multiband RF energy harvesting. The rectenna gives maximum RF-DC measured conversion efficiency of 63% and 69% at 1.95 GHz and 2.5 GHz, respectively. it also operates over a wide range of the input power; it covers the range of 14 and 15.5 dBm at f_1 and f_2, respectively for a conversion efficiency higher than 50% with load resistance (R_L) = 1K. The rectenna is simulated, fabricated and measured. The simulated and measured results show good agreement.

Author details

Mohamed Aboualalaa* and Hala Elsadek Electronics

Research Institute, Cairo, Egypt

*Address all correspondence to: mohamed.ali@ejust.edu.eg

References

[1] Kim D, Park J, Park HH, Ahn S. Generation of magnetic propulsion force and torque for microrobot using wireless power transfer coil. EEE Transactions on Magnetics. 2015;51(11): 1-4

[2] Ahn D, Kim S-M, Kim S-W, Moon J, Cho I-K. Wireless power transfer receiver with adjustable coil output voltage for multiple receivers application. IEEE Transactions on Industrial Electronics. 2018;66(5): 4003-4012

[3] Kim T, Yun G, Lee WY, Yook J. Asymmetric coil structures for highly efficient wireless power transfer systems. IEEE Transactions on Microwave Theory and Techniques. 2018;66(7):3443-3451

[4] Hekal S, Abdel-Rahman AB, Jia H, Allam A, Pokharel RK, Kanaya H. Strong resonant coupling for short- range wireless power transfer applications using defected ground structures. In: IEEE Wireless Power Transfer Conference (WPTC), Colorado, USA. 2015

[5] Hekal S, Abdel-Rahman AB, Jia H, Allam A, Barakat A, Pokharel RK. A novel technique for compact size wireless power transfer applications using defected ground structures. IEEE Transactions on Microwave Theory and Techniques. 2017;65(2):591-599

[6] Tahar F, Barakat A, Saad R, Yoshitomi K, Pokharel RK. Dual-band defected ground structures wireless power transfer system with independent external and inter-resonator coupling. IEEE Transactions on Circuits and Systems II: Express Briefs. 2017;64(12):1372-1376

[7] Sasaki S, Tanaka K, Maki K. Microwave power transmission technologies for solar power satellites. Proceedings of the IEEE. 2013;101(6): 1438-1447

[8] McSpadden P, Jaffe J. Energy conversion and transmission modules for space solar power. Proceedings of the IEEE. 2013;101(6):1424-1437

[9] Li X, Duan B, Song L, Zhang Y, Xu W. Study of stepped amplitude distribution taper for microwave power transmission for SSPS. IEEE Transactions on Antennas and Propagation. 2017;65(10):5396-5405

[10] Nguyen VT, Kang SH, Choi JH, Jung CW. Magnetic resonance wireless power transfer using three- coil system with single planar receiver for laptop applications. IEEE Transactions on Consumer Electronics. 2015;61(2): 160-166

[11] Low ZN, Chinga RA, Tseng R, Lin J. Design and test of a high-power high- efficiency loosely coupled planar wireless power transfer system. IEEE Transactions on Industrial Electronics. 2009;56(5):1801-1812

[12] Wireless Charging Pad, Black Sapphire. Samsung Mobile, Samsung Electronics America. [Online]. Available from: http://www.samsung.com/us/ mobile/cell-phones-accessories/eppg 920ibugus

[13] Wang G, Liu W, Sivaprakasam M, Kendir GA. Design and analysis of an adaptive transcutaneous power telemetry for biomedical implants. IEEE Transactions on Circuits and Systems I: Regular Papers. 2005;52(10):2109-2117

[14] Jou AY, Azadegan R, Mohammadi S. High-resistivity CMOS SOI rectenna for implantable applications. IEEE Microwave and Wireless Components Letters. 2017;27(9):854-856

[15] Loubet G, Takacs A, Dragomirescu D. Implementation of a battery-free wireless sensor for cyber-physical systems dedicated to structural health monitoring applications. IEEE Access. 2019;7:24679-24690

[16] Awais Q, Jin Y, Chattha HT, Jamil M, Qiang H, Khawaja BA. A compact rectenna system with high conversion efficiency for wireless energy harvesting. IEEE Access. 2018;6: 35857-35866

[17] Shen S, Chiu C-Y, Murch RD. A dual-port triple-band L-probe microstrip patch rectenna for ambient RF energy harvesting. IEEE Antennas and Wireless Propagation Letters. 2017; 16:3071-3074

[18] Khemar A, Kacha A, Takhedmit H, Abib G. Design and experiments of a dual-band rectenna for ambient RF energy harvesting in urban environments. IET Microwaves, Antennas and Propagation. 2018;12(1): 49-55

[19] Valenta CR, Durgin GD. Harvesting wireless power: Survey of energy- harvester conversion efficiency in far- field, wireless power transfer systems. IEEE Microwave Magazine. 2014;15(4): 108-120

[20] Nie M, Yang X, Tan G, Han B. A compact 2.45-GHz broadband rectenna using grounded coplanar waveguide. IEEE Antennas and Wireless Propagation Letters. 2015;14:986-989

[21] Assimonis SD, Daskalakis S, Bletsas A. Sensitive and efficient RF harvesting supply for batteryless backscatter sensor networks. IEEE Transactions on Microwave Theory and Techniques. 2016;64(4):1327-1338

[22] Haboubi W, Takhedmit H, Luk J- DLS, Adami S-E, Allard B, Costa F, et al. An efficient dual-circularly polarized rectenna for RF energy harvesting in the 2.45 GHz ISM band. Progress In Electromagnetics Research. 2014;148: 31-39

[23] Chou J, Lin D, Weng K, Li H. All polarization receiving rectenna with harmonic rejection property for wireless power transmission. IEEE Transactions on Antennas and Propagation. 2014; 62(10):5242-5249

[24] Collado A, Georgiadis A. Optimal waveforms for efficient wireless power transmission. IEEE Microwave and Wireless Components Letters. 2014; 24(5):354-356

[25] Song C, Huang Y, Carter P, Zhou J, Joseph SD, Li G. Novel compact and broadband frequency-selectable rectennas for a wide input-power and load impedance range. IEEE Transactions on Antennas and Propagation. 2018;66(7):3306-3316

[26] Okba A, Takacs A, Aubert H, Charlot S, Calmon P-F. Multiband rectenna for microwave applications. Comptes Rendus Physique. 2017;18(2): 107-117

[27] Pham BL, Pham A. Triple bands antenna and high efficiency rectifier design for Rf energy harvesting at 900, 1900 and 2400 MHz. In: IEEE MTT-S International Microwave Symposium Digest (MTT), USA. 2013

[28] Liu J, Zhang XY. Compact triple- band rectifier for ambient RF energy harvesting application. IEEE Access. 2018;6:19018-19024

[29] Sarma SS, Chandravanshi S, Akhtar MJ. Triple band differential rectifier for RF energy harvesting applications. In: Asia-Pacific Microwave Conference (APMC), New Delhi. 2016

[30] Lopez-Yela A, Segovia-Vargas D. A triple-band bow-tie rectenna for RF energy harvesting without

matching network. In: IEEE Wireless Power Transfer Conference (WPTC), Taipei. 2017

[31] Chandravanshi S, Sarma SS, Akhtar MJ. Design of triple band differential rectenna for RF energy harvesting. IEEE Transactions on Antennas and Propagation. 2018;66(6): 2716-2726

[32] Niotaki K, Kim S, Jeong S, Collado A, Georgiadis A, Tentzeris MM. A compact dual-band rectenna using slot-loaded dual band folded dipole antenna. IEEE Antennas and Wireless Propagation Letters. 2013;12:1634-1637

[33] Sun H, Guo Y, He M, Zhong Z. A dual-band rectenna using broadband Yagi antenna array for ambient RF power harvesting. IEEE Antennas and Wireless Propagation Letters. 2013;12: 918-921

[34] Collado A, Georgiadis A. Conformal hybrid solar and electromagnetic (EM) energy harvesting rectenna. IEEE Transactions on Circuits and Systems I: Regular Papers. 2013;60(8):2225-2234

[35] Noghabaei SM, Radin RL, Savaria Y, Sawan M. A high-efficiency ultra-low- power CMOS rectifier for rf energy harvesting applications. In: IEEE International Symposium on Circuits and Systems (ISCAS), Florence. 2018

[36] Yadav RK, Das S, Yadava RL. Analysis and modelling of a novel compact rectenna for indoor applications. IET Communications. 2014;8(15):2642-2651

[37] Xie F, Yang G, Geyi W. Optimal design of an antenna array for energy harvesting. IEEE Antennas and Wireless Propagation Letters. 2013;12:155-158

[38] Monti G, Tarricone L, Spartano M. X-band planar rectenna. IEEE Antennas and Wireless Propagation Letters. 2011; 10:1116-1119

[39] Mavaddat A, Armaki SHM, Erfanian AR. Millimeter-wave energy harvesting using 4x4 microstrip patch antenna array. IEEE Antennas and Wireless Propagation Letters. 2015;14: 515-518

[40] Huang F, Lee C, Chang C, Chen L, Yo T, Luo C. Rectenna application of miniaturized implantable antenna design for triple-band biotelemetry communication. IEEE Transactions on Antennas and Propagation. 2011;59(7): 2646-2653

[41] Lu P, Yang XS, Li JL, Wang BZ. A compact frequency reconfigurable rectenna for 5.2 and 5.8 GHz wireless power transmission. IEEE Transactions on Power Electronics. 2015;30(11): 6006-6010

[42] Lu P, Yang X-S, Li J-L, Wang B-Z. A dual-frequency quasi-pifa rectenna with a robust voltage doubler for 2.45 and 5.8 GHz wireless power transmission. Microwave and Optical Technology Letters. 2015;57(2):319-322

[43] Liu Z, Zhong Z, Guo Y. Enhanced dual-band ambient RF energy harvesting with ultra-wide power range. IEEE Microwave and Wireless Components Letters. 2015;25(9): 630-632

[44] Xu J, Ricketts DS. An efficient, watt-level microwave rectifier using an impedance compression network (ICN) with applications in outphasing energy recovery systems. IEEE Microwave and Wireless Components Letters. 2013; 23(10):542-544

[45] Aboualalaa M, Abdel-Rahman AB, Allam A, Elsadek H, Pokharel RK. Design of a dual-band microstrip antenna with enhanced gain for energy harvesting applications. IEEE Antennas and Wireless Propagation Letters. 2017; 16:1622-1626

[46] Balanis CA. Modern Antenna Handbook. New York: John Wiley & Sons; 2011

[47] Ahn D, Park J, Kim C, Kim J, Qian Y, Itoh T. A design of the low-pass filter using the novel microstrip defected ground structure. IEEE Transactions on Microwave Theory and Techniques. 2001;49(1):86-93

[48] Abdel-Rahman AB, Verma AK, Boutejdar A, Omar AS. Control of bandstop response of Hi–Lo microstrip low-pass filter using slot in ground plane. IEEE Transactions on Microwave Theory and Techniques. 2004;52(3): 1008-1013

[49] Maktoomi MA, Gupta R, Hashmi MS. A dual-band impedance transformer for frequency-dependent complex loads incorporating an L-type network. In: Asia-Pacific Microwave Conference (APMC), Nanjing. 2015

[50] Liu X, Liu Y, Li S, Wu F, Wu Y. A three-section dual-band transformer for frequency-dependent complex load impedance. IEEE Microwave and Wireless Components Letters. 2009; 19(10):611-613

[51] Monzon C. Analytical derivation of a two-section impedance transformer for a frequency and its first harmonic. IEEE Microwave and Wireless Components Letters. 2002;12(10): 381-382

[52] Aboualalaa M, Mansour I, Bedair A, Allam A, Zahhad MA, Elsadek H, et al. Dual-band rectenna using voltage doubler rectifier and four-section matching network. In: IEEE Wireless Power Transfer Conference (WPTC), Montreal, QC, Canada. 2018

5

Solar Rectennas: Analysis and Design

Ahmed M.A. Sabaawi and Oras Ahmed Al-Ani

Abstract

There is a growing interest in recent years on developing solar cells and increasing their conversion efficiency. This interest was motivated by the demand on producing clean and inexpensive energy, where the current solar cell technology failed to fulfill the market demand due to its low efficiency obtained. Thus, an efficient alternative is highly required to overcome the drawbacks of current photovoltaic technologies. In this chapter, the concept and operation of solar rectennas will be introduced as an efficient energy-harvesting technology and as a better alternative to conventional solar cells. Nanoantennas are used for receiving solar radiation at both visible and infrared regions as AC electromagnetic signals. The received power is then passed to a nanodiode that acts as a rectifier to convert the power from AC to DC form. Nanoarrays are utilized often to increase the captured energy and decrease the number of rectifiers of the entire system. The biggest challenge is how to design an efficient nanoantenna integrated efficiently into a nanodiode in order to maximize the overall efficiency. State-of-the-art designs for nanoantennas and nanodiodes will be highlighted in this chapter mentioning the figure of merits used to compare between one design and another.

Keywords: energy harvesting, solar rectennas, nanoantennas, nanorectifiers, THz detection

Introduction

The increasing demand on clean and inexpensive energy has led to the emergence of solar cells in the early 1950s, where the main source of the world's power is fossil fuels. The creation of photovoltaics (PV) has opened a new era on exploiting solar radiation for the production of electricity. However, the development of the PV industry is not sufficient to cover the market demand on solar panels due to their low efficiency. Therefore, cheaper and higher-efficiency technologies are required for the solar power market. These requirements have induced the researchers to find an alternative solution by replacing the current solar cells with optical antennas integrated to diodes forming a rectifying antenna (rectenna) using the wave nature of light [1, 2]. Most of the recent researches are focused on developing solar rectennas to convert the visible region of solar spectrum efficiently to electric power and exploiting the unused portion of solar radiation (i.e., infrared region) [3]. The proposed solar rectennas are expected to exhibit higher efficiency (theoretically 100% for monochromatic illumination) than current solar cells [4]. Rather than the low efficiency, solar rectennas overcome the other drawbacks of PVs which include the dependence on the bandgap energy and the narrowband operation (visible region only).

However, several challenges contribute to make the actual conversion efficiency much lower than expected such as the poor coupling between the optical antenna and the diode [5].

Each photon in semiconductor solar cells produces electron hole pair to generate electrical power. However, the device absorbs only those photons that have energy higher than the band gap energy. This limits the conversion efficiency to 44% or even less in real devices. On the other hand, classical rectifiers receive the electromagnetic energy and convert it into DC power with a conversion efficiency reaching 100%. Solar rectennas are designed to operate in a similar way with the expectation to obtain very high efficiencies at a wide range of the electromagnetic spectrum. The field of solar rectennas appears to be promising and attractive due to the fact that high efficiency is theoretically obtainable and the material used is inexpensive and available.

Why solar rectennas?

- Solar rectennas can achieve as high as the efficiency of solar cells or even higher.

- The material of solar rectennas is widely available in the form of thin films, and the fabrication process is inexpensive compared to conventional solar cells.

- Solar rectennas demonstrate versatility over PV devices by exceeding efficiency during the day.

- Other forms of infrared such as waste heat can also be harvested by solar rectennas rather than the solar irradiation.

In contrast, there are several drawbacks and challenges associated with solar rectennas such as [6]:

- When converting visible light, the time constant must be in the range of 0.1 fs, which is hard to achieve using the planar MIM diodes.

- The leakage current of the diode must be as small as 1 µA, which is quite challenging.

- A strong matching between the antenna impedance and the diode's to ensure maximum power transfer and hence higher efficiency.

It is obvious that the technology of solar rectennas is still young in the early stage of research and faces numerous challenges and limitations. Thus, in this chapter, the theoretical understanding is presented highlighting the development of each part of a solar rectenna.

History of rectennas

In the last century, the story of solar rectenna begun when electrical power has been transferred without the use of wires. This technique is called wireless power transmission (WPT). It is worth to mention that all the rectenna systems conceived at that time were working at microwave frequencies with efficiencies exceeding 80% at a single frequency.

A brief historical background on this technique is presented here:

- Early experiments on WPT return to the work of Hertz and Tesla which was implemented by exploiting a giant coil and a 3-ft-diameter copper ball to transport the electromagnetic wave with low frequency from one point in space to another one. Later, the idea of power transmission has been developed by researchers particularly after the significant progress that witnessed in microwave technology [7].

- In 1963, the first rectenna has been invented by Raytheon Co., which was constructed from 28 half-wave dipole antennas. Each one terminated with a bridge rectifier. The overall efficiency of this design was 40%. The rectenna has then been developed by the same company to use as a power source for a microwave-powered helicopter.

- In 1972, Bailey proposed an idea to use the rectennas to generate electricity from solar power. This idea was based on using a pair of pyramids or cones as a modified dipole, which is similar to rod antennas. The pair is connected to a load via a diode (half-wave rectifier) [1].

- In 1984, arrays of crossed dipoles (**Figure 1**) have been proposed by Marks, where an insulating sheet with fast full-wave rectification is used [9].

- In contrast, Bailey proposed a conventional broadside array antenna, in which the output signal is collected after passing in several dipoles. The latter is used to feed a transmission line in which the signals are transferred to a rectifier. Combined signals are used in that approach to add in-phase.

- In 1996, Lin et al. achieved the first experimental work [10] that based on the absorption of light by fabricated metallic resonant nanostructures and rectification at light frequency. The device that used this technique uses dipole antenna array that connected in parallel and constructed on a silicon substrate. The device components also include a p-n diode as a half-wave rectifier.

- In 2003, infrared (IR) rectenna structure-based metal-insulator-metal (MIM) diodes have been designed by Berland [11]. It has been designed using dipoles, operating at 10 μm wavelength. The overall recorded efficiency, however, was very low (<1%) [11].

- In 2010, spiral nanoantenna for solar energy has been designed and fabricated to collect energy at mid-IR region [12]. Kotter et al. demonstrated the progress related to this technique.

- In 2011, a monopole antenna has been designed by Midrio et al., where nickel is used as the main material to fabricate the reception of thermal radiation. This type of antenna is overlapping with the ground plane. MIM that consists of nickel-nickel oxide-nickel diode is used to convert terahertz fields into electrical current. Furthermore, other research studies [13] are interested to study the impacts of geometrical parameters on the antenna performance.

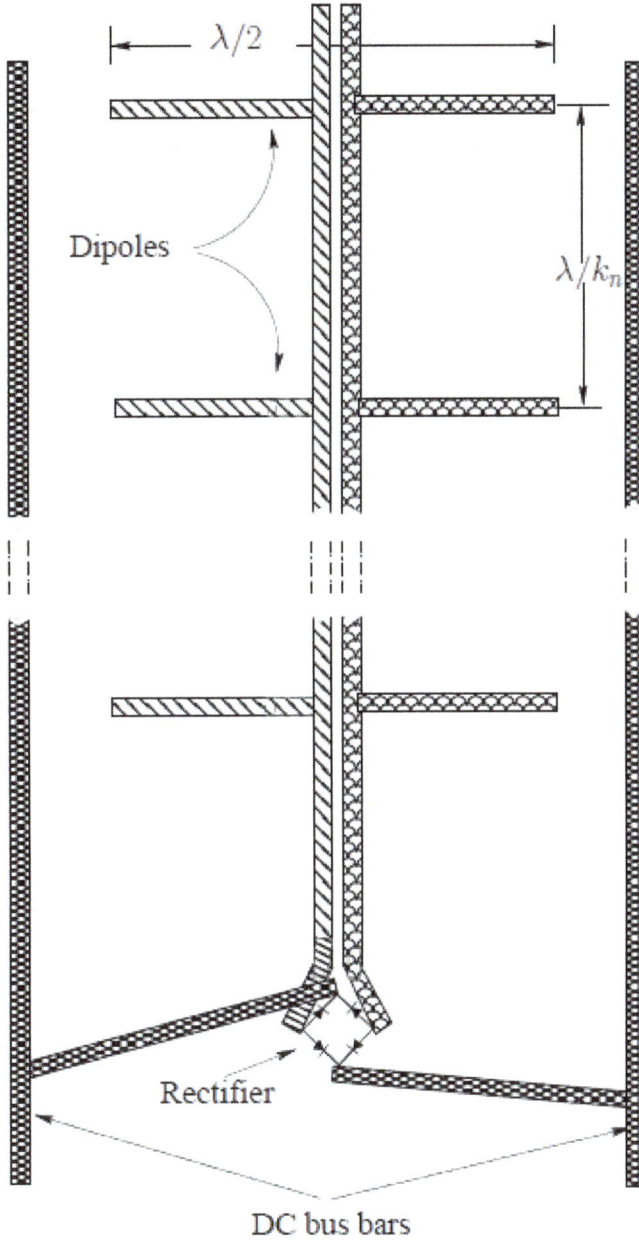

Figure 1. *The first optical rectenna proposed by mark [8].*

After that, there was a significant interest by researchers to study nanoantennas coupled to MIM diode for solar power-harvesting applications or THz sensing, which cannot be covered here due to space limitations.

Basics of solar rectennas

The structure and the operation theory of nanoantennas have been presented in this section. The same as the response of the conventional RF antenna to the electromagnetic wave, nanoantenna responds to the visible light and IR. Induced AC current, which is formed on the surface of the antenna, interacts with the incident wave and oscillates with it in the same frequency. The presence of a feeding gap in the antenna can help to collect the solar power, and then DC power is produced by rectifying the oscillated AC current with the aid of a specific diode-based rectifier.

Based on the theory of boundary conditions, the tangential electric field vanishes on the antenna surface and is equal to zero (Et = 0). This is fundamental to the traditional RF antenna, where metals are considered to have ideal electrical conductivity. In other words, Es = −Ei, where Es and Ei are scattered electric and incident electric fields, respectively.

In contrast, the operation of nanoscale antennas is based on the optical and IR regimes. In this case, metals are considered to be non-ideal conductors since they exhibit lower conductivity. Thus, the expression Et has to be taken into account. This expression can be presented by multiplying the value of surface impedance by the value of the surface current.

Figure 2 shows the block diagram of a typical optical rectenna, in which the solar antenna receives the electromagnetic wave within a proper frequency band to deliver it to the low-pass filter (LPF) [8]. The latter, which is placed between the antenna and diode (rectifier), is used to prevent the reradiation of the higher harmonics that generated from the rectification process by the nonlinear diode. Generally, power losses result in from this reradiation.

Furthermore, the LPF matches the impedance between the antenna and the subsequent circuitry. The DC LPF smoothly delivers the rectified signal to DC and then passes it to the external load. In general, MIM diode is considered being the most common rectifier in the solar rectenna system; based on the electron tunneling process, the rectification is generally occurring through the insulator layer.

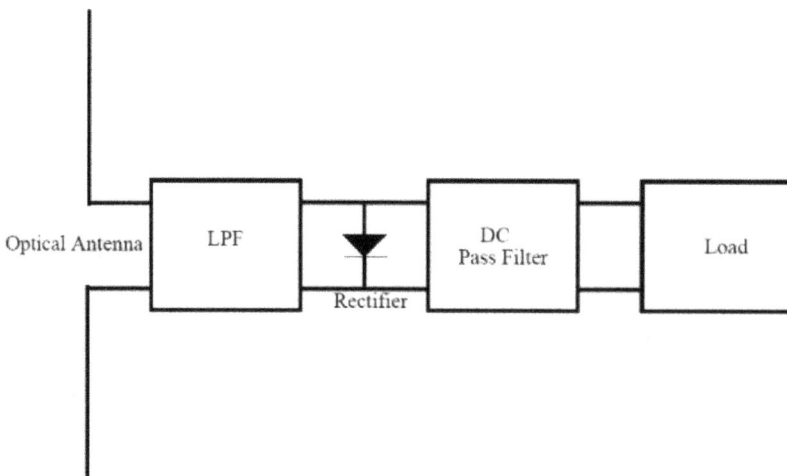

Figure 2. *Block diagram of optical rectenna [8].*

Nanoantennas

Mirrors and lenses are usually utilized to control light propagation. However, they are unable to concentrate the light in a tiny area (smaller than λ/2), whereas antennas can easily confine the electromagnetic wave in subwavelength (beyond the diffraction limit). The urgent need to local-ize the light beyond the diffraction limit has motivated the researchers and helped toward the development of nanoantennas. With the rapid growth of nanotechnology techniques, scientists are now able to fabricate nanoantennas in the order of 10 nm using E-beam lithography [14, 15]. The dimensions of nanoantenna must be in the order of the incident light wavelength to ensure efficient performance. Light/matter interaction has been exploited extensively in many applica-tions such as photovoltaics, microscopy, and THz sensing.

The main role that nanoantenna plays in solar rectennas is to receive external fields and con-fine the energy at its feed gap to be rectified by a nanodiode. The technological advances in the development of a new generation of nanodiodes such as point-contact diodes have contributed significantly to the emergence of solar rectennas in its modern form [16]. **Figure 3** demonstrates numerous fabricated nanoantennas for various applications.

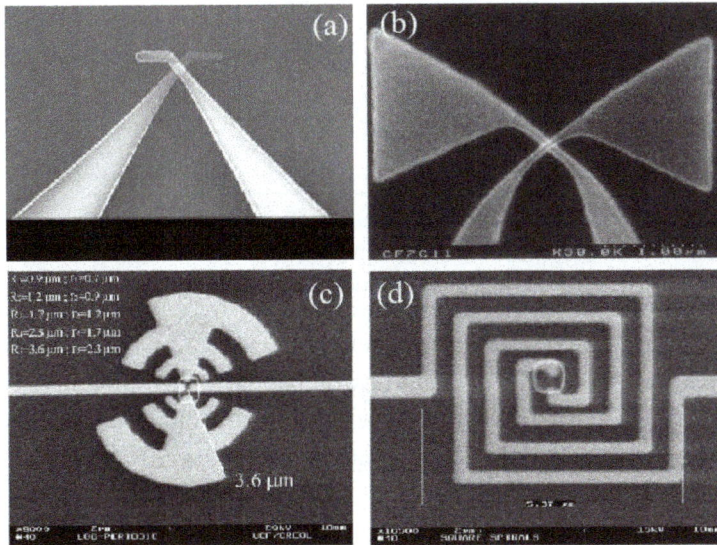

Figure 3. *Fabricated nanoantennas: (a) dipole, (b) bowtie, (c) log-periodic, and (d) spiral.*

The performance of nanoantennas in solar rectennas is measured by their ability to efficiently concentrate the received solar energy at the feed gap of the antenna.

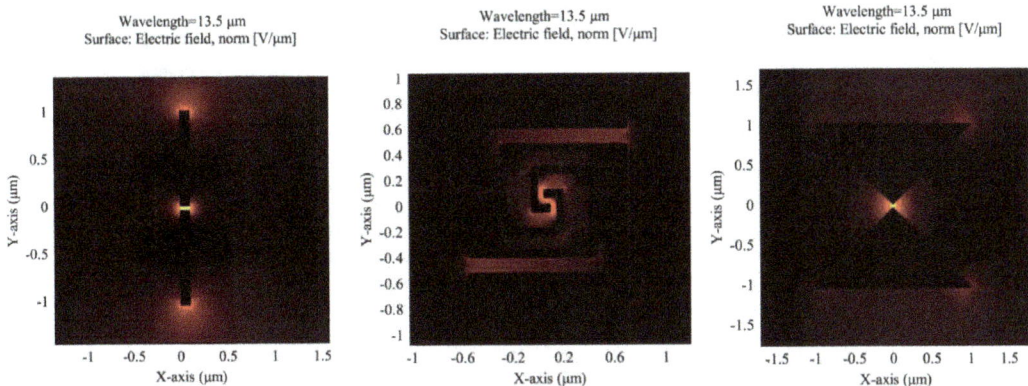

Figure 4. *Concentration of the electric field at the feed gap of different nanoantennas [17].*

The electric field generated at the feed gap varies from one type of antenna to another depending on the characteristics of the antenna itself. Thus, the confined electric field can be enhanced by choosing the proper antenna type for this application or by gathering a number of antennas in one rectenna system forming an antenna array. A comparison between different types of nano-antennas is presented in **Figures 4** and **5**, where the figure of merit is the value of the received electric field at the antenna's gap [18].

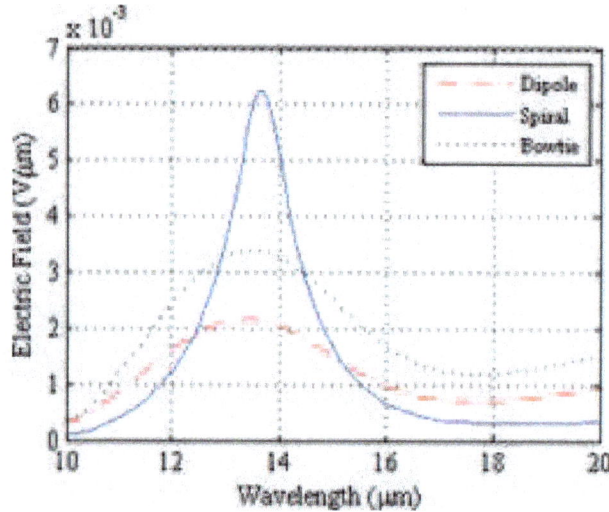

Figure 5. *Electric field variation versus wavelength for different nanoantennas [17].*

Another way to increase the captured electric field is to arrange several antenna elements in an array form. **Figure 6** shows an eight-element bowtie nanoarray as suggested in [19], where the concentration in the feed gap is also illustrated, while **Figure 7** shows the variation of the electric field with increasing the wavelength. The nanoarray exhibits multiple resonances with maximum capturing at longer wavelengths.

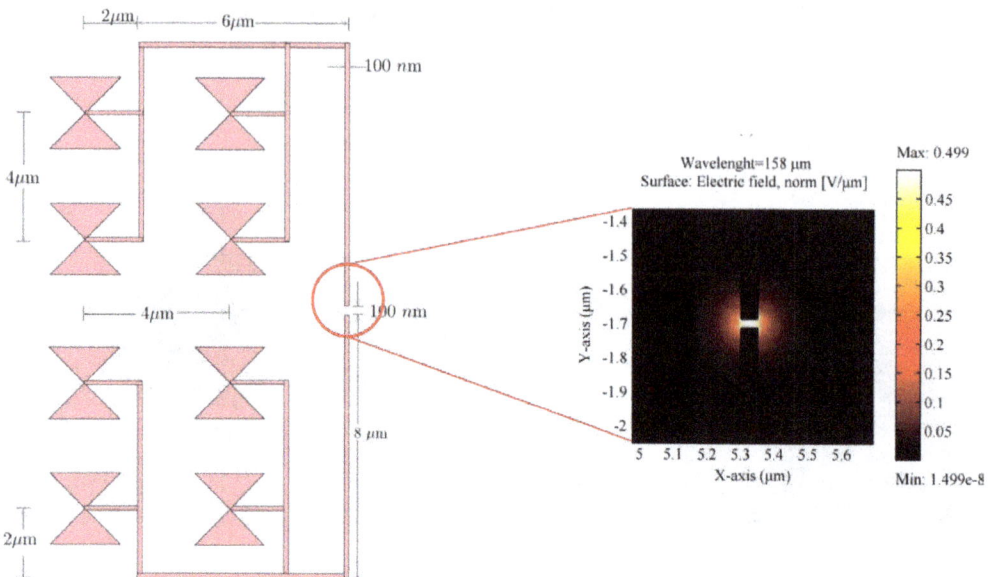

Figure 6. *Bowtie nanoarray configuration [19].*

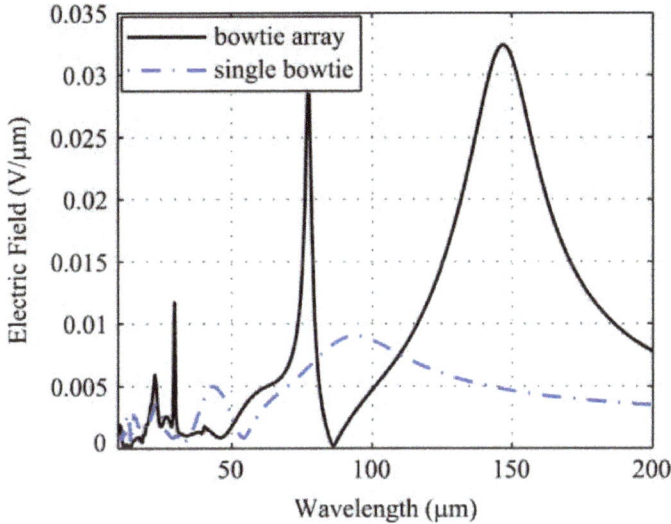

Figure 7. *Electric field variation with wavelength for bowtie nanoarray and single bowtie of the same footprint area [19].*

Nanodiodes

The most commonly used nanodiodes in solar rectennas are metal-insulatormetal diodes, which act as a promising rectifying element in solar rectennas. MIM diodes are made of thin insulator layer sandwiched between metal electrodes and depend on the tunneling mechanism. Work functions of metals and the electron affinity of insulators play an important role in MIM diodes by making a barrier at the interface between metal and insulator. **Figure 8** shows a typical MIM diode where a difference between metal work function is clearly indicated to ensure efficient electron transport across the insulator. The quantum-mechanical tunneling of electrons governs the charge transport mechanism through the barrier. Electron tunneling in MIM diodes is ultrafast, and this makes them operate at THz frequencies. A thin insulator layer (few nanometers) is required to ensure the tunneling of electrons through the diode layers.

Recent years have witnessed tremendous lithographical efforts to reduce the size of MIM diodes. To this end, the insulator layer is grown by oxidizing metal films to achieve the desired thickness. The second metal is then deposited, where this method helps to avoid vacuum break at the barrier and reduce the contamination. It is worth mentioning that controlling the roughness of the insulator layer as well as the metal films is very important during the fabrication process.

MIM diode characterization

The major obstacle in using MIM at optical frequencies is the high RC time constant. The diode resistance and capacitance must be well controlled through the fabrication techniques and processes in ordered to reduce it. In this section, the most important parameters of the MIM diode will be discussed:

- **Resistance:** The diode resistance (R_D) can be obtained directly from the I-V characteristics of the diode. Since the antenna impedance is low (in the order of 100 Ohm), the diode impedance must be low as well to achieve a reasonable impedance matching and hence ensure a maximum power transfer between the antenna and the diode.

- **Responsivity:** The responsivity of MIM diodes is a measure of the diode rectification efficiency. It is the second derivative of the diode's I-V curve over the first derivative. The responsivity represents the DC power generated by the incident AC power on the diode. The larger I-V curvature, the higher responsivity obtained, and hence the higher DC power generated. High curvature is associated directly with high barrier diodes.

- **Asymmetry:** The ratio of the forward current to the reverse current represents the diode asymmetry, which is another measure of the diode's rectification efficiency. High asymmetry can be obtained by employing different metals on both sides of the diode with a difference in their work functions.

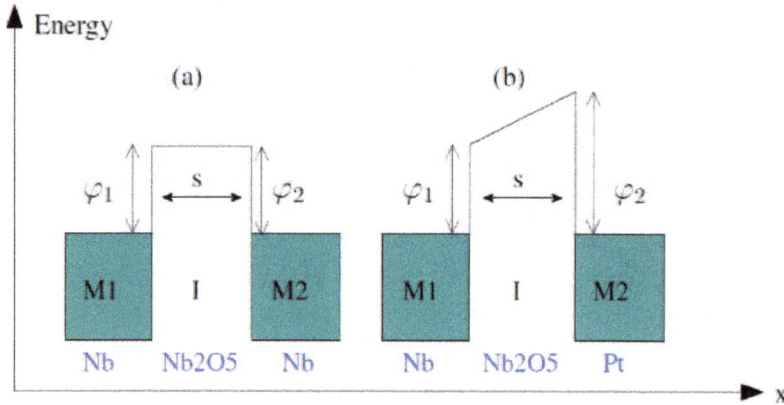

Figure 8. *Equilibrium band diagram of (a) symmetric Nb/Nb2O5/Nb diode and (b) asymmetric Nb/Nb2O5/Pt diode [20].*

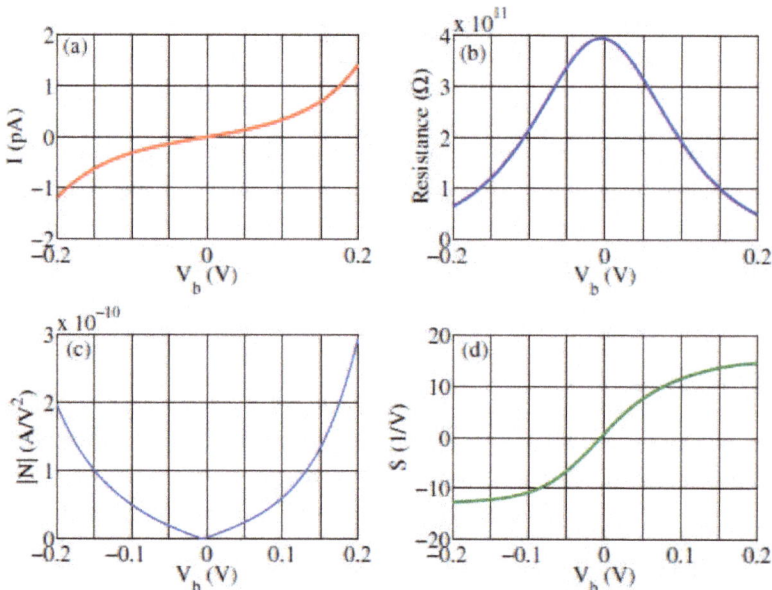

Figure 9.*Current versus biasing voltage for the asymmetric MIM diode with insulator thickness s = 5 nm and barrier heights $\varphi 1 = 0.4\ eV$ and $\varphi 2 = 1.75\ eV$ [20].*

Most of the MIM diode parameters are extracted directly from the I-V characteristics, which is the key factor in the characterization of MIM diodes. **Figure 9** demonstrates typical MIM diode parameters.

Semiconductor nanoantennas

In this section, a comparison between the performance of nanoantennas fabricated by different materials will be presented. The characteristics of the designed dipole nano-antennas have been obtained by solving Hallen's integral equation numerically. Obtained results show that carbon exhibits very low conductivity compared with other types of proposed semi-conductors like Si and Ge.

This is because of the fact that carbon has a relatively wide energy gap, which is the main reason to enhance carbon nanoantenna performance. In contrast, creating extra defect states by phos-phor or iron doping in the narrow band gap of Si and Ge can increase the conductivity and, thus, the efficiency of the host material.

The calculated efficiencies of these heavily doped semiconductor nanoantennas are unity. This is because of the high conductivity of these materials. Moreover, obtained results show that these materials behave like a perfect electric conductor at the wavelength range of interest. In addition, the performance of these semiconductor nanoantennas is compared with nanoantennas made of gold that showed approximately similar performance.

To investigate the impact of the conductivity (σ) on the antenna parameters, pure and heavily doped semiconductors materials are used instead of metal in designing nanoantennas. Since plasmonic materials like gold are being used to fabricate metallic nanoantenna, a modeling com-parison between the metallic and heavily doped semiconductor antenna is proposed to study the impact of the material on the performance of nanoantenna to exploit the mid-IR to generate presentable power.

Furthermore, the mid-IR radiation provides very low penetration depths for the electromagnetic fields. Generally, most studies on this area were focused on operating system with 10 μm wave-lengths, which may provide a wide range of energies [12].

To solve Hallen's integral equation, which is numerically used to evaluate the input impedance of the cylindrical dipole nanoantennas [21], method of moments (MOM) is generally used for this purpose. A study has been conducted to investigate the effect of replacing gold in plasmonic na-noantennas at mid-IR by heavily phosphorus-doped germanium on the antenna operation [22]. In this study, however, carbon nanotube semiconductor material is extended to heavily doped silicon by iron as common unavoidable contamination in Si. In both cases, the characteristics of the gold center-fed cylindrical dipole antenna that is used in this study include L = 0.47λ, N = 51, and a = 50 nm for λ = 10 μm, where L is the total length of the dipole, N is the number of segments, and a is the radius of dipole. In addition to that, for the delta-gap source, MoM is used to solve Hallen's integral equation. An approximate kernel can only be utilized since the ratio a/λ $\leqslant$ 0.01, and in this example, a ratio of 0.005 is used which gives an acceptable approximation.

One of the methods used to increase the efficiency of nanoantenna is by developing the quality of materials that are used to fabricate the nanoantenna. In this work, heavily doped Ge with phosphorous and heavily doped Si with Fe have been proposed as an alternative to carbon. The value of the frequency-dependent dielectric constant of heavily doped Ge with a doping concentration of 2.23×1019 cm^{-3} has been given somewhere else [23]. On the other hand, the values of heavily doped Si with a doping concentration of 1×1020 cm^{-3} have been obtained by another study [24]. The interaction between Fe and Si has been studied and reported in [25].

The dielectric constant (ε_r) has frequency-dependent real and imaginary parts, in which the metal conductivity (σ) at IR wavelengths can be obtained as the following equation [26]:

$$\sigma = i\omega\, \varepsilon_0 (\varepsilon r - 1) \tag{1}$$

The complex form of material conductivity at IR wavelengths is illustrated as real and imaginary parts in **Figures 10** and **11**, respectively. Both figures show that heavily doped semiconductors exhibit considerably high conductivity at a range of wavelength between 5 and 15 μm. Consequently, both of the heavily doped semi-conductors behave like perfect electric conductor [22].

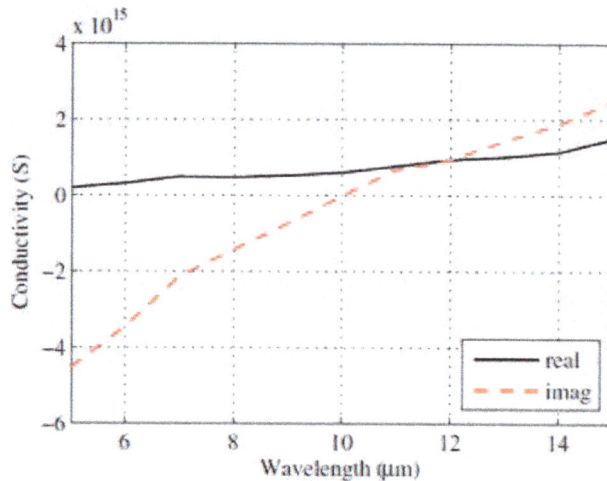

Figure 10. *Real and imaginary parts of Ge conductivity versus wavelength [22].*

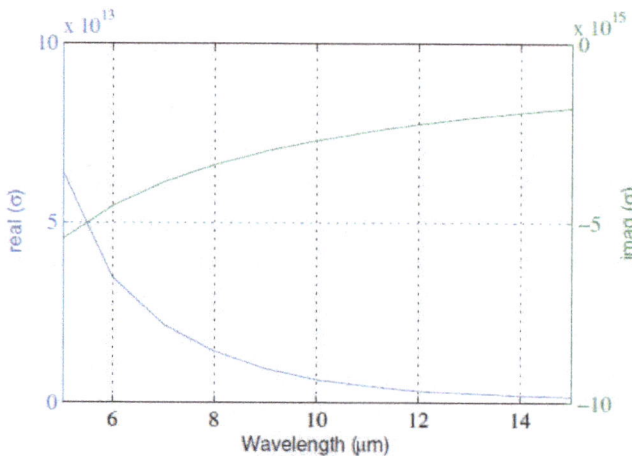

Figure 11. *Real and imaginary parts of Si conductivity versus wavelength [22].*

It is found that the conduction-dielectric efficiencies at the wavelength 10 μm for both Ge and Si are 100% as what is expected to having the same behavior as the perfect electric conductor. In contrast, the relatively low conductivity of gold yield decreases in the efficiency to around 90% at wavelengths of interest.

Conversion efficiency

The figure of merit in solar rectennas is the conversion efficiency, which depends on several factors related to both the antenna and the MIM diode. The conversion efficiency, η_t, of a solar rectenna can be described as [27].

$$\eta_t = \eta_r \eta_s \eta_q \eta_c \tag{2}$$

where η_r is the antenna radiation efficiency, η_s is the efficiency that related to the losses inside the antenna, η_q is the quantum efficiency that is responsible for the rectification of the received power, and η_c is the coupling efficiency between the antenna and the diode. It is worth noting that the term $\eta_r \eta_s$ in (2) depends on the antenna type and its characteristics and is referred, in this chapter, to as antenna-dependent efficiency of solar rectenna. On the other hand, the term $\eta_q \eta_c$ relates strongly to the diode parameters and is referred to as the diode-dependent efficiency.

For solar energy conversion, each efficiency factor is required to be optimized and maximized. Recent works have focused on improving only the quantum efficiency [28] or the diode-dependent efficiency by assuming a perfect antenna (i.e., do not include antenna efficiency limits) [4]. The analysis of the complete conversion efficiency in one single work gives the reader a close physical insight on how the IR solar rectenna works, including the parameters that affect its performance.

In the following sections, we will investigate each term of (2) individually with a detailed description of its main parameters and how to compute them. After finding the optimum values of each efficiency term in (2), the overall conversion efficiency will then be calculated and plotted.

Antenna-dependent efficiency

As mentioned in Section 7, the antenna-dependent efficiency is represented by the term $\eta_r \eta_s$. This section demonstrates how to find this efficiency numerically, which depends totally on antenna parameters. The calculation of antenna efficiency should take into account the losses that relates to reflection, conduction, and dielectric inside the antenna. The reflection losses will be represented by the coupling efficiency, ηc, and will be discussed in details in the following section. Thus, this section will be dedicated to the calculation of the conduction and dielectric losses inside the antenna structure. Since it is very difficult to compute and separate these losses individually, they will, therefore, be lumped together to form the conduction- dielectric efficiency, cd, which can be defined as [29].

$$\eta_{CD} = \frac{R_r}{R_r + R_l}; \tag{3}$$

where R_r is the radiation resistance of the antenna and R_l represents the conduction-dielectric resistance, which can be written as [29]

$$R_l = 2\frac{L}{P}R_s \; ; \tag{4}$$

where L is the antenna length, P is the cross-section perimeter of the wire antenna of radius a, and Rs is the conductor surface resistance that can be calculated as follows

$$R_s = \sqrt{\frac{\omega\mu_0}{2\sigma}} \; ; \tag{5}$$

where ω is the angular frequency, μ_o is the free-space permeability, and σ is the metal conductivity. It is worth mentioning here that Eq. (4) is valid for the case of a uniform current distribution.

Before starting the calculation of the conduction-dielectric efficiency, it is important to recall that metals are no longer perfect electric conductors at optical and infrared frequencies [30]. Consequently, the DC bulk conductivity of metal cannot be utilized in (5). Instead, the frequency-dependent conductivity at optical frequencies should be calculated.

Diode-dependent efficiency

The two terms of the diode-dependent efficiency are the coupling efficiency, η_c, and quantum efficiency, η_q. In this chapter, we will set $\eta_q = S\,h\omega/e$, where S is the MIM diode responsivity, which will be explained more explicitly later in this section, h is Plank's constant, and e is the charge of the electron. In contrast, the coupling efficiency can be written as [5].

$$\eta_c = \frac{4R_aR_D/(R_a+R_D)^2}{1 + \left(\omega\left(\frac{R_aR_D}{R_a+R_D}\right)C_D\right)^2} \; ; \tag{6}$$

R_a is the antenna resistance, ω is the angular frequency, R_D is the diode resistance, and C_D is the diode capacitance. For simplicity of analysis, the reactance of the antenna was assumed to be negligible; however, for the antenna in this work, this approximation is more realistic for the wavelength between 8 and 12 μm where a low reactance part of the impedance is noticed. The value of R_D depends on the I-V characteristics of the MIM diode, whereas the C_D can be given by

$$C_D = \frac{\varepsilon_0\varepsilon_r A}{s} \; ; \tag{7}$$

where ε_r represents the relative permittivity of the insulator layer of the MIM diode, ε_o is the free-space permittivity, A is the diode junction area (overlapping area), and s is the thickness of insulator layer. It is clearly evident that the MIM diode parameters play a significant role in determining the entire conversion efficiency of solar rectennas.

Figure 12 shows the total conversion efficiency (solid line) for an IR solar rectenna versus the wavelength. Moreover, we have added the diode-dependent efficiency (dashed line) to the same graph to show the role that the antenna plays in shaping the conversion efficiency. The total conversion efficiency has been calculated based on the terms of (2), where every single term is calculated individually and all terms are then combined together. The main reason behind this low efficiency is the mismatch between the resistance of the designed MIM diode, R_D, and the antenna resistance, R_a. This mismatch led to a lower coupling efficiency in (6), where one of the conditions to achieve unity efficiency is to have Rd = Ra. Although the diode characteristics have been optimized, the coupling efficiency still needs further improvement. However, this value of efficiency demonstrates an enhancement to recently reported conversion efficiencies of (η_t ~ 10^{-9}–η_t ~ 10^{-12}) [31]. It is worth mentioning that the antenna efficiency is very high and the diode responsivity is acceptable; however, the total conversion efficiency is quite low due to the poor coupling efficiency between the antenna and the diode. Recent studies are paying attention and efforts to increase coupling efficiency. Once the coupling is improved, we would expect a high conversion efficiency, which makes solar antennas a promising alternative to conventional solar cells and a great addition to the renewable energy sector.

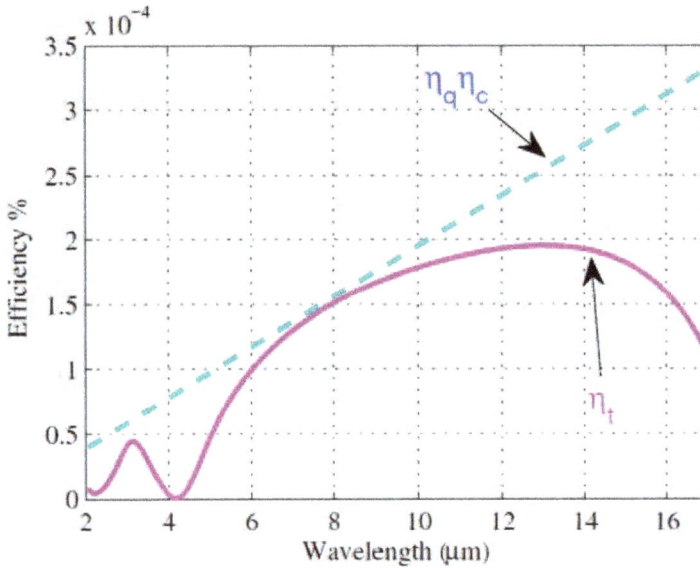

Figure 12. *Total conversion efficiency and the diode-dependent efficiency of a typical IR solar rectenna [8].*

The promising features of solar rectennas have motivated the researcher recently to come up with new approaches and ideas in order to improve the total conversion efficiency. Examples of these approaches include improving the impedance matching and the coupling between the antenna and rectifier [32, 33]. Another approach is to use metasurface absorbers to enhance the performance of solar rectenna [34].

In addition, light concentrators represented by adding a layer of micro lenses lead to increase the captured electric field as demonstrated in [35] or design dual-polarized nanoantennas [36] and/or multiband nanoantennas [37] to get benefits of all received spectrum. The approach was even extended to include harvesting thermal energy at infrared wavelengths from hot bodies [38], which sometimes focuses on preselected narrow frequencies in the infrared region [39].

Conclusions

Researchers worldwide pay attention and effort to reduce the cost of conventional solar cells and increase their efficiency by using new materials and different approaches. However, there is no significant improvement in their conversion efficiency, which is still quite low. Breakthroughs in designing efficient nanoantennas led to rapid development in solar rectenna for harvesting solar radiation. Efficient nanoantennas were designed for receiving the solar energy as an AC signal and coupling it to a nanodiode to convert it to DC power.

The focus of this chapter was to highlight different types of nanoantennas that are commonly used in this application. The design and simulation results of four types of nanoantennas have been presented, and a comparison is made to find the best candidate. The figure of merit in the selection process was the captured electric field at the feed gap of the antenna, which is a key factor in calculating the harvested energy. As a result of the comparison, it was found that the spiral nanoantenna exhibited better performance at resonance. Furthermore, it was found that the captured electric field at the feed gap could be increased by coupling many elements in one structure.

Finally, this chapter highlighted the most important factors that influence the conversion efficiency of solar rectennas with the aim to improve and optimize it. It was shown that even when optical antennas couple thermal radiation efficiently, the total conversion efficiency is still low. This is due to the poor matching between the diode and the antenna, where a very high diode resistance is obtained compared to the low antenna resistance, albeit the diode characteristics have been optimized.

As a summary, solar rectennas are an attractive option to replace PV cells in harvesting solar energy; however, this technique requires further developments in the rectification process.

Author details

Ahmed M.A. Sabaawi[1*] and Oras Ahmed Al-Ani[2]

College of Electronics Engineering, Ninevah University, Mosul, Iraq

College of Electrical Engineering Techniques, Middle Technical University, Baghdad, Iraq

*Address all correspondence to: ahmed.sabaawi@uoninevah.edu.iq

References

[1] Corkish R, Green MA, Puzzer T. Solar energy collection by antennas. Solar Energy. 2002;73(6):395-401

[2] Berland B. Photovoltaic technologies beyond the horizon: Optical rectenna solar cell. Final report, NREL/SR-520- 33263. National Renewable Energy Laboratory (NREL); 2003

[3] Biagoni P, Huang JS, Hecht B. Nanoantennas for visible and infrared radiation. Reports on Progress in Physics. 2012;75:1-76

[4] Joshi S, Moddel G. Efficiency limits of rectenna solar cells: Theory of broadband photon-assisted tunneling. Applied Physics Letters. 2013;102:083901

[5] Grover S, Moddel G. Applicability of metal/insulator/metal (MIM) diodes to solar rectennas. IEEE Journal of Photovoltaics. 2011;1(1):78-83

[6] Nozik AJ, Conibeer G, Beard MC, editors. Advanced Concepts in Photovoltaics. Cambridge, UK: Royal Society of Chemistry; 2014

[7] Brown WC. The history of power transmission by radio waves. IEEE Transactions on Microwave Theory and Techniques. 1984;32(9):1230-1242

[8] Sabaawi AMA. Nanoantennas for solar energy harvesting [PhD dissertation]. Newcastle University; 2015

[9] Marks AM. Device for conversion of light power to electric power. US Patent 4,445,050. April 24, 1984

[10] Lin GH, Abdu R, Bockris JOM. Investigation of resonance light absorption and rectification by sub nanostructures. 1996;**80**:565

[11] Berland B. Photovoltaic Technologies beyond the Horizon: Optical Rectenna Solar Cell. Golden, CO, US: National Renewable Energy Laboratory; 2003

[12] Kotter D, Novack S, Slafer W, Pinhero P. Theory and manufacturing processes of solar nanoantenna electromagnetic collectors. Journal of Solar Energy Engineering. 2010;**132**(1):011014

[13] Midrio M, Romagnoli M, Boscolo S, De Angelis C, Locatelli A, Modotto D, et al. Flared monopole antennas for 10-μm radiation. IEEE Journal of Quantum Electronics. 2011;**47**(1):84-91

[14] Bharadwaj P, Deutsch B, Novotny L. Optical antennas. Advances in Optics and Photonics. 2009;**1**(3):438-483

[15] Bareiß M, Kälblein D, Krenz PM, Zschieschang U, Klauk H, Scarpa G, et al. Large-area fabrication of antennas and nanodiodes. In: Rectenna Solar Cells. Springer; 2013. pp. 297-311

[16] Fumeaux C, Herrmann W, Kneubühl F, Rothuizen H. Nanometer thin-film Ni–NiO–Ni diodes for detection and mixing of 30 THz radiation. Infrared Physics & Technology. 1998;**39**(3):123-183

[17] Sabaawi AMA, Tsimenidis CC, Sharif BS. Infra-red spiral nano- antennas. In: Loughborough Antennas and Propagation Conference (LAPC); November 12-13, 2012. pp. 1-4

[18] Sabaawi AM, Tsimenidis CC, Sharif BS. Infra-red nano-antennas for solar energy collection. In: 2011 IEEE Loughborough Antennas & Propagation Conference. November 14, 2011. pp. 1-4

[19] Sabaawi AM, Tsimenidis CC, Sharif BS. Planar bowtie nanoarray for THz energy detection. IEEE Transactions on Terahertz Science and Technology. 2013;**3**(5):524-531

[20] Sabaawi AM, Tsimenidis CC, Sharif BS. Characterization of coupling and quantum efficiencies in solar rectennas. In: 2013 Loughborough Antennas & Propagation Conference (LAPC); November 11, 2013; IEEE. pp. 363-368

[21] Sabaawi AM, Tsimenidis CC, Sharif BS. Analysis and modeling of infrared solar rectennas. IEEE Journal of Selected Topics in Quantum Electronics. 2013;**19**(3):9000208

[22] Sabaawi AMA, Al-Ani OA. Heavily- doped semiconductor infrared antennas for solar energy harvesting. In: 2015 Loughborough Antennas & Propagation Conference (LAPC). IEEE; 2015

[23] Baldassarre L et al. Mid-infrared plasmonic platform based on heavily doped epitaxial Ge-on-Si: Retrieving the optical constants of thin Ge epilayers. In: 39th IEEE International Conference on Infrared, Millimeter, and Terahertz Waves; IRMMW-THz; 2014

[24] Al-Ani OA, Sabaawi AMA, Goss JP, Cowern NEB, Briddon PR, Rayson MJ. Investigation into efficiency-limiting defects in mc-Si solar cells. Solid State Phenomena. 2016;**242**:96-101

[25] Oras A, Al-Ani JP, Goss NEB, Cowern PR, Briddon M, Al- Hadidi RA-H, et al. A density functional study of iron segregation at ISFs and Σ5-(001) GBs in mc-Si. Solid State Phenomena. 2016;**242**:224-229

[26] Hanson GW. On the applicability of the surface impedance integral equation for optical and near infrared copper dipole antennas. IEEE Transactions on Antennas and Propagation. 2006;**54**:3677-3685

[27] Grover S, Dmitriyeva O, Estes MJ, Moddel G. Traveling wave metal/ insulator/metal diodes for improved infrared bandwidth and efficiency of antenna-coupled rectifiers. IEEE Transactions on Nanotechnology. 2010;**9**(6):716-722

[28] Dagenais M, Choi K, Yesilkoy F, Chryssis AN, Peckerar MC. Solar spectrum rectification using nanoantennas and tunneling diodes. Proceedings of SPIE. 2010;**7605**: 76050E. 11 p

[29] Balanis CA. Antenna Theory: Analysis and Design. New Jersey: Wiley; 2005

[30] Gonzalez FJ, Alda J, Simon J, Ginn J, Boreman G. The effect of metal dispersion on the resonance of antennas at infrared frequencies. Infrared Physics and Technology. 2009;**52**(1):48-51

[31] Briones E, Alda J, González FJ. Conversion efficiency of broadband rectennas for solar energy harvesting applications. Optics Express. 2013;**21**(103):A412-A418

[32] Di Garbo C, Livreri P, Vitale G. Optimal matching between optical rectennas and harvester circuits. In: 2017 IEEE International Conference on Environment and Electrical Engineering and 2017 IEEE Industrial and Commercial Power Systems Europe (EEEIC/I&CPS Europe); IEEE; 2017

[33] da Costa KQ , Souza JL, Dmitriev V. Impedance matching analysis of cylindrical plasmonic nanoantennas fed by optical transmission lines. In: Barbillon G, editor. Nanoplasmonics: Fundamentals and Applications. Intech; 2017. p. 267

[34] Almoneef T, Ramahi OM. Dual- polarized multiband infrared energy harvesting using H-shaped metasurface absorber. Progress In Electromagnetics Research. 2017;**76**:1-10

[35] Kashif MF, Rakos B. A Nanoantenna-MIM diode-lens device concept for infrared energy harvesting. In: International Conference on Global Research and Education. Cham: Springer; 2018

[36] Eltresy NA et al. Dual-polarized nanoantenna so-

lar energy collector. In: 2016 33rd National Radio Science Conference (NRSC). IEEE; 2016

[37] Chekini A, Sheikhaei S, Neshat M. Multiband plasmonic nanoantenna structure for infrared energy harvesting based on electron field emission rectification. Microwave and Optical Technology Letters. 2017;**59**(10):2630-2634

[38] Sandeep R et al. Design of Nanoantennas for harvesting waste thermal energy from hot automobile exhaust system. In: 2018 IEEE Indian Conference on Antennas and Propagation (InCAP). IEEE; 2018

[39] Fountain MW, Saffold GL. Thermal rectifying antenna complex (trac). US Patent Application No. 15/976,167; 2018

6

An Overview on Synthesis Techniques for Near-field Focused Antennas

*Marcos R. Pino, Rafael G. Ayestarán, Paolo Nepa
and Giuliano Manara*

Abstract

Microwave and millimeter-wave antennas focused in their radiative near-field (NF) region, which are usually named as near-field focused (NFF) antennas, are becoming increasingly popular. Indeed, when compared to conventional far-field focused antennas, they can guarantee performance improvement at a relatively limited implementation cost, in short-range communication systems, wireless power transfer arrangements, remote nondestructive sensing setups, and radio- frequency identification apparatus, among many others. In this chapter, application fields and metrics for NFF antennas will be briefly summarized. Most of the chapter is then devoted to the description, classification, and critical review of the many synthesis techniques that go beyond the simple, yet effective and with a clear physical insight, conjugate-phase approach.

Keywords: near-field focusing, focused antennas, focused arrays, near-field shaping, near-field synthesis, focused array synthesis, antenna focusing, phase conjugate technique

Introduction

Many wireless systems require directive antennas for proper operation. These are antennas that are able to radiate most of their input power into a limited angular sector. Indeed, above feature can improve either the spatial resolution of a localization system or the sensitivity of a remote measurement system, reduce the interference level with other wireless systems, and increase either the transfer efficiency of wireless power transfer systems or the signal-to-noise ratio in radio communication links. As it is well known to antenna designers, antenna directivity can be increased at the expense of electrically larger antennas, namely antennas that are large in terms of the free-space wavelength. Indeed, electrically large antenna arrays are becoming popular in several industrial applications, indoor wireless communication systems, and even in consumer electronic goods.

Let us consider a specific wireless system whose frequency band is assigned, as well as the typical expected distance between the transmitting antenna and the target (transponder, receiving antenna, material sample under analysis, scatterer,

etc.). Then, when enlarging the transmitting antenna size to increase its directivity, it must be considered that the boundary of the antenna far-field region moves far from the antenna itself and the target may end up with belonging to the antenna radiative near-field region, where conventional antenna far-field parameters, as radiation pattern, array factor, half-power beamwidth and gain, all become meaningless. Therefore, different design criteria must be considered when synthetizing the excitations of a large array or the layout of a large aperture antenna. It is worth noting that above phenomenon may not be rare in real-world short-range wireless systems with operating frequencies larger than a few GHz (microwave frequencies and beyond).

In above situations, the most evident design criterion consists in exploiting the well-known optical focusing concept. Indeed, by simply controlling the phase of the array element currents (or equivalent currents on the antenna aperture), it is possible to achieve a constructive combination, namely an in-phase summation, of the fields radiated by the array basic radiators at an assigned point in the antenna radiative near-field region, i.e., the focal point, where the target may be located. It is worth noting that when the assigned focal point moves far from the antenna, beyond the boundary of the far-field region, the above phase profile smoothly converges to that of a conventional far-field focused antenna (simply named as unfocused antenna in the following), i.e., a constant phase profile for a broadside antenna or a linear phase profile for a phased array, as expected.

At the focal region, the power density radiated by an NFF antenna is larger than that radiated by an unfocused antenna with equal size and input power. Equivalently, the NFF antenna can guarantee at the targeted focal point an assigned power density as for the unfocused antenna, but with a lower far-field radiation, thereby limiting the coupling with nearby wireless systems. In the following section, a planar array of printed patches operating at the 2.4 GHz ISM band will be used to quantify above advantages as well as to introduce the parameters that can be used to characterize NFF antennas.

The usefulness of an NFF antenna has been demonstrated in several applications [1]: remote sensing of material sample properties with high sensitivity and high spatial resolution, wireless power transfer at distances larger than those achieved with magnetic coupling, reliable radio-frequency identification systems, effective biomedical devices, dedicated short-range communication (DSRC) systems, among many others.

Almost any antenna technology can be used to realize NFF antennas, as it is just required to modify the layout of a conventional unfocused antenna to get the proper phase profile of the radiation currents on the array/aperture surface. An overview of the different technologies that have been applied to prototype NFF

antennas can be found in [2]. Among them, it is worth mentioning reflectarrays and transmitarrays, linear and planar arrays (made of dipoles, patches, dielectric resonator antennas), planar slotted waveguide antennas, lens antennas, dielectrically loaded horn antennas, elliptical reflectors, substrate integrated waveguide (SIW) antennas, leaky wave antennas, and Fresnel Zone Plate Lens antennas. A number of more recent papers deal with the design of focused metasurface antennas [3, 4]. With respect to the layouts of conventional unfocused antennas, the implementation of the focusing phase profile for the array currents only requires small layout modifications, as the adjustment of the feeding network in microstrip arrays, or the tuning of the geometrical parameters of the quasi-periodic cells in reflectarrays/transmitarrays, or the tapering of the guiding structure in leaky-wave antennas. The required array excitation profile can be derived by either the basic phase-conjugate approach or *ad hoc* synthesis techniques.

Near-field focused antennas: design criteria and metrics

Let us consider a planar array of 8 x 8 linearly polarized inset-fed patches resonating at 2.4 GHz, which are realized on a thin grounded dielectric layer (FR-4 substrate with a relative dielectric constant of $\epsilon_r = 4.4$, loss tangent $\tan \delta = 0.02$ and thickness of 1.52 mm). The array layout is shown in **Figure 1** and its geometrical parameters are listed in **Table 1**. The array surface coincides with the z-plane of a rectangular coordinate system, and the assigned focal point is at $\vec{r}_{focal} = z_f \hat{z}$.

To focus the field radiated by the patches at the assigned focal point, the phase of the array element excitations can be simply derived by imposing a phase shift that compensates for the distance between the assigned focal point $\vec{r}_{focal}$ and each array element position $\vec{r}_{mn}$ (conjugate-phase – CP – approach):

$$\phi_{mn} = \frac{2\pi}{\lambda} \left\| \vec{r}_{focal} - \vec{r}_{mn} \right\|_2, \tag{1}$$

where λ is the free-space wavelength and $\| \cdot \|_2$ stands for the Euclidean distance.

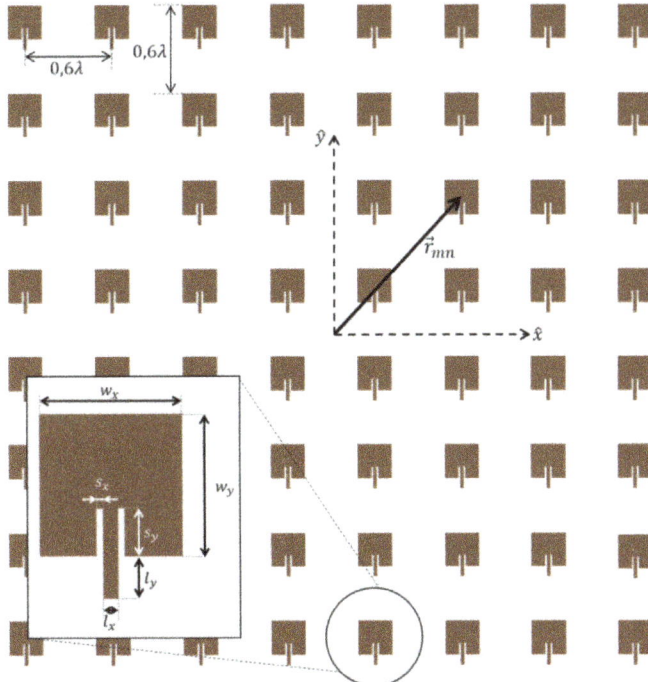

Figure 1. *8 x 8 planar array layout. The inset shows the dimensions of the array elements separated at 0.6λ in both directions (0.6λ = 7.5 cm at 2.4 GHz).*

Table 1. *Array element geometrical parameters.*

Dimensions of the array element (mm)			
w_x	30	w_y	29.5
s_x	1.5	s_y	10
l_x	3.0	l_y	9.0

If the focal point is assumed at $z_f = 4\lambda = 50$ cm, all the ports are fed with a voltage of $A_{mn} = A_o e^{j\phi_{mn}}$ and the obtained E-field distribution radiated by the NFF array along the z-axis is shown in **Figure 2**. In the same figure, the field amplitude radiated by a similar unfocused array ($A_{mn} = A_o$) is also plotted. As a reference, the electric field amplitude radiated by an ideal isotropic antenna has been added in the same figure. All three antennas radiate with an equal effective isotropic radiated power (EIRP) level, which is here assumed of 32 dBm. It is worth noting that the far-field region of the arrays starts at around $46\lambda = 5.75$ m, as confirmed by the fact that all three curves become parallel after around 6 m. The realized gain of the unfocused array is 20.1 dB; meanwhile, the gain of the NFF array is only 8.1 dB. Due to its lower gain, the maximum power transmitted by the NFF can be higher than the one transmitted by the unfocused array, for an assigned EIRP level. This fact, together with the near-field focusing effect, causes that the NFF antenna generates in the neighborhood of the focal region, a field level that is up to 21 dB larger than that achieved by the conventional unfocused array.

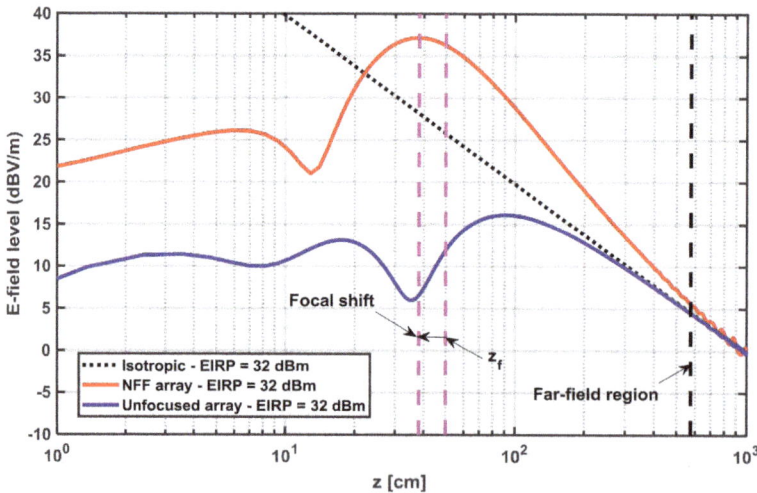

Figure 2. *E-field distribution along z-axis comparing the 8 x 8 NFF array (red) with a broadside unfocused array (blue). The dot line represents the radiated field amplitude for an ideal isotropic radiator transmitting 32 dBm.*

As verified in many NFF antennas whose synthesis uses the CP approach [1], the field amplitude peak does not coincide with the assigned focal point, it being located between the focal point and the antenna aperture. Indeed, for an assigned focal point, the CP is the approach that maximizes the field amplitude at that point, although it does not correspond to the field peak in the antenna near-field region. This well-known phenomenon (focal shift) is simply due to the fact that the field radiated by each array element increases as the observation point moves close to the antenna surface. For the planar array considered here, the focal shift is around 12 cm. It is apparent that it is possible to increase the focal distance needed to evaluate the phase profile in Eq. (1) until the field amplitude peak moves at the desired location.

The field focusing boosts a focal spot around a field amplitude peak, whose -3 dB size along three orthogonal axes is used to quantify the achieved focusing effect [1]: the -3 dB width along the focal point direction (depth of focus, DoF); the -3 dB width of the focal spot along two orthogonal axes perpendicular to the focal point direction (width of focus, WoF). For the NFF planar array considered here, a DoF of 40 cm and a WoF of 10 cm are obtained, as shown in **Figure 3** where the field distribution over different planes is given. In **Figure 3**(a), the field distribution at a plane parallel to the array, $z = 38$ cm, is shown and the WoF = 10 cm along x- and y-axes can be verified. In **Figure 3**(b), the field distribution is plotted over the $y = 0$ plane and the DoF = 40 cm is shown along the z-axis.

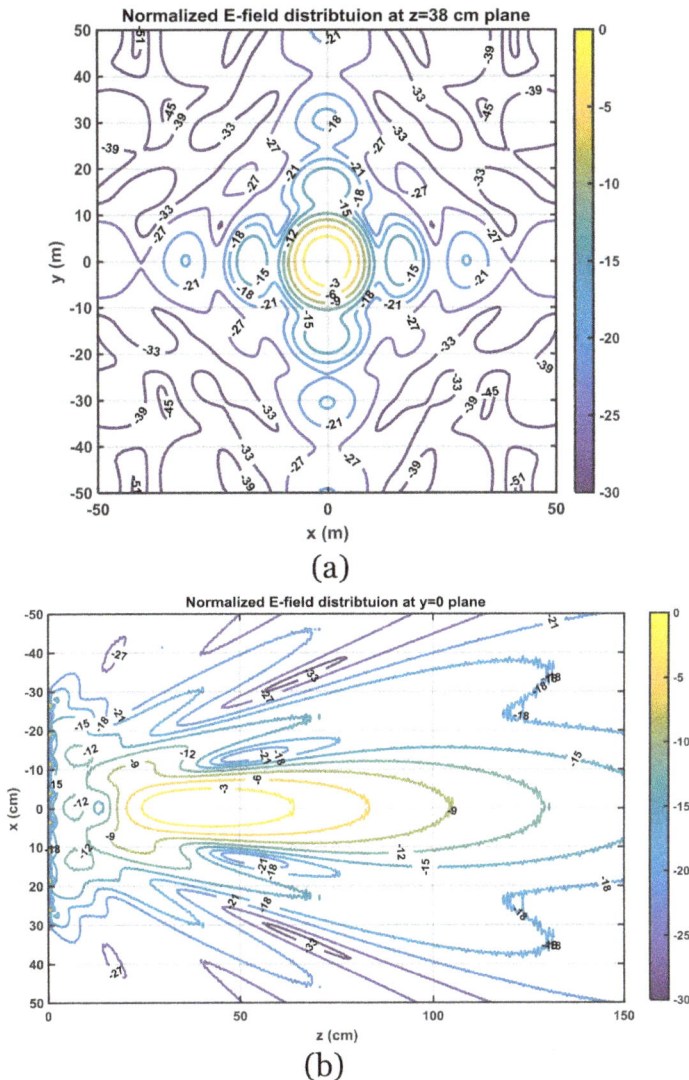

(a)

(b)

Figure 3. *8 x 8 NFF array normalized field distribution over different planes: (a) XY plane with z = 38 cm, (b) XZ plane with y = 0. In case of multifocus NFF antennas, the CP approach can still be applied by a simple superposition of the excitations as derived from Eq. (1) for each single focus, as long as the foci are not too close to each other.*

As shown in [1], DoF and WoF are related to the normalized geometrical parameters z_j/λ and D/λ, where D is the array size, and they can be used as the design requirements for the NFF antenna, depending on the specific short-range radio links at hand.

Power transfer efficiency optimization

In the previous section, the CP approach has been introduced, where a simple closed-form expression for the phase profile allows to obtain a limited focal spot by in-phase addition of all array-element contributions at a given point located in the array NF region. One issue inherent to the CP approach is that no coupling between array elements is taken into account. Indeed, the geometrical distribution of the array elements and the focal point position determine on their own the feeding phase of each array element (see Eq. (1)).

The mutual coupling between the array elements is included if a synthesis method based on the optimization of the power transmission efficiency (PTE) between the transmitting array antenna and a test receiving antenna is adopted [5]. When the testing antenna is located into the radiative NF region or far-field region of the transmitting array, the above optimization problem naturally yields to an NFF array or a conventional unfocused array, respectively. The technique is based on using the scattering matrix $\mathbf{S}$ of the $(N + 1)$-port network defined by the N-ports of the NFF transmitting array and the port of the testing antenna:

$$\begin{bmatrix} \mathbf{b_{tx}} \\ \mathbf{b_{rx}} \end{bmatrix} = \mathbf{S}_{(N+1)\times(N+1)} \begin{bmatrix} \mathbf{a_{tx}} \\ \mathbf{a_{rx}} \end{bmatrix} \tag{2}$$

where $\mathbf{a_{tx}}$ and $\mathbf{b_{tx}}$ are the incident and reflected waves, respectively, at the transmitting array ports (1 port for each array element), and $\mathbf{a_{rx}}$ and $\mathbf{b_{rx}}$ are the incident and reflected waves, respectively, at the testing/receiving antenna port. The complete scattering matrix can be obtained with full-wave electromagnetic software in order to consider a proper electromagnetic analysis of the radiation of each array element, including the mutual coupling between them. Any obstacle nearby the two antennas can be considered too.

The optimization parameter PTE is defined as the ratio between the power delivered to the load connected to the testing antenna and the power transmitted by the array:

$$\text{PTE} = \frac{|\mathbf{b_{rx}}|^2 - |\mathbf{a_{rx}}|^2}{|\mathbf{a_{tx}}|^2 - |\mathbf{b_{tx}}|^2}. \tag{3}$$

Expressing the vectors $\mathbf{b_{tx}}$ and $\mathbf{b_{rx}}$ as functions of $\mathbf{S}$ and $\mathbf{a_{tx}}$ and assuming that the whole system is well matched, a matrix equation can be derived to obtain the incident waves at each NFF array port:

$$\mathbf{A}\,\mathbf{a_{tx}} = \text{PTE}\,\mathbf{a_{tx}} \tag{4}$$

where $\mathbf{A} = \mathbf{S_{rt}}^T \mathbf{S_{rt}}$, $\mathbf{S_{rt}}$ being the scattering matrix taking into account the coupling between all receiving and transmitting ports. Eq. (4) corresponds to an eigenvalue problem for getting the optimal solution of the array excitation. Applying the obtained solution to a 4x4 array working at 2.45 GHz, an optimal PTE close to 30% is achieved, with a maximum E-field spot located at 10 cm from the array surface (less than one wavelength) [5].

It is also possible to apply the proposed technique to a circularly polarized (C-Pol) array by considering a C-Pol single-port microstrip patch as an array element, to obtain the scattering matrix as presented in [6]. Here, a pin-fed rectangular patch with two squares cutting off two opposite corners is used to obtain a C-Pol near field. The authors have also presented an in-line-fed patch with an L-shape slot and cutoff corners to obtain a single-port C-Pol patch. These two different elements have been used to implement two different 4x4 NFF array prototypes. After applying the PTE technique, the corresponding excitations generate an NFF spot with an axial ratio below 1 dB.

The PTE approach can be generalized to consider multiple receiving ports when the receiving antenna is an array. In [7], the authors extend the application of the technique to optimize the PTE between two linearly polarized arrays. The mutual coupling between the two arrays is included through the above matrix **A**. Using two arrays (for both transmitting and receiving antennas) allows to increase the transmission efficiency, reaching up to a maximum of 42% for a system of a 6 x 6 transmitting antenna and a 4 x 4 receiving antenna, which are 40 cm separated, at 5.8 GHz. For the same frequency band and using two equal 8 x 8 arrays that are 100 cm separated, the measured PTE reaches up to 46.9%.

Another interesting feature of the PTE technique is that it is able to determine the optimal feeding values of a transmitting array even when the testing antenna radiates in presence of an unknown medium. This medium can be around the testing antenna in its nearby scenario, or the antenna can be inside the unknown medium. This application is presented in [8], where a dielectric slab is placed between the transmitting and receiving antennas. In this case, the scattering matrix S is obtained from the measurements of the S-parameters for the $N + 1$ ports, by using a vector network analyzer (VNA). Using measurements may substitute the full-wave electromagnetic analysis when the latter cannot be used since the electromagnetic parameters of the surrounding media are unknown.

In all previous examples, the optimal feeding excitation of the transmitting antenna leads to a single focal spot in the NF region. Recently, in [9], the authors have proposed this technique to get a flat-top radiation pattern, for both NF and far-field (FF) applications. In particular, for the NF application, it is necessary to define a number of testing ports and their location in order to impose the flat-top E- field distribution in the targeted zone. The flat-top distribution is obtained by imposing equal values for the power of the outgoing waves at all testing ports; under this, Eq. (3) cannot be solved as an eigenvalue problem, as yielding to an optimization problem in which both the objective function and the constraints are quadratic functions (quadratically constrained quadratic program). By imposing some simplifications under FF assumptions and using weighting coefficients to reduce the ripple presented in the NF region, the authors are able to derive linearly constrained quadratic programming:

$$\text{maximize}_{\mathbf{x}} \mathbf{x}^H \mathbf{A} \mathbf{x}$$
$$\text{subject to } \mathbf{S}_{rt}\mathbf{x} = \mathbf{W}\mathbf{y} \tag{5}$$

where the vector x will provide the optimal excitations and **Wy** stands for the weighting coefficients at the testing ports. The authors have applied this technique to obtain a uniform field distribution along a bookshelf for an RFID detection application. Using a five-element transmitting antenna for the RFID reader and 11 testing points along the bookshelf, an almost flat-top distribution (ripple <3 dB) is obtained at 22 cm from the reader antenna. The flat-top distribution is imposed in a 73-cm long area in the direction parallel to the transmitting antenna.

As the PTE maximization approach is based on the S matrix evaluation, it works for any value of the focal distance. When the array is in free space and the receiving antenna lies in the array radiative near-field region, it is expected that the PTE solution approaches that predicted by the CP technique.

More recently, Cicchetti et al. [10, 11] presented an interesting and flexible NF synthesis procedure based on the eigenfields of the radiation matrix of the antenna array. Once a surface close to the antenna has been assigned, the active power passing through that surface can be optimized or a specific field pattern can be synthesized. The synthesis surface can be either opened or closed and surrounding the antenna; the field synthesis can be based on both the electric and magnetic fields, or the electric field only. In [11], a set of different complex NF distributions have been considered for a numerical validation of the proposed synthesis procedure: tilted Bessel beams, orbital angular momentum (OAM) Bessel beams and Airy beams.

Synthesis algorithms for an assigned NF distribution

The most popular approaches to the design of NFF antennas are based on assigning a required field distribution in the antenna NF region, and not only the targeted focal point positions. In some cases, such field distribution is completely assigned (it is usually referred to as a shaped field distribution or a 3D field distribution), while it is only partially specified in other cases (as for example at certain locations, such as the assigned focal points or volumes). Moreover, in some cases, a realistic field distribution is required, while some schematic values suffice for the algorithms to converge for other cases.

By specifying a target field distribution, some of the limitations of previous methods are overcome, as far as any requirement may be included for the field: multiple focal spots, arbitrary focusing volumes, a vanishing field at certain positions, a complete 3D field distribution. This fact leads to a wider range of potential applications and scenarios where NFF antennas might be useful, or even considered as a subset of NF problems. The price to pay is the complexity of the algorithms, usually associated to higher computational costs too.

The way in which specifications are given or treated differs from one method to another. However, the most significant differences between the methodologies are given by the type of algorithms used to get a solution, and not by the potential applications for each method. Most of the methods make use of different optimization schemes in order to minimize a cost function designed according to the requirements, which accounts for the distance between the targeted field distribution and that associated to a given array excitation vector. In some of the methods, additional constraints are included in the optimization in order to deal with different types of requirements, in many cases related to technological issues such as implementation or manufacturing simplifications.

Although this is a general framework, differences in the statement of the problem arise between methods solving different implementation problems or facing with different applications.

Direct optimization is not the only way to deal with NF focusing based on a given field distribution. Other techniques have been proposed to achieve the desired distribution through iterative methods using intersection approach (IA), machine learning, genetic algorithms (GA), or compressive sensing frameworks, among others. Although all of them include the minimization of the distance between target and achieved distributions, the statement of the problem is different from direct optimization, so they are studied separately in the following.

Optimization algorithms based on cost function minimization

In a general optimization problem, a cost function is designed and minimized so that the resulting solution fulfills the specifications. This means that the key point is the definition of a proper cost function that must account for the requirements. In case of NFF antennas, different sets of requirements may lead to different procedures. For example, the number of assigned focal points might vary depending on the method, or even a complete 3D field distribution might be specified. Even more, other requirements might be accounted for, as those regarding manufacturing issues, radiation properties, etc.

As a pioneer work, [12] presented a method intended to synthesize an assigned shaped NF distribution. Although it is not strictly a focusing method, it is based on ideas that became the bases for next NFF optimization-based methods. In particular, a certain NF distribution is assigned through a set of field samples, and the relation between these samples and the excitation weights to be applied to the elements of the array is represented in a matrix formulation, then solving the optimization least-squares problem given by

$$argmin_{\mathbf{w}} \left\| E_{ref} - E_{NF}(\mathbf{w}) \right\|_2 \tag{6}$$

where E_{ref} is the target or reference field distribution, and $E_{NF}(\mathbf{w})$ is the achieved distribution, which is a function of the set of weights $\mathbf{w}$ applied to the elements of the array. This is a well-known problem in the scientific literature that can be efficiently solved through a direct derivation (as it is in that paper) or using any iterative scheme. This is not a strict NFF method, but it may be considered the original approach for the following methods.

Let us consider first those methods aimed at getting a single focal spot. A direct optimization of the aperture field of an antenna to obtain a targeted focal spot is proposed in [13, 14], where the scalar case is considered. A focal point in the antenna broadside direction (z-axis) is assigned, at a given distance z_f, and the real part of the radiated field is maximized at that point. The formulation relates the near-field distribution at the plane $\vec{r}_{focal} = z_f\hat{z}$ and the aperture field (field at $z = 0$) through a plane-wave decomposition based on the Fourier transform (FT). In [13], the result of the optimization is the required aperture field, and the optimization includes a constraint to keep the field level below a bound outside the region of interest (the side lobe level, SLL). This problem can be formulated as:

$$\text{maximize}_{E_{AP}} \, \Re\left\{ E_{NF}\left(\vec{r}_{focal} \right) \right\}$$
$$\text{subject to } \left| E_{NF}\left(\vec{r} \right) \right| \leq \text{SLL}, \forall \vec{r} \in R_{SLL} \tag{7}$$

where it is stated that the aperture field, E_{AP}, is the expected result; the real part of the field is maximized at the assigned focal point; and the field level is bounded by the SLL value at any other position in the region R_{SLL}. This is a convex optimization problem that may be efficiently solved, for example, using the proposed toolbox CVX [15], which is a very popular choice specifically developed to deal with this type of problems [16]. The main feature in convex problems relies on the existence of a unique solution, with improved convergence avoiding local minima. Although it is not strictly a requirement for optimization problems, it is an interesting property as far as it simplifies implementation and reduces computational costs.

In [14], the focal spot is allowed to be designed in positions out of the broadside axis, and a further cost function to be minimized is defined as the infinite norm of the field samples in each aperture zone, where N zones are defined, which enforces the field at each zone to have a constant amplitude. It is formulated as:

$$\text{minimize}_{E_{AP}} \sum_{n=1}^{N} \left\| E_{AP}(R_{AP}^n) \right\} \right\|_{\infty}$$
$$\text{subject to}$$
$$\Re\left\{ E_{NF}\left(\vec{r}_{focal} \right) \right\} \geq 1 \tag{8}$$
$$\left| E_{NF}\left(\vec{r} \right) \right| \leq \text{SLL}, \forall \vec{r} \in R_{SLL}$$

plus other constraints in the aperture.

The field at the focal point $\vec{r}_{focal}$ is included as a constraint enforcing it to be greater than or equal to 1 (normalized value), while additional constraints are included to limit the field level below an assigned SLL at any other position (recall that it is enforced to be constant by the cost function), and to deal with lossy stratified media and the geometric properties of the aperture. The resulting problem is still convex and may be solved making use of CVX tool [15].

An interesting example is presented in [17], where a complete set of techniques are combined to obtain the weights to be applied to an array so that a designed focal spot is achieved with a minimum number of elements. This method may be included in the general class of *compressive sensing* (CS) techniques [18], where sparse information (the array elements in this case, resulting in a *sparse array*) is linearly related to a compressed set of data. The recovery of the sparse information from the compressed data is done through an optimization problem where the cost function is the $l0$ norm of the solution (i.e., the number of non-null elements). It is this recovery that is used to perform NFF synthesis in this work. The core of the method is the optimization scheme, although the distance between the target field distribution and the achieved one is considered a constraint and limited by an assigned tolerance. The minimized function is, interestingly, the $l0$

norm of the weights (i.e., the number of non-null weights, and hence the number of active radiating elements) so that the resulting set of weights represents an array with a minimum number of elements. It represents a good example of how to modify the optimization formulation to account for technological/implementation issues, and making use of other existing approaches such as CS. The schematic formulation of the resulting problem is:

$$\text{minimize}_{\mathbf{w}} \|\mathbf{w}\|_0$$
$$\text{subject to } E_{ref}\left(\overrightarrow{r}\right) - E_{NF}\left(\overrightarrow{r}\right) < e, \forall \overrightarrow{r}$$

(9)

where the parameter e is related to the pattern matching tolerance. The optimization itself takes advantage of the CS general framework, based on convex formulation, and hence resulting in a convex problem that might be efficiently solved. In [17], the popular toolbox CVX is used together with some probabilistic estimation methods.

Additionally, an important effort is made both in defining the target field distribution that better suits the given radiating structure, and taking into account the desired control over SLL. Also, in postprocessing, the results are checked to avoid *clusters* of elements concentrated at certain locations (a secondary effect of the proposed formulation, which established a grid of possible element locations closely spaced). To complete the formulation, mutual coupling between array elements may also be included in the formulation via an FEKO-computed impedance matrix that may be used to relate the weights for the coupling and noncoupling cases.

Once the optimization framework is developed to deal with a single focal spot problem, the straightforward extension is using it to include other requirements such as multiple focal spots (*multifocusing*) or a combination of spots and radiation nulls. A cost function based on a template or mask for the field distribution in the NF region is used in [19] to calculate the weights that must be applied to an antenna array for multifocusing applications. The idea behind the method is relatively simple: a set of bounds for maximum and minimum field levels at each position is specified, so that higher values are required in the focal spots, and lower levels are requested at any other position. A typical cost function used for FF synthesis is adapted to the NFF case:

$$\text{minimize}_{\mathbf{w}} \sum_{p=1}^{P} \left[\left(G^2\left(\overrightarrow{r}_p\right) - \left|E_{NF}\left(\overrightarrow{r}_p\right)\right|^2 \right) \left(g^2\left(\overrightarrow{r}_p\right) - \left|E_{NF}\left(\overrightarrow{r}_p\right)\right|^2 \right) \right.$$
$$\left. + \left|G^2\left(\overrightarrow{r}_p\right) - \left|E_{NF}\left(\overrightarrow{r}_p\right)\right|^2\right| \left|g^2\left(\overrightarrow{r}_p\right) - \left|E_{NF}\left(\overrightarrow{r}_p\right)\right|^2\right| \right]$$

(10)

where $G\left(\overrightarrow{r}_p\right)$ and $g\left(\overrightarrow{r}_p\right)$ are the maximum and minimum field levels at the p-th position of the NF region defined by $\overrightarrow{r}_p$.. This cost function penalizes field values $E_{NF}\left(\overrightarrow{r}_p\right)$ out of the specified limits. The Levenberg-Marquardt algorithm (LMA) is used to reach the solution, although any other minimization algorithm might be used by adapting the formulation in [20]. The same formulation, also using the LMA, may be found in [21] for the case of reflectarray antennas and the design of an antenna for generating an assigned quiet radiation zone.

The use of different constraints or the optimization of different parameters present in the formulation as arguments to be calculated allow accounting for technological issues: considering both the amplitude and phase of the weights or only the phase [22], accounting for coupling effects between the elements of the array through an impedance matrix [23], or allowing for nonuniform structures where the location of the elements of the array may also be optimized [20].

Another direct optimization approach is presented in [24, 25]. Although [24] is intended for a scalar case in multifocusing applications and [25] is presented as a method for a complete optimization of the radiated field, accounting for its three components, both methods are based on the maximization of the real part of the field radiated at the focal points with additional constraints to vanish the imaginary part of the field and to control the field level at positions apart from the focal points. The choice of a real value field at the focal point does not reduce the degrees of freedom of the problem as far as it represents a simple change of the overall phase reference. For the single-spot focusing problem, the maximization problem becomes:

$$\text{maximize}_{\mathbf{w}} \; \Re\left\{ E_{NF}\left(\vec{r}_{focal} \right) \right\}$$
$$\text{subject to}$$
$$\Im\left\{ E_{NF}\left(\vec{r}_{focal} \right) \right\} = 0 \tag{11}$$
$$\left| E_{NF}\left(\vec{r} \right) \right|^2 \leq B, \forall \vec{r} \neq \vec{r}_{focal}$$

where B is a non-negative function that represents a field-level bound at positions outside the focal spot (i.e., a mask for the field level). Notice that this problem is quite similar to that in Eq. (7). This approach is referred to as focusing via optimal constrained power (FOCO). The main contribution of this work is its extension to multiple targets. This extension results in an NP-hard problem that is interestingly transformed into several different convex problems, allowing a much easier implementation. For example, for two targets, the problem is formulated as:

$$\text{maximize}_{\mathbf{w}} \; \Re\left\{ E_{NF}\left(\vec{r}_{focal\_1} \right) \right\}$$
$$\text{subject to}$$
$$\Im\left\{ E_{NF}\left(\vec{r}_{focal\_1} \right) \right\} = 0$$
$$\Re\left\{ E_{NF}\left(\vec{r}_{focal\_2} \right) \right\} = \Re\left\{ E_{NF}\left(\vec{r}_{focal\_1} \right) \right\} \cos\phi \tag{12}$$
$$\Im\left\{ E_{NF}\left(\vec{r}_{focal\_2} \right) \right\} = \Re\left\{ E_{NF}\left(\vec{r}_{focal\_1} \right) \right\} \sin\phi$$
$$\left| E_{NF}\left(\vec{r} \right) \right|^2 \leq B, \forall \vec{r} \neq \vec{r}_{focal\_1}, \vec{r}_{focal\_2}$$

with $\phi \in [-\pi, \pi]$ being an auxiliary value that represents the phase shift between the fields at the two target positions. The resulting formulation corresponds to a convex problem, where the phase shift has to be also explored to get a convenient value. The price to pay is, obviously, computational complexity, as the problem has to be repeated several times, depending on the number of targets and sampling points of the auxiliary variable ϕ.

In case of the methods presented in [24, 25], it is also important to point out the differences between the scalar and vector cases, as the scalar case seems to represent a simplified formulation perfectly suitable for focusing applications, while the vector case fits a complete shaping of the 3D NF distribution, accounting for the three components of the radiated field.

Both multifocusing and 3D field shaping are addressed in [26], where multitarget time reversal is proposed as the foundation of a method also involving optimization. This interesting work shows how multifocusing and 3D NF shaping are closely linked, so that the statement of the problem in both cases is equivalent, simply selecting a different number of specified positions in the near-field region. Another example is presented in [27], where the cost function is minimized by resorting to the Steepest-Descent method, with two particularities: only the phase of the excitations is calculated, and the CP approach is used to establish an initial guess of the solution in the iterative procedure, thereby improving the speed of convergence.

A step beyond is given by [28, 29], where both near- and far-field specifications are considered, simultaneously. Although synthesizing an FF pattern with NF constraints is not a novel topic, considering simultaneous synthesis for both regions is an interesting improvement from the NFF perspective, as in this case, some applications such as wireless power transfer, may benefit from constraints in the FF region, while it is actually the NF region that represents the main subject of interest. Optimization is again the basis of the proposed methodology, which begins with the definition of a cost function balancing near- and far-field requirements. The NF term is based on the same principles as in the previous methods, while the FF term is defined as in typical FF synthesis algorithms. A trade-off between both terms is controlled using a multiplier that properly balances both requirements. The general formulation proposed in this work is:

$$\text{minimize}_{\mathbf{w}} f_{NF}(\mathbf{w}) + \gamma f_{FF}(\mathbf{w}) \tag{13}$$

where $f_{NF}(\mathbf{w})$ and $f_{FF}(\mathbf{w})$ are functions of the weight vector $\mathbf{w}$ accounting for NF and FF requirements, balanced by the trade-off parameter γ. Depending on the application, different functions may be specified in Eq. (13). For example, for given NF and FF distributions, the problem becomes:

$$argmin_{\mathbf{w}} \sum_{p=1}^{P} \left| E_{NF_{ref}}\left(\vec{r}_p\right) - E_{NF}\left(\vec{r}_p\right) \right|^2 + \gamma \sum_{q=1}^{Q} \left| E_{FF_{ref}}\left(\vec{r}_q\right) - E_{FF}\left(\vec{r}_q\right) \right|^2 \tag{14}$$

minimizing the distance between the target or specified distributions and the achieved ones in P-assigned positions of the NF region and Q-assigned directions of the FF region. Other examples are given for other problems. For example, if minimum transmitted power is required, the NF term is defined as

$$f_{NF}(\mathbf{w}) = \|\mathbf{w}\|^2 + \sum_{p=1}^{P} \left| E_{NF_{ref}}\left(\vec{r}_p\right) - E_{NF}\left(\vec{r}_p\right) \right|^2 \tag{15}$$

so that the applied excitations are reduced.

By using all the previously mentioned ideas, multifocusing combined with any type of FF requirement may be addressed, resulting in a set of weights to be applied to the array so that the resulting radiated field distribution fulfills all the specifications at one time.

Iterative algorithms using IA

Iterative algorithms represent an alternative method to obtain a field distribution E_{AP} over an antenna aperture that generates a given E_{ref} field distribution (or some given field specifications) inside a focusing area. Iterative algorithms seek to obtain an $E_{NF}^{(n)}$ over the focusing area at the n-th iteration complying with the given specification or with a controlled error-level respect to the given reference field E_{ref}. Above approach is usually referred as IA where a set $\mathcal{R}$ of the fields that can be generated by the antenna aperture must present an intersection with the set $\mathcal{M}$ that includes the fields complying with the specifications or the field E_{ref} plus the tolerable errors. At each iteration, two operations known as projectors are performed, to obtain the final solution EAP that radiates an E_{NF} field that belongs simultaneously to both sets $\mathcal{R}$ and $\mathcal{M}$.

The algorithms are initialized establishing a starting solution, generally an initial guess of $E_{AP}^{(0)}$ is used. Then, the i-th iteration can be resumed in four main steps: in the first step, $E_{NF}^{(i)}$ is calculated as a forward propagation of $E_{AP}^{(i)}$. In general, $E_{NF}^{(i)}$ will not belong to $\mathcal{M}$ so different constrains are imposed on the field to adapt it to $\mathcal{M}$ obtaining $\tilde{E}_{NF}^{(i)}$ (known as *forward projection* of $E_{NF}^{(i)}$ on set) and to gradually converge to the desired solution. As a third step, the constrained $\tilde{E}_{NF}^{(i)}$ is backward propagated to obtain a new field at the aperture $\tilde{E}_{AP}^{(i)}$. In this situation, additional constrains may be imposed on $\tilde{E}_{AP}^{(i)}$ to obtain $E_{AP}^{(i+1)}$ to adapt it to the possible fields at the aperture before proceeding with the forward propagation in the next iteration. The combination of these last three steps (backward propagation, applied constrains on $\tilde{E}_{AP}^{(i)}$, and forward propagation) can be seen as *backward projection* of $\tilde{E}_{NF}^{(i)}$ on set $\mathcal{R}$. Once the computed $E_{NF}^{(n)}$ belongs to $\mathcal{M}$ with no need to apply a new *forward projection*, it is considered that it complies with the desired specifications and a solution is met.

The different versions using this technique mainly differ in the following aspects: the regions where E_{ref} is imposed, the applied constrains to E_{NF} in order to get a convergent solution, and the expressions and techniques used to related E_{AP} and E_{NF} in the forward and backward propagation (direct integral evaluation, physical optics – PO, etc.). Some contributions also may apply constrains on E_{AP} once it is updated from the backward propagation of E_{NF} when considering additional restrictions mainly related to the antenna physics.

One common characteristic of these techniques is that all of them relate E_{AP} with E_{NF} in terms of an FT, in order to apply the fast FT (FFT) to accelerate the numerical computation.

In this contest, [30] applies above iterative procedure to obtain an NFF antenna with a radial line slot array (RLSA). In this case, both the forward and backward propagations are performed in terms of plane wave spectrum (PWS) propagation and the FT is used to compute the PWS from each field distribution and vice versa. The field spectra at the two planes ($\mathbb{E}_{NF}$ and $\mathbb{E}_{AP}$) can be related using:

$$\mathbb{E}_{NF}(k_x, k_y, d) = \mathbb{E}_{AP}(k_x, k_y, 0)e^{-jk_z d} \tag{16}$$

where $k = \sqrt{k_x^2 + k_y^2 + k_z^2}$ is the free-space propagation constant and d is the distance between the aperture and focusing planes (supposing them parallel oriented). The algorithm requires four FFTs evaluations: two of them to compute ENF in the forward propagation of the field and other two to compute EAP in the backward propagation.

Once E_{NF} is computed at each iteration, the mask defining E_{ref} is applied to trim the field amplitude $|E_{NF}|$ to keep it within the mask values. Specifically, Ettorre et al. [30] propose limiting the minimum field value in the focal spot area and define a maximum value for the rest of the region, to keep a low SLL. After convergence, the radial component of E_{AP} is used to synthesize an RLSA antenna compliant with the E_{ref} desired distribution.

More recently, [31] applies a similar procedure to the synthesis of a millimeterwave reflector antenna. In this case, PO is used to calculate the E_{NF} along the axis perpendicular to the antenna aperture (z-axis). Exploiting radial symmetries together with a Taylor expansion of the field and proper mathematical transformations, the authors come to an FT-like expression to relate E_{AP} and E_{NF} so the FFT and inverse FFT can be used to speed up the synthesis algorithm.

In this case, the proposed technique just considers a phase-only synthesis (POS) so that the computed E_{NF} along the z-axis is constrained to maintain the phase variation while the amplitude is replaced by $|E_{ref}|$. In the *backward projection*, the E_{AP} is updated with the new phase calculated keeping the original amplitude distribution at the aperture.

The previous contributions have presented results to obtain a focal spot at a given distance along the aperture broadside direction. In [32], a 3D shaping of the NF distribution is presented, which can be seen as a generalization of the previous techniques when defining a different mask at multiple planes along z-direction. These multiple masks allow to maintain the focal spot along the transversal planes generating a 3D spot along the z-axis. The technique requires the definition of m masks along z, the computation of $E_{m,NF}$ at all the z-planes where the masks have been defined, trimming them considering each mask and then individually perform the backward projection to obtain $E_{m,AP}$ related to $E_{m,NF}$. Considering that the iterative procedure requires only one E_{AP}, the multiple $E_{m,AP}$ obtained are averaged as:

$$E_{AP} = \frac{1}{N} \sum_{i=1}^{m} E_{i,AP} \tag{17}$$

An RLSA of 12 cm radius working at 30 GHz is synthesized and prototyped, obtaining a uniform spot along z between 6 and 18 cm away from the antenna.

IA combined with optimization techniques

In [33], the use of optimization techniques is proposed to perform one of the above-mentioned projectors inside an iterative algorithm of IA. The combination of an optimization technique inside the IA iterative algorithm can be seen as a generalization of the IA to the problem of antenna FF radiation pattern synthesis, by taking advantage of the optimization technique properties applied to given steps of the iterative procedure. In particular, in [34], the LMA is proposed as an optimization technique to perform the backward projector of the IA to synthesize the far-field radiation pattern of a reflectarray antenna. At each iteration of the IA, only a few iterations of

LMA are performed to get the new distribution $E_{AP}^{(i+1)}$ from the constrained far-field radiation pattern (through a local search around $E_{AP}^{(i)}$) to gradually achieve the convergence of the IA algorithm.

The application of this generalized IA to an NF focusing problem has been presented in [35] to synthesize a reflectarray antenna with a quite zone in its NF region ($5.3 \times 5.3 \ \lambda^2$ zone at 20 λ from the reflector aperture). The LMA is used inside each iteration of IA algorithm as part of the *backward* projector to obtain a new field distribution $E_{NF}^{(i+1)}$ from the constrained $\tilde{E}_{NF}^{(i)}$. At the i-th iteration of the IA, the LMA starts from $E_{AP,LMA}^{(0)} = E_{AP}^{(i)}$ and performs a local search through l iterations to get $E_{AP,LMA}^{(l)}$ that minimizes the defined cost function. The cost function is usually defined in terms of the field distribution over the NF focusing region $E_{NF,LMA}^{(l)}$ trying to minimize its distance to the constrained $\tilde{E}_{NF}^{(i)}$:

$$\text{minimize} \sum_{p=1}^{P} \left[\left(\left| \tilde{E}_{NF}^{(i)} \left(\vec{r}_p \right) \right|^2 - \left| E_{NF,LMA}^{(l)} \left(\vec{r}_p \right) \right|^2 \right) \right]^2 \tag{18}$$

noticing that $E_{NF,LMA}^{(l)}$ is obtained from $E_{AP,LMA}^{(l)}$ at the l-th iteration of the LMA applying the forward propagation used in the IA operations. After a few iterations of the LMA, the $(i+1)$-th iteration of the IA starts using the solution obtained with LMA, that is $E_{AP}^{(i+1)} = E_{AP,LMA}^{(l)}$ that generates $E_{NF}^{(i+1)} = E_{NF,LMA}^{(l)}$ to be forward projected on the set M.

The results achieved with this hybrid technique clearly improves those presented in [21] where only the LMA is directly applied as optimization technique using Eq. (10) without considering the IA iterative algorithm. The quite zone achieved using IA with LMA is $7.3 \times 7.3 \ \lambda^2$ fulfilling the specifications of a maximum amplitude ripple of 1.5 dB and the phase distribution at the aperture presents a smoother variation, which leads a physical realizable antenna in comparison with the results obtained by direct application of LMA where an almost random phase distribution is achieved.

The main drawback of this technique is the high computational cost required by the LMA to evaluate the gradient, as far as it performs multiple evaluations of the NF at the focusing region from the known field at the aperture. This drawback can be overcome by applying the differential technique presented in [36] where the new $E_{NF,LMA}$ evaluations are calculated in terms of the differential variations produced at each iteration without requiring the whole evaluation of the field in the NF region. This gradient evaluation in terms of differential contributions (DFC) is faster than using FFT for the FF problems. The application of the DFC to speed up the NF computation inside the IA-LMA has been presented in [37] where a reflectarray has been designed to generate a uniform coverage area of 100×30 cm^2 at a distance of 2.2 m in its NF region using an aperture of 15×15 cm^2 at a central frequency of 28 GHz, for 5G applications.

Further techniques

Additional methods have been proposed to solve NF problems from an assigned field distribution but by using approaches that cannot be considered into previous categories.

As a first example, in the pioneer work presented in [38], a unified approach for the synthesis of an assigned NF distribution is based on the decomposition of both the amplitude and phase reference patterns in terms of the coefficients of spherical vector wave functions. This formulation leads to a set of linear equations that may be solved to calculate the complex excitations to be applied to certain array structures. This method is intended for NF synthesis problems, but it is also the back- ground for the already cited work [12], which tries to overcome the limitations of this method by becoming also the pioneer method in optimization through the specification of a reference NF distribution.

Another approach, based on machine learning, was proposed in [39], where a trained neural network (NN) is used to relate a targeted near-field distribution to the corresponding radiating element excitations. In an NN, a mapping between input and output data is established through a previous training process where the system *learns* from previously known data (the so-called *training patterns*). The internal weights in the synapsis between the neurons are optimized so that the overall behavior of the network approximates the unknown function relating inputs and outputs. To use an NN approach in NFF antenna design, different sets of excitations are applied to determine the resulting NF distribution for a given radiating structure (through electromagnetic analysis or numerical simulations). During the training process, the NN discovers the relation between excitations and NF so that it may provide the array excitations required to achieve certain field distributions, typically with a minimum square error criterion. Once the NN is trained and ready for operation, a distribution consisting on constant real values at the focal points and zero-values at any other position is presented to the network to get the corresponding weights. It is interesting to point out that this approach allows for specifying any shaped 3D near-field distribution, taking advantage of the specification of field values at the assigned positions of the specified NF region. One of the main advantages of this approach is given by the fast computation time of the trained NN, which may be considered a real-time process. Exhaustive computation is required during the training process, but it is done only once.

That in [39] is not the first technique making use of machine learning methods. In [23], support vector machines (SVMs), and in particular, the so-called support vector regression framework is used together with the previously mentioned optimization scheme. However, it is not used to solve the NF synthesis problem but to calculate a model of the radiating system accounting for coupling effects, individual radiation patterns of the array elements, nonuniformities, nonidealities or any realistic effect that might affect the global behavior of the antenna. SVMs are able to extract such model from known patterns consisting on effective feedings applied to the array and samples of the resulting field distribution. These patterns may be obtained through measurements or simulations, provided that they are able to account for all above effects. The obtained model is incorporated into the optimization so that it becomes much more accurate and realistic.

It is interesting to relate [39] with [40]. At first glance, the two methods appear as being completely different. However, the internal structure of an NN is an approximation of the represented function as a combination of local or global basis functions, just as it is proposed in [40] to represent the field distribution for multiple focusing points. In [40], the field close to the antenna aperture is calculated from the expansion of the field distribution in these basis functions, although through analytical formulation instead of training data.

Although it is related to optimization, the approach in [41, 42] is based in a leastsquares calculation, to generate a quiet zone, i.e., a local plane-wave field generated in an assigned volume of the NF region. The oscillation of the amplitude and phase of the field distribution in such volume is minimized, as well as the field outside the quiet zone in order to reduce multipath interferences. The proposed formulation of the problem is a quite standard LS case whose solution is well known and directly applied.

In [43], an unconventional approach combining NF-FF transformation and a GA is proposed. A desired NF distribution is assigned and its corresponding FF distribution is calculated. Then, an FF synthesis method based on a GA approach is used to determine the excitations of an array of short dipoles. GAs are known to be excellent global optimizers, as they are based on the so-called *random search* principle. Although inspired by biological evolution, any GA is actually composed of different stages that perform an exhaustive search on the space of solutions with different mechanisms to avoid local minima. Their performance is usually excellent provided that a proper *fitness function* is defined (a function playing a role equivalent to the cost function in direct optimization), although the computation time is typically larger than using other techniques. In [43], the proposed method takes advantage of the maturity of FF synthesis methods based on GA, which have been proven to be successful, although a direct application to the NF distribution might have been explored.

Although most of the papers here referenced present simulation and experimental results for planar arrays, it is worth mentioning that in many cases, the proposed synthesis technique can be easily extended to conformal arrays. Indeed, the phase compensation required to focus the near-field radiated by an antenna array can also be obtained by controlling the physical distance between each array element and the assigned focal point, namely using a conformal array. In this context, Otera [44] showed that a curved slotted waveguide array is effective to get a focusing effect if the phase variation along the guiding structure is properly combined with the waveguide curve shape. More recently, a detailed analysis of an NFF leaky wave array antenna based on a curved slotted SIW has been presented in [45]. In that paper, the authors use the further degree-of-freedom represented by the array linear shape (namely, the spatial location of each slot) to improve the control of the array focus steering when the latter is achieved through frequency variation. The shape of the linear slotted SIW is approximated with an m-order polynomial, whose $m + 1$ coefficients are synthesized through a conventional least- squares method that minimizes the discrepancy between the achieved position of the field amplitude peak and the its assigned position, for a number of frequency values of the frequency scanning range. In the field evaluation, for each hypothetical shape, the slot position along the SIW is derived by using the CP-based approach.

As already introduced through [45], another interesting aspect when designing NFF antennas is the analysis of the synthetized near-field pattern vs. frequency, especially when such a dependence is exploited on purpose to implement a low-cost focus steering. The frequency dependence of the involved traveling wave phase constant (as in series-fed arrays, leaky wave antennas, etc.) and free-space phase constant can be easily introduced in most of the previous synthesis techniques, if the higher computational costs can be tolerated. Not surprisingly, most of the papers presenting NFF antennas with frequency focus steering use a CP-based synthesis approach. In this context, a NFF planar array is proposed in [46] to implement a two-dimensional microwave imaging system, where the focal point steering is obtained by using frequency scan (from 8.5 to 11.5 GHz) and phase shift between 8 series-fed parallel linear arrays, in the two planes orthogonal to the array surface, respectively. The details of the array design process are given in [47], where the phase synthesis is based on the quadratic approximation of the phase profile calculated by the robust and valuable CP approach. The quadratic phase approximation helps to control the focus steering independently along the two principal planes, as far as the required focus displacement from the broadside direction is relatively small.

Finally, it is worth noting some recent investigations on focused antennas implementation for wireless power transfer by exploiting frequency-diverse arrays or time-modulated arrays [48].

Conclusions

Focusing and shaping the field radiated by an antenna in its radiative near-field region may be important to improve the performance of several wireless systems based on short-range radio links at microwave and millimeter-wave frequency bands, as for instance radio links for effective wireless power (and data) transfer. In this chapter, it has been shown that the large number of available synthesis techniques can efficiently solve many real-world focusing problems, when different requirements on the antenna near field are assigned.

Future work could deal with comparing and merging the presented synthesis techniques with the well-known focusing techniques developed at the optical region, as well as with those techniques usually applied when the target is closer to the antenna, in its reactive near-field region. Interest is also devoted to potential radio link architectures where both the transmitting and receiving antennas are relatively large with respect to their separation distance, and a simultaneous syn- thesis of the coupled NFF antennas is required.

Finally, a larger and larger utilization of near-field focused antennas in practical scenarios is desirable, as they can improve the performance of the short-range wireless systems with a very limited increase of the antenna layout complexity and realization costs.

Acknowledgements

This work was supported in part by the Ministerio de Ciencia, Innovación y Universidades under project TEC2017-86619-R (ARTEINE) and by the Gobierno del Principado de Asturias/ FEDER under project GRUPIN-IDI/2018/000191.

Author details

Marcos R. Pino[1], Rafael G. Ayestarán[1], Paolo Nepa[2]* and Giuliano Manara[2]

1Department of Electrical, Electronics, Computers and Systems Engineering, University of Oviedo, Gijón, Spain

2 Department of Information Engineering, University of Pisa, Pisa, Italy

*Address all correspondence to: paolo.nepa@unipi.it

References

[1] Nepa P, Buffi A. Near-field focused microwave antennas. IEEE Antennas and Propagation Magazine. 2017;59(3): 42-53

[2] Nepa P, Buffi A, Michel A, Manara G. Technologies for near-fielsd focused microwave antennas. International Journal of Antennas and Propagation. 2017;2017:17

[3] Yu S, Liu H, Li L. Design of near-field focused metasurface for high-efficient wireless power transfer with multifocus characteristics. IEEE Transactions on Industrial Electronics. 2019;66(5): 3993-4002

[4] Yurduseven O, Smith DR, Fromenteze T. Design of a reconfigurable metasurface antenna for dynamic near-field focusing. In: 2018 IEEE International Symposium on Antennas and Propagation. Boston, MA,USA; 2018. pp. 8-13

[5] Shan L, Geyi W. Optimal design of focused antenna arrays. IEEE Transactions on Antennas and Propagation. 2014;62(11):5565-5571

[6] Jiang YH, Geyi W, Yang LS, Sun HC. Circularly-polarized focused microstrip antenna arrays. IEEE Antennas and Wireless Propagation Letters. 2016;15: 52-55

[7] Yang XD, Geyi W, Sun H. Optimum design of wireless power transmission system using microstrip patch antenna arrays. IEEE Antennas and Wireless Propagation Letters. 2016;16:1824-1827

[8] Sun HC, Geyi W. Optimum design of wireless power transmission systems in unknown electromagnetic environments. IEEE Access. 2017;5: 20198-20206

[9] Cai X, Geyi W. An optimization method for the synthesis of flat-top radiation patterns in the near-field and far-field regions. IEEE Transactions on Antennas and Propagation. 2019;67(2): 980-987

[10] Cicchetti R, Faraone A, Testa O. Energy-based representation of multiport circuits and antennas suitable for near- and far-field synthesis. IEEE Transactions on Antennas and Propagation. 2019;67(1):85-98

[11] Cicchetti R, Faraone A, Testa O. Near field synthesis based on multi-port antenna radiation matrix Eigenfields. IEEE Access. 2019;7:62184-62197

[12] Narasimhan MS, Philips B. Synthesis of near-field patterns of arrays. IEEE Transactions on Antennas and Propagation. 1987;35(2):212-218

[13] Iliopoulos I, Fuchs B, Sauleau R, Pouliguen P, Potier P, Ettorre M. On the use of convex optimization for electromagnetic near-field shaping. In: 2017 11th European Conference on Antennas and Propagation (EUCAP). Paris, France; 2017. pp. 1013-1016

[14] Iliopoulos I, Fuchs B, Sauleau R, Pouliguen P, Potier P, Ettorre M. Scalar near-field focusing in lossy media. In: 2017 International Conference on Electromagnetics in Advanced Applications (ICEAA). Verona, Italy; 2017. pp. 718-721

[15] Grant M, Boyd S. CVX: Matlab software for disciplined convex programming, version 2.1. 2014. Available from: http://cvxr.com/cvx

[16] Boyd S, Vanderberghe L. Convex Optimization. Cambridge, UK: Cambridge University Press; 2004

[17] Huang ZX, Cheng YJ. Near-field pattern synthesis for sparse focusing arrays based on Bayesian compressive sensing and convex optimization. IEEE Transactions on Antennas and Propagation. 2018;66(10):5249-5257

[18] Candès E, Walkin MB. An introduction to compressive sampling. IEEE Signal Processing Magazine. 2008; 25(2):21-30

[19] Álvarez J, Ayestarán RG, Las Heras F. Design of antenna arrays for near-field focusing requirements using optimization. Electronics Letters. 2012; 48:1323-1325

[20] Ayestarán RG, Álvarez J, Las Heras F. Design of non-uniform antenna arrays for improved near-field multifocusing. Sensors. 2019;19(3):645

[21] Prado DR, Vaquero AF, Arrebola M, Pino MR, Las-Heras F. General near field synthesis of reflectarray antennas for their use as probes in CATR. Progress In Electromagnetics Research. 2017;160:9-17

[22] Álvarez J, Ayestarán RG, León G, Herrán LF, Arboleya A, López JA, et al. Near field multifocusing on antenna arrays via non-convex optimization. IET Microwaves, Antennas and Propagation. 2014;8(10):754-764

[23] Muñiz JA, Ayestaran RG, Laviada J, Las Heras F. Support vector regression for near-field multifocused antenna arrays considering mutual coupling. International Journal of Numerical Modelling: Electronic Networks Devices and Fields. 2016;29(2):146-156

[24] Bellizzi GG, Iero DAM, Crocco L, Isernia T. Three-dimensional field intensity shaping: The scalar case. IEEE Antennas and Wireless Propagation Letters. 2018;17(3):360-363

[25] Iero DAM, Crocco L, Isernia T. Constrained power focusing of vector fields: An innovative globally optimal strategy. Journal of Electromagnetic Waves and Applications. 2015;29(13): 1708-1719

[26] Bellizzi GG, Bevaqua MT, Crocco L, Isernia T. 3-D field intensity shaping via optimized multi-target time reversal. IEEE Transactions on Antennas and Propagation. 2018;66(8): 4380-4385

[27] Chou H-T, Hung K-L, Chou H-H. Design of periodic antenna arrays with the excitation phases synthesized for optimum near-field patterns via steepest descent method. IEEE Transactions on Antennas and Propagation. 2011;59(11):4342-4345

[28] Ayestarán RG, León G, Pino MR, Nepa P. Wireless power transfer through simultaneous near-field focusing and far-field synthesis. IEEE Transactions on Antennas and Propagation. 2019;67:5623-5633

[29] Ayestarán RG, Pino MR, Nepa P. Synthesis of near field focused arrays including far-field constraints. In: 2017 International Applied Computational Electromagnetics Society Symposium (ACES). Florence, Italy; 2017. pp. 1-2

[30] Ettorre M, Casaletti M, Valerio G, Sauleau R, Le Coq L, Pavone SC, et al. On the near-field shaping and focusing capability of a radial line slot array. IEEE Transactions on Antennas and Propagation. 2014;62(4):1991-1999

[31] Xu G, Zeng R, Chen S, Shi M, Yu C. A novel synthesis method for millimeter-wave antenna with contoured-beam at near-field region. In: 2017 47th European Microwave Conference (EuMC). 2017. pp. 264-267

[32] Iliopoulos I, Casaletti M, Sauleau R, Pouliguen P, Potier P, Ettorre M. 3-D shaping of a focused aperture in the near field. IEEE Transactions on Antennas and Propagation. 2016;64(12):5262-5271

[33] Bucci OM, D'Elia G, Mazzarella G, Panariello G. Antenna pattern synthesis: A new general approach. Proceedings of the IEEE. 1994;82(3):358-371

[34] Prado DR, Arrebola M, Pino MR, Florencio R, Boix RR, Encinar JA, et al. Efficient crosspolar optimization of shaped-beam dual-polarized reflectarrays using full-wave analysis for the antenna element characterization. IEEE Transactions on Antennas and Propagation. 2017;65(2):623-635

[35] Vaquero AF, Prado DR, Arrebola M, Pino MR, Las-Heras F. Near field synthesis of reflectarrays using intersection approach. In: 2017 11th European Conference on Antennas and Propagation (EUCAP). 2017. pp. 3644-3648

[36] Prado DR, Vaquero AF, Arrebola M, Pino MR, Las-Heras F. Acceleration of gradient-based algorithms for array antenna synthesis with far-field or near-field constraints. IEEE Transactions on Antennas and Propagation. 2018; 66(10):5239-5248

[37] Vaquero AF, Prado DR, Arrebola M, Pino MR. Reflectarray Antennas for 5-G Indoor Coverage. In: 2019 13th European Conference on Antennas and Propagation (EUCAP). Krakow, Poland; 2019

[38] Narasimhan MS, Varadarangan K, Christopher S. A new technique of synthesis of the near- or far-field patterns of arrays. IEEE Transactions on Antennas and Propagation. 1986;34(6):773-778

[39] Ayestarán RG. Fast near-field multifocusing of antenna arrays including element coupling using neural networks. IEEE Antennas and Wireless Propagation Letters. 2018;17(7):1233-1237

[40] Chou H-T, Wang N-N, Chou H-H, Qiu J-H. An effective synthesis of planar array antennas for producing near-field contoured patterns. IEEE Transactions on Antennas and Propagation. 2011; 59(9):3224-3233

[41] Hill DA. A numerical method for near-field array synthesis. IEEE Transactions on Electromagnetic Compatibility. 1985;27(4):201-211

[42] Hill DA, Koepke GH. A near-field array of Yagi-Uda antennas for electromagnetic-susceptibility testing. IEEE Transactions on Electromagnetic Compatibility. 1986;28(4):170-178

[43] Clauzier S, Mikki SM, Antar YMM. Design of near-field synthesis arrays through global optimization. IEEE Transactions on Antennas and Propagation. 2015;63(1):151-165

[44] Othera I. Focusing properties of a microwave radiator utilizing a slotted rectangular waveguide. IEEE Transactions on Antennas and Propagation. 1990;38(1):121-124

[45] Wu YF, Cheng YJ. Proactive conformal antenna array for near-field beam focusing and steering based on curved substrate integrated waveguide. IEEE Transactions on Antennas and Propagation. 2019;67(4):2354-2363

[46] Li P-F, Qu S-W, Yang S. Two-dimensional imaging based on near- field focused array antenna. IEEE Antennas and Wireless Propagation Letters. 2019;18(2):274-278

[47] Li P-F, Qu S-W, Yang S, Nie Z-P. Microstrip array antenna with 2-D steerable focus in near-field region. IEEE Transactions on Antennas and Propagation. 2017;65(9):4607-4617

[48] Mani F, Masotti D, Costanzo A. Exploitation of time modulated arrays for multisine power transmission. In: IEEE International Symposium on Personal, Indoor and Mobile Radio Communications (PIMRC). Bologna; 2018. pp. 9-12

7

Voltage-Doubler RF-to-DC Rectifiers for Ambient RF Energy Harvesting and Wireless Power Transfer Systems

Abdul Quddious, Marco A. Antoniades,
Photos Vryonides and Symeon Nikolaou

Abstract

Wireless Power Transfer (WPT) is promoted as a key enabling technology (KET) for the widespread use of batteryless Internet of Things (IoT) devices and for 5G wireless networks. RF-to-DC rectifiers are essential components for the exploitation of either ambient RF power or wireless transmitted power from a dedicated source. There are several alternative rectifier topologies which can be selected depending on the desired wireless charging scenario and may include one or more diodes. For full rectification, a minimum of two diodes are needed. The current chapter discusses various implementations of voltage-doubler designs, which revolve around the basic topology of two diodes and two capacitors. Schottky diodes are usually used, in combination with lumped capacitors. Off-the-shelf diodes include both separate diodes and integrated voltage-doubler topologies in a single package. Rectifiers are inherently narrowband, non-linear devices, and the RF-to-DC efficiency, which is usually the figure of merit, depends non-linearly on both the termination load and the received RF power. The bandwidth of the rectifier depends on the preceding matching network.

Keywords: wireless power transfer (WPT), voltage-doubler rectifier, rectenna, RF energy harvesting and DC-to-DC booster

Introduction

In the past few decades, substantial research efforts have been devoted to wireless power transfer (WPT) applications from both dedicated [1] and ambient RF sources [2, 3]. The sensitivity of the receiver and the energy transfer efficiency are the main concerns in any WPT system, and they eventually dictate the RF energy harvesting (EH) system's performance, its range, and the overall reliability. The sensitivity depends on the minimum power that is required in order to power up the semiconductor devices used [4] in the EH circuit. On the other hand, RF-to-DC efficiency depends on the receiving antenna performance, the impedance matching network between the antenna and the rectifying circuit, and the overall power conversion efficiency of the rectifier's subsequent stage, that is, the voltage multiplier. Traditionally, the efficiency of any rectifying circuit is controlled and improved by optimizing the circuit design [5, 6]. Additionally, it can be further improved by using a high peak-to-average power ratio (PAPR) multi-sine signals that have demonstrated improved efficiency performance compared to the conventional sinusoidal signals [7, 8]. Usually, the available ambient RF power is very low and therefore the harvested power is not sufficient to support any immediate application, although, it can be utilized to directly control the external supply dynamically [9–11]. Recently, one of the key developments in wireless communication is the exploitation of the same RF signal for information and energy harvesting. This refers to the concept of simultaneous wireless information and power transfer (SWIPT) [12] systems. Finally, RF EH is widely used in RFID tags that are used for tracking and identification [13]. With the development and integration of a low cost compact, energy harvesting circuit on the RFID tag, the conventional passive tags [14–16] can be converted into "active-equivalent" tags with improved life time and increased read range through RF energy harvesting coming from designated power beacons without relying on the reader to activate the RFID oscillating circuit [17].

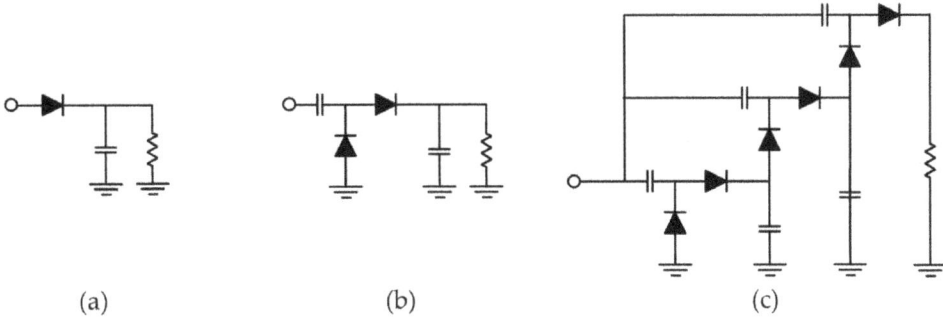

Figure 1. *Rectifier topologies, (a) single diode (single-stage), (b) voltage doubler, and (c) voltage multiplier (multi-stage).*

To convert RF to DC power, the RF energy harvesting circuits are generally implemented using semiconductor-based rectifying elements such as CMOS diodes with transistors [18] or Schottky diodes, due to their low cost and low power requirements [4]. The most common Schottky diodes include Skywork's SMS [19] and Broadcom's HSMS [20] series of surface mount devices (SMD). **Figure 1** illustrates the basic RF-to-DC rectifier topologies. The first one (**Figure 1a**) is a single diode half-wave rectifier (envelope detector) that consists of a series diode with shunt capacitor and performs half-wave rectification by passing either the negative or the positive half of the AC current while the remaining half is blocked. Single diode, half-wave rectifiers have large rectified voltage ripples compared to fullwave rectifiers, hence, additional filtering is required to remove the harmonics from the DC output. The use of Schottky diodes with lower built-in threshold voltage, such as the SMS7630, contributes toward improved RF-to-DC conversion efficiency for the lower input power signals. This happens because a fixed amount of power is consumed for biasing the diodes, which for lower power signals, it is high percentage of the original RF power. The second rectifier topology (**Figure 1b**) is the voltage doubler, or single stage voltage multiplier, that works as a full-wave rectifier and converts the incident AC signal to a constant polarity voltage at its output. Compared to the half-wave rectifier, it results in a higher average DC voltage. It consists of a one-stage clamper with a pumping capacitor at the input of a shunt diode and a series rectifying diode along a shunt capacitor at the output stage, which works as a low pass filter to smoothen the output DC voltage. The third topology (**Figure 1c**) is the voltage multiplier (multi-stage), which is a full-wave rectifier that further increases the output voltage with a network of capacitors and diodes. It has a similar operation principle to the voltage doubler. The voltage of each stage is used as a reference for the next stage, and the maximum output voltage depends on the overall number of stages. The undesired effect of the multi-stage rectifiers is the decrease in the overall efficiency which degrades for every additional stage since the efficiency of every single stage is a multiplying factor smaller than one (1). It is important to note that the overall RF-to-DC conversion efficiency depends directly on the selected topology. Where the available AC power is high, the use of multi-stage rectifiers results in an increased DC voltage level, although the overall power efficiency is decreased. For RF EH systems though, the available input power is usually very low. For relatively low input power situations where the available power is comparable to the amount of power that is required to switch on the rectifying diodes, the lower the number of diodes is, the higher the overall RF-to-DC efficiency is.

Voltage doubler

The use of the voltage doubler circuit implemented with Schottky diodes appears often in the literature. The voltage doubler circuit consists of a combination of two diodes and two capacitors connected in the topology presented in the schematic of **Figure 2a**. Several types of diodes can be used for the implementation of a voltage doubler. **Figure 2b** presents the equivalent circuit for the Skyworks SMS7630 Schottky diode, which was used for many of the rectifiers which are discussed in this chapter.

Figure 2a illustrates the schematic of a voltage double topology with the preceding antenna and the required matching network. The antenna and rectifier, including the intermediate matching network, are commonly referred to as a rectenna. The presented voltage doubler implementation consists of the clamper stage formed by the capacitor C1 and the diode D1, the rectifying diode (D2), and the RC low pass filter (C2 and RL). The preceding matching network illustrated here as a box, is usually part of the rectifier circuit and it assures the maximum power transfer by reducing the impedance mismatch loss between the preceding antenna and the rectifier circuit. Often, the matching network is built with reactive lumped or distributed components, optimized for the intended operation frequency and the expected power levels, since rectifiers are non-linear devices. Several different matching topologies are discussed in the subsequent sections.

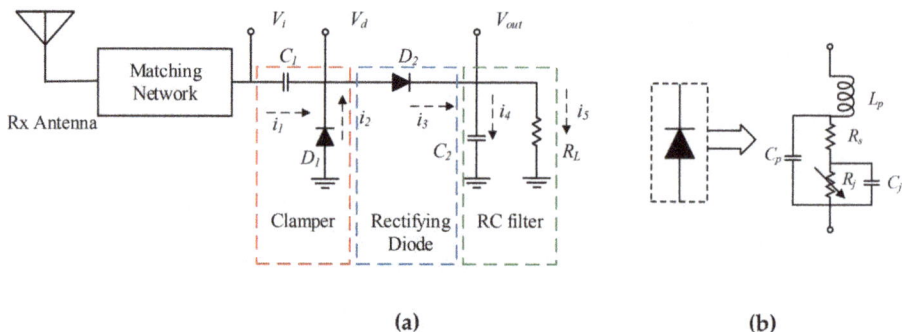

(a)   (b)

Figure 2. RF energy harvesting systems (a) voltage doubler circuit (b) equivalent model for Schottky diodes in voltage doubler topology.

Voltage doubler operation principle

As the first voltage doubler stage following the matching network, the clamper is used to enhance the input voltage to the rectifying diode (D_2), as can be seen in **Figure 2a**. Clamping circuits are mostly used to implement the voltage multiplier [8]. For a time-varying sinusoidal input signal, the clamping circuit uses the peak of the negative voltage of the input signal to produce a positive shift in the signal during its positive part, as explained below.

From **Figure 2a**, V_i is the voltage at the input of the capacitor C_1, i_1 is the current passing through C_1, V_d is the voltage on diode D_1, and i_2 is the current on D_1. The clamper is activated during the negative cycle of V_i, and C_1 is charging with D_1 in ON state. During the positive cycle, C_1 acts as a voltage source of $-V_{c1}$ with D_1 in OFF state. According to Kirchhoff's law at the relevant loop, V_d is associated with V_{C1} and V_i through (1)

$$V_i = V_{C1} + V_d \tag{1}$$

$$i_1 = C_1 \frac{dv_{C1}}{dt} = -i_2 = -I_s\left(e^{V_d/nV_T} - 1\right) \tag{2}$$

where $V_d = V_{D1} \times I_s$, V_T, I_s, and n show the thermal voltage, the saturation current, and the ideality factor, respectively. The voltage enhancement factor is defined as the ratio of input voltage to rectifying diode voltage V_d with peak input voltage to clamper circuit V_i [8]. For the ideal continuous wave excited voltage doubler, the voltage enhancement factor is two, if the clamper moves perfectly the input signal to the DC offset, equal to the peak of V_i during the negative input cycle.

It can be seen in **Figure 2a** that the rectifying circuit consists of the diode D_2 and the low pass RC filter with capacitor C_2 and termination load R_L. V_d is the input voltage and i_3 is current through the diode D_2, while the current though the capacitor C_2 and load resistor R_L is denoted by i_4 and i_5, respectively, and the output voltage is V_{out}. According to Kirchhoff's law, the voltage across diode D_2 is:

$$V_{D2} = V_d - V_{out} \tag{3}$$

where $\quad i_3 = i_4 + i_5 \tag{4}$

$$i_3 = I_s\left(e^{V_d/nV_T} - 1\right) \tag{5}$$

$$i_1 = C_2 \frac{dV_{out}}{dt}, i_5 = V_{out}/R_L \tag{6}$$

The ordinary differential equation (ODE) by proper transformation of nonlinear behavior of the diode described in Eqs. (5) and (6) is:

$$\frac{dV_{out}}{dt} = i_5/C_2 = I_s\left(e^{V_d - V_{out}/nV_T} - 1\right) - V_{out}/R_L C_2 \tag{7}$$

Equation (7) is a nonlinear form of an ODE, and a closed form solution for the general case is not available. Though, its numerical solution is possible using the ODE solver [8], the calculated DC output voltage is

$$V_{DC} = \frac{1}{T-t}\int_t^T V_{out}(t)dt \tag{8}$$

where t is the steady state charging starting time, and $T = n\tau$ is the total number of n periods of the output signal [8].

The non-linear response of the diode is represented by a varying ohmic load R_j (see **Figure 2b**). A real diode requires a combination of parasitic capacitors and inductors, which are needed to accurately model the package of the semiconductor device. The non-linear nature of the Schottky diodes used is reflected in the matching of the rectifier device ($|S_{11}|$) and also on the resulting RF-to-DC efficiency. **Figure 3a** presents the simulated reflection coefficient for a single-port voltage doubler rectifier. The reflection coefficient resonance shifts when the input power is varied and therefore a wideband matching network is needed in order to maintain the $|S_{11}|$ below -10 dB, despite the shift of the resonance. The achieved RF-to-DC efficiency also depends non-linearly on the termination load and the input RF power, as can be seen in the simulated efficiency surface presented in **Figure 3b**.

Implemented rectifier circuits

A thorough study of voltage-doubler rectifiers in the UHF frequency range is presented in this section, considering possible variations on (a) the diode model the substrate type, and (c) several alternative matching configurations. In particular, the Broadcom HSMS2850 and the Skyworks SMS7630 Schottky diodes were investigated. The use of low-cost, lossy FR4 substrate ($\varepsilon_r = 4.3$, $\tan\delta = 0.016$) was compared with the higher-cost and less lossy Duroid 5880 substrate ($\varepsilon_r = 2.2$, $\tan\delta = 0.0009$). Finally, a wide range of matching circuits that consisted of one or more lumped inductors, one or more radial and custom shaped open and shorted stubs, and the use of tapered microstrip lines for bandwidth enhancement were considered in an attempt to enhance the bandwidth and the bandwidth stability for various power levels of the UHF rectifiers investigated.

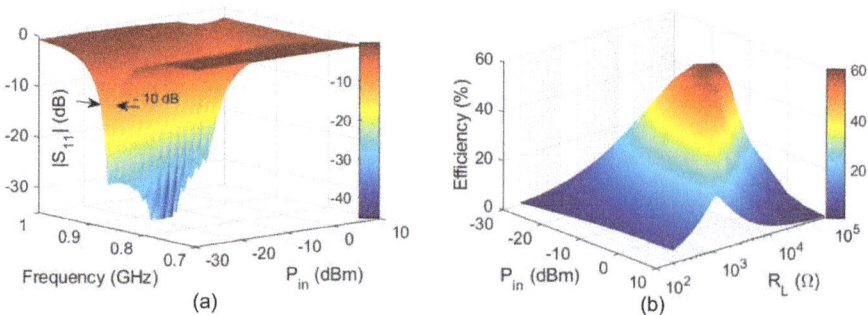

Figure 3. *Non-linearity of the rectifying circuit (a) simulated |S11| versus frequency and input power (b) efficiency percentage vs. input power and termination load.*

For this study, 11 rectifier variations were fabricated and measured. The fabricated rectifiers labeled A to K with a short description of their characteristics, matching networks, and diodes used are presented in **Table 1**. The first comparison between a voltage-doubler rectifier (Rectifier A) and a single-stage rectifier (Rectifier B) presented in **Figure 4** suggests that for input power levels higher than -15 dBm, the voltage doubler topology exhibits a higher RF-to-DC efficiency. For most wireless power transfer applications and due to the unavoidable free-space loss, the available power at the receiver is rather low. However, the received power level is random and cannot be predicted, and the efficiency plots for Rectifiers A and B cross at the -15 dB input power mark (**Figure 4d**), therefore a direct comparison of the efficiency performance is not straight forward. The improved average efficiency performance over the entire power range is the main reason why the voltage doubler was preferred over the single stage rectifier and was further studied in detail. The free-space loss is inversely proportional to $\lambda 2$ and this is the reason why the UHF frequency (relatively low frequency) was preferred for the presented study.

Table 1. *Implemented rectifiers labeled A to K.*

Rect.	Schematic	Description (matching network)	Diode
A		Consists of low-loss distributed elements, which include shorted stub, two-stage tapered line, radial stub followed by distributed inductor	SMS7630
B		Consists of low-loss distributed elements, which include shorted stub, two-stage tapered line, radial stub followed by distributed inductor	SMS7630
C		Consists of distributed elements, which include a shorted stub and a radial stub at the beginning and at the end of a U-shaped microstrip line	HSMS2850
D		Consists of distributed elements, which include a shorted stub and a radial stub	SMS7630
E		Consists of a hybrid matching network, which includes a radial stub, a shorted stub, and a series inductor	HSMS2850
F		Consists of a hybrid matching network, which includes a radial stub and two series inductors	SMS7630
G		Consists of low-loss distributed elements, which include a shorted stub, a radial stub followed by a distributed inductor	HSMS2850
H		Consists of low-loss distributed elements, which include a shorted stub, a radial stub followed by a distributed inductor	HSMS2850
I		Consists of low-loss distributed elements, which include a shorted stub, two-stage tapered line, a radial stub followed by a distributed inductor	SMS7630
J		Consists of a hybrid matching network, which includes a radial stub and two series inductors	SMS7630
K		Consists of a hybrid matching network, which includes a radial stub and two series inductors	HSMS2850

Voltage doubler matching considerations

Rectifiers C, D, E, and F were designed on an FR-4 substrate, and are used to discuss the effect of the matching circuit on the stability of the $|S_{11}|$ resonance that defines the matching, and on the

-10 dB bandwidth of the $|S_{11}|$. As can be seen from Figures 5 and 6, the matching depends on the input power level. For different power levels, the exact position of the resonance that defines the matching ($|S_{11}| < $ -10 dB) shifts, and for certain values, the rectifier is mismatched, and as a result, the RF-to-DC efficiency degrades. Rectifiers C and D, which are compared in **Figure 5**, use different diode types, (Broadcom and Skyworks, respectively), and similar matching networks that consist of shorted and radial matching stubs.

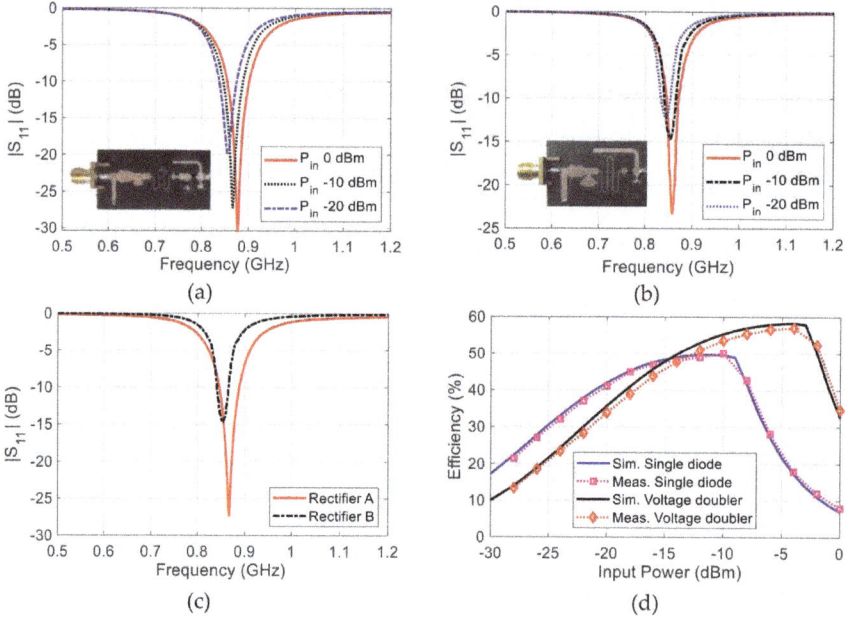

Figure 4. *Voltage doubler (A) and single stage rectifier (B) comparison; (a) Rectifier A $|S_{11}|$, (b) Rectifier B $|S_{11}|$, (c) $|S_{11}|$ at -10 dBm comparison, and (d) RF-to-DC efficiency.*

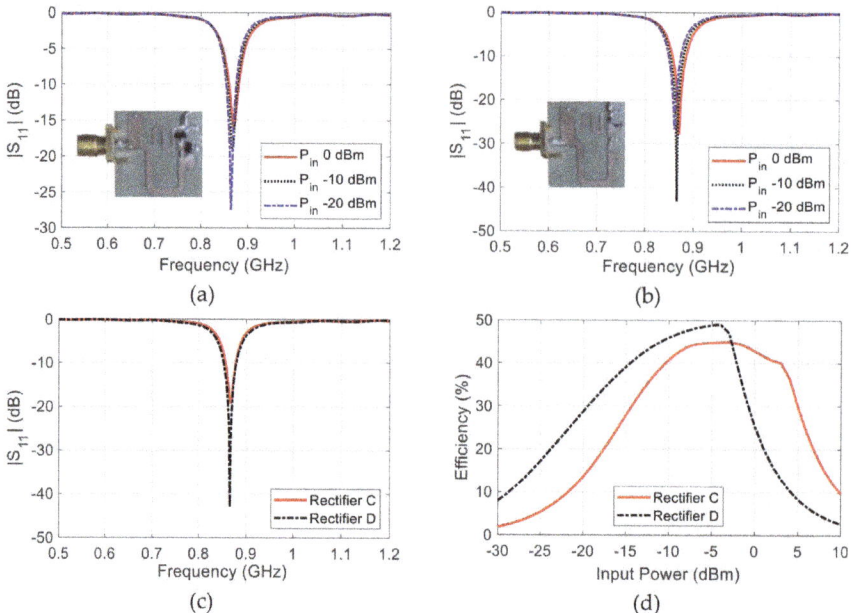

Figure 5. *Rectifier (C) and rectifier (D) comparison; (a) Rectifier C $|S_{11}|$, (b) Rectifier D $|S_{11}|$, (c) $|S_{11}|$ at -10 dBm comparison, and (d) RF-to-DC efficiency.*

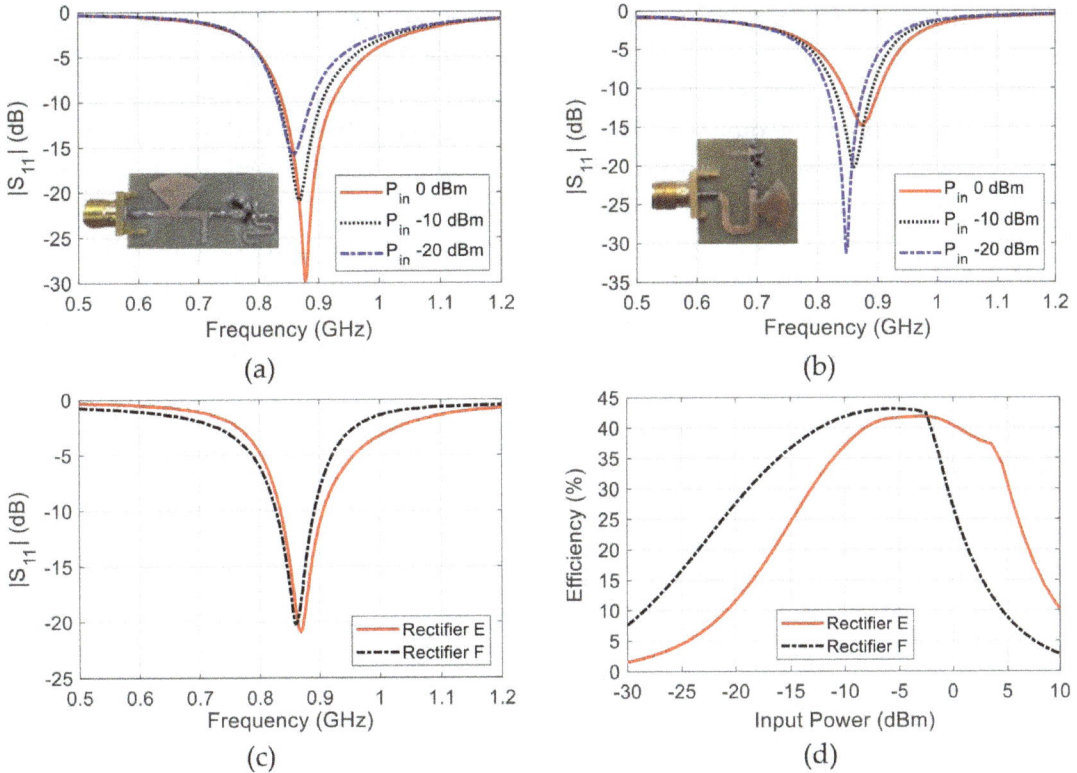

Figure 6. *Rectifier E and rectifier F comparison; (a) Rectifier E $|S_{11}|$, (b) Rectifier F $|S_{11}|$, (c) $|S_{11}|$ at -10 dBm comparison, and (d) efficiency.*

The $|S_{11}|$ resonance remains rather stable as the input power varies, but it is narrowband (rv25 MHz). The different diode models have minimum effect on the matching, but they affect the efficiency as will be discussed subsequently. For Rectifiers E and F, presented in **Figure 6**, that use a combination of radial stubs with series inductors, the $|S_{11}|$ shift is much more evident. For this hybrid matching circuit, what overcomes the mismatch problem in the UHF frequency is the considerably wider bandwidth (greater than 65 MHz) for both rectifiers. Apparently, the combination of radial stubs with lumped inductors and the resulting hybrid matching network enhances the bandwidth and makes the matching more tolerant to frequency shifts.

RF-to-DC efficiency

For any RF-to-DC rectifier, the figure of merit is the RF-to-DC efficiency, which has been shown to depend non-linearly on both the input power and the termination load, as can be verified from **Figure 3b**. In an attempt to identify the design parameters that affect the efficiency, several rectifier prototypes were fabricated and tested. Parameters such as the diode type, the use of distributed or lumped inductors for the matching network and their associated losses and quality factor, or the substrate losses of the preferred fabrication board were investigated experimentally.

The Skyworks SMS7630 [19] and Broadcom HSMS2580 [20] diodes were chosen for the high-efficiency voltage doubler topologies that were investigated at low input power, because they have been the most commonly used diodes. In the recent years, the Broadcom HSMS2852 type has been a popular diode for many RF-to-DC rectifiers used for WPT systems. For low input power levels, the higher series resistance (compared to, e.g., the Broadcom HSMS-286x family) does not seem to degrade the measured efficiency. In addition to their specifications, an important role for the selection of the diodes had to do with the availability (or not) of the diode model for the Keysight - Advanced Design System (ADS) simulator. The Spice model of the Skyworks SMS7630 diode was characterized at 1.8 GHz, and the same model was used for the UHF rectifier simulations. S-parameter measurements of the implemented rectifier indicated a shift in the $|S_{11}|$ resonance. The resonance shift could be remedied with the matching network, and especially by modifying the inductance values of the inductors that were used for the matching. For the Broadcom HSMS2580 diode, an ADS library model was available and the measurements indicated that the S-parameter simulations using the ADS library model were more accurate. Similar rectifier prototypes with similar matching networks using either type of Schottky diodes (C and D or E and F) were implemented.

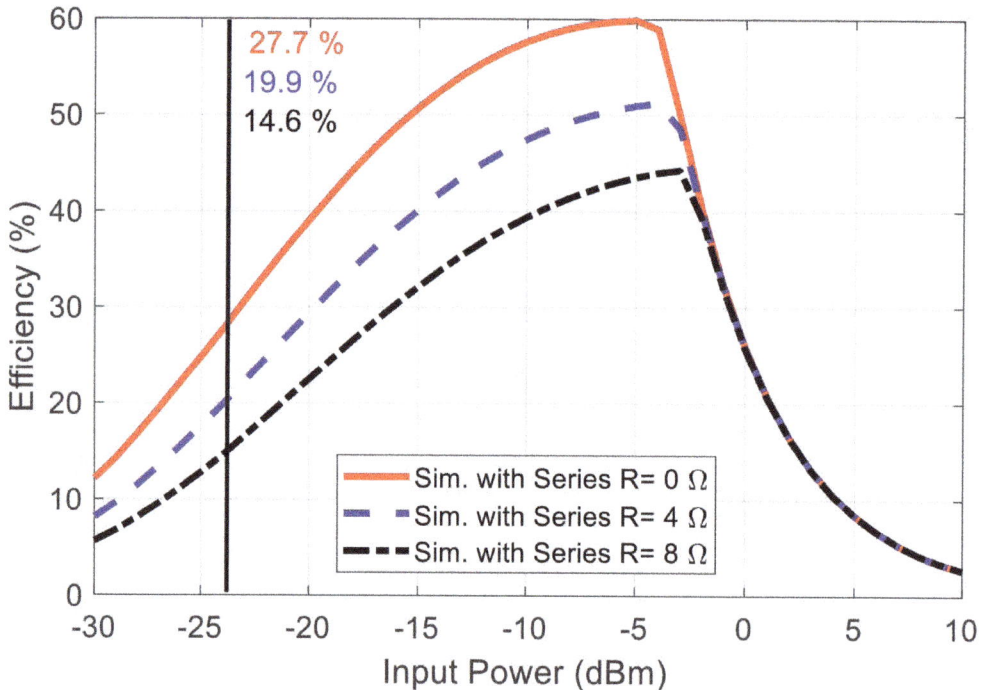

Figure 7. *Effect of the ohmic losses of the lumped inductor modeled as series resistor, on the simulated efficiency.*

Considering the direct comparison between the efficiency of rectifier C (Broadcom HSMS2580) and D (Skyworks SMS7630) presented in **Figure 5d**, the general observation is that the rectifier with the Skyworks diodes has better efficiency for input power less than -3 dBm, and the diode with the HSMS2580 diodes has better efficiency for input power higher than -3 dBm. The same behavior is verified by inspecting **Figure 6d**, where the efficiencies of rectifiers E and F are presented. This can be explained by noting that the SMS7630 diode requires a forward bias voltage between 60 and 120 mV, and it has a breakdown voltage of 2 V, while the HSMS2850 diode has a forward bias voltage between 150 and 250 mV, and it has a breakdown voltage of 3.8 V. For the implementation of highefficiency UHF rectifiers with lower input power levels, the Skyworks

SMS7630 diodes are preferred, since improved RF-to-DC efficiency can be achieved as a result of the model's lower saturation current and lower junction capacitance.

By comparing the efficiencies for rectifiers C, D, E, and F, another important observation can be made. The use of series lumped inductors for the matching circuit seems to degrade the RF-to-DC efficiency, something verified by comparing the maximum efficiency of C (46%) with E (42%) and the efficiency of D (50%) with F (42%). For a simulated voltage doubler circuit, a real inductor was modeled as an ideal inductor in series with an ohmic resistor to model the inductor's losses. A small variation in the ohmic resistance has a direct significant impact on the simulated efficiency, as can be verified in **Figure 7**. Based on this observation, Rectifier H was fabricated using one printed distributed inductor (meander shaped thin line in the inset of **Figure 8b**) instead of lumped inductors in the matching circuit. The comparison between Rectifier H and Rectifier K is presented in **Figure 8**. Although the matching bandwidth for the rectifier with the distributed inductor decreased, the lack of ohmic losses for the packaged lumped inductor caused a small improvement in the efficiency, which can be verified in **Figure 8d**. This is also verified in the comparison between Rectifier A and Rectifier J presented in **Figure 9**, that were both fabricated on Rogers 5880 material. The efficiency of Rectifier A is consistently better than the efficiency of Rectifier J (**Figure 9d**), while the use of tapered lines for Rectifier A improved the bandwidth as well. Comparing the maximum efficiency of K (46%) from **Figure 8d** and the efficiency of J (56%) from **Figure 9d**, which have similar designs with the same diode models, it is obvious that the substrate losses are important in the resulting efficiency.

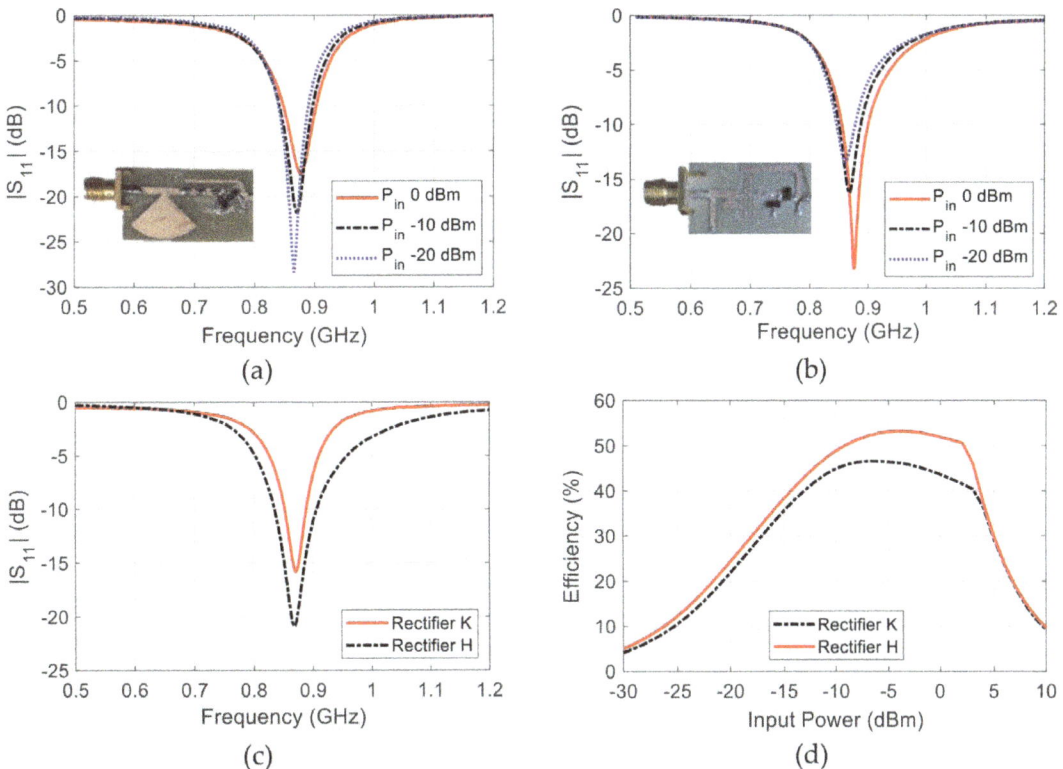

Figure 8. *Rectifier K and rectifier H comparison; (a) Rectifier E $|S_{11}|$, (b) Rectifier F $|S_{11}|$, (c) $|S_{11}|$ at -10 dBm comparison, and (d) efficiency.*

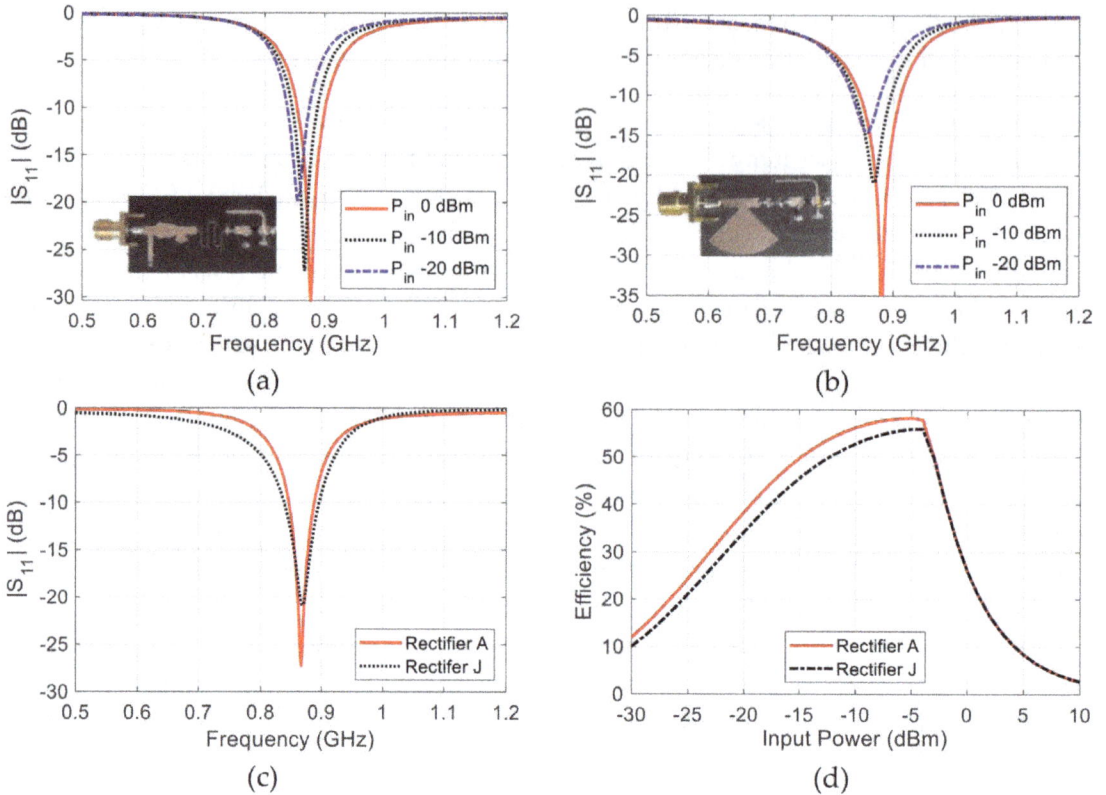

Figure 9. *Rectifier (A) and rectifier (J) comparison; (a) Rectifier A $|S_{11}|$, (b) Rectifier J $|S_{11}|$, (c) $|S_{11}|$ at -10 dBm*

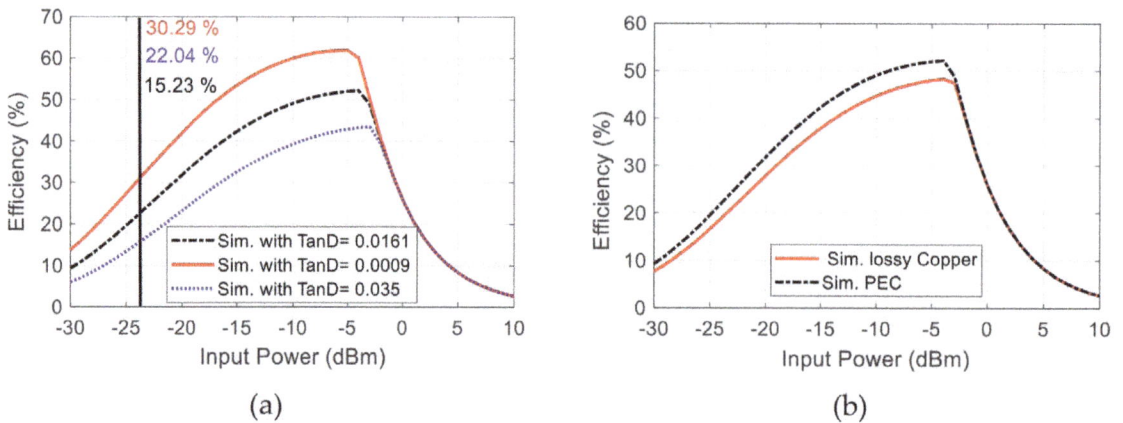

Figure 10. *Effect of the substrate loss on the simulated efficiency (a) with variations in the loss tangent (tanδ) and (b) variations in the copper's conductivity.*

When distributed inductors are used instead of lumped inductors, the substrate losses become even more critical for the implemented efficiency. Low loss substrates limit the quality factor of the implemented inductors and consequently degrade the efficiency. **Figure 10a** shows the simulated rectifier efficiency when the loss tangent (tanδ) of the substrate was varied. For the lower end of the input power, around -24 dB, the low loss of the Rogers 5880 (tanδ = 0.009) results in 30% efficiency, while for the same design, tanδ = 0.035 reduces the efficiency to approximately 15%. For low input power, even small variations in the copper's conductivity affect the efficiency, as can be verified in **Figure 10b**. The importance of a low-loss substrate is evident in the measured efficiencies for the implemented rectifiers presented in **Figure 11**, where the lossy FR-4 (tanδ = 0.016) was used for Rectifier H, and the more expensive low loss (tanδ = 0.0009) Rogers 5880 was used for Rectifier G. **Figure 11d** verifies that the efficiency of rectifier G fabricated on the low-loss material is consistently higher than the efficiency of the similar design Rectifier H.

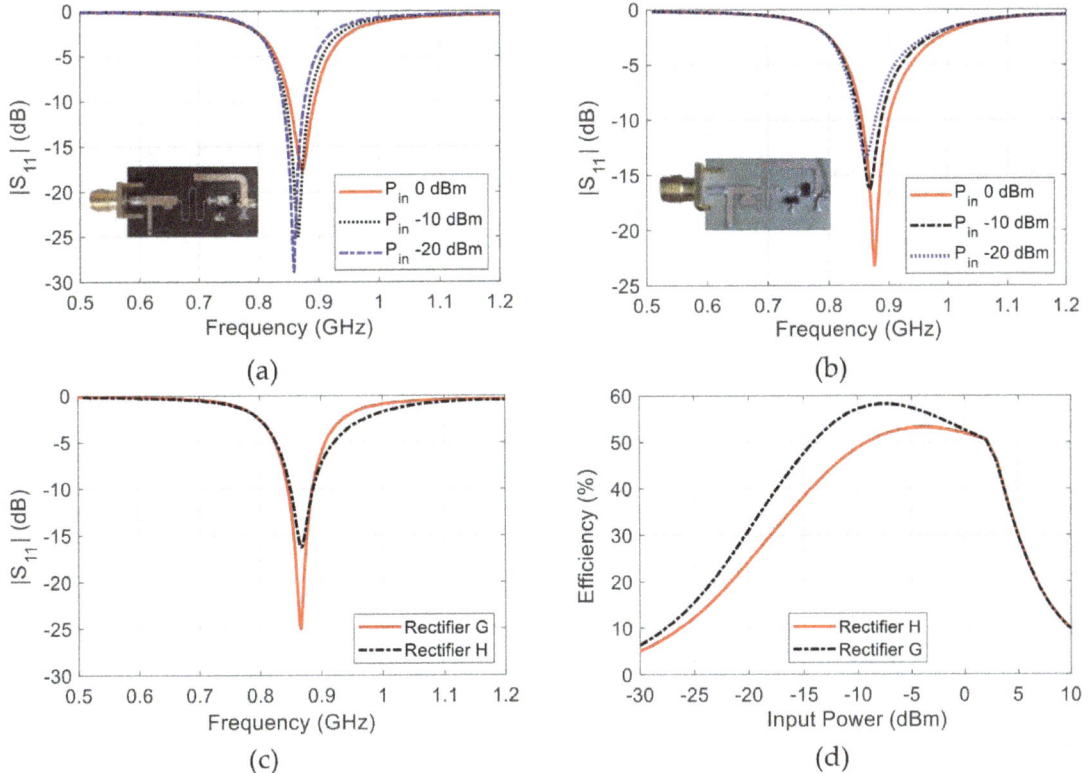

Figure 11. *Rectifier (G) and rectifier (H) comparison; (a) Rectifier E $|S_{11}|$, (b) rectifier F $|S_{11}|$, (c) $|S_{11}|$ at -10 dBm*

Size considerations

Considering that the fundamental voltage doubler topology that consists of two diodes and two capacitors is rather inflexible, the overall dimensions of any rectifier depend on the matching network. When a compact size is a priority, as for example in the case of implantable devices, the use of lumped components is preferred. **Figure 12** shows how the overall size of a rectifier can be significantly reduced with the use of lumped components instead of distributed ones.

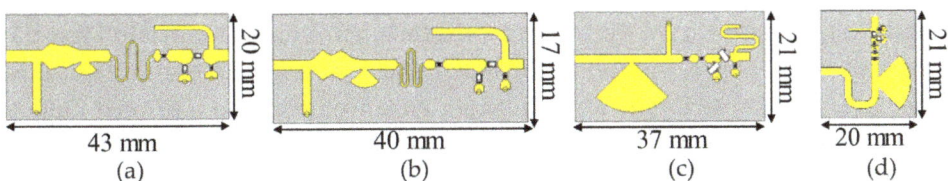

Figure 12. *Size comparison (a) Rectifier A, (b) Rectifier I, (c) Rectifier E, and (d) Rectifier F.*

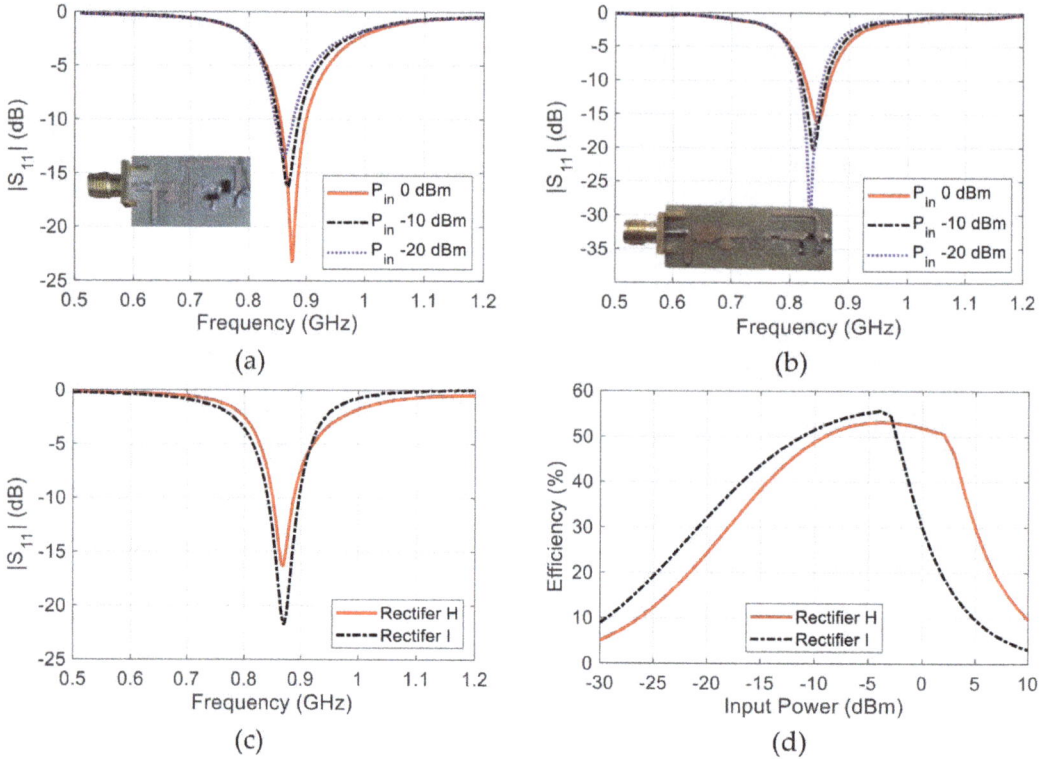

Figure 13. *Rectifier H and Rectifier I comparison; (a) Rectifier E $|S_{11}|$, (b) Rectifier F $|S_{11}|$, (c) $|S_{11}|$ at -10 dBm comparison, and (d) efficiency.*

Using lumped inductors for the matching network is often preferred because they can also result in a more wideband design. In order to exploit the low losses of distributed printed inductors and further improve the bandwidth of rectifiers with distributed inductors, the use of multi-stage tapered microstrip lines was considered. Rectifier I was implemented on FR-4 and used a two-stage tapered microstrip line. The performance of Rectifier I is compared with the performance of Rectifier H, which uses conventional microstrip lines, in **Figure 13**. The bandwidth of Rectifier I is 40 MHz and the bandwidth of I is 36 MHz. The 10% bandwidth enhancement (**Figure 13c**) in addition to the improved efficiency, as can be seen in **Figure 13d**, implies that if there are not strict size constraints, the best efficiency rectifier should have the following characteristics: (a) should use Skyworks 7630 diodes, (b) should use tapered microstrip lines with distributed inductors for the matching circuit, and (c) should be fabricated on low-loss material. From the implemented rectifiers A to K, Rectifier A has the aforementioned characteristics.

Proposed UHF rectifier – Rectifier A

The detailed design parameters of the implemented Rectifier A are presented in **Figure 14a** and are summarized in the figure caption. The matching network consists of a shorted linear stub, a two-stage tapered microstrip line for bandwidth enhancement, and a printed inductor. The maximum measured efficiency is almost 60% at -3 dB. For input power higher than -3 dB, the voltage saturates close to 2 V and as a result the RF-to-DC efficiency degrades, since only the denominator RF power increases while the DC power remains saturated. The simulated results presented in **Figure 14b** indicate the non-linear dependence of the efficiency with the termination load. Although the peak efficiency per power level is shifted, the rectifier parameters were optimized for a termination load equal to 13.5 kΩ. The combined effect of the input power and the termination load can be seen in the chart presented in **Figure 14c**, which presents the achieved efficiency when both input power and termination load are varied logarithmically.

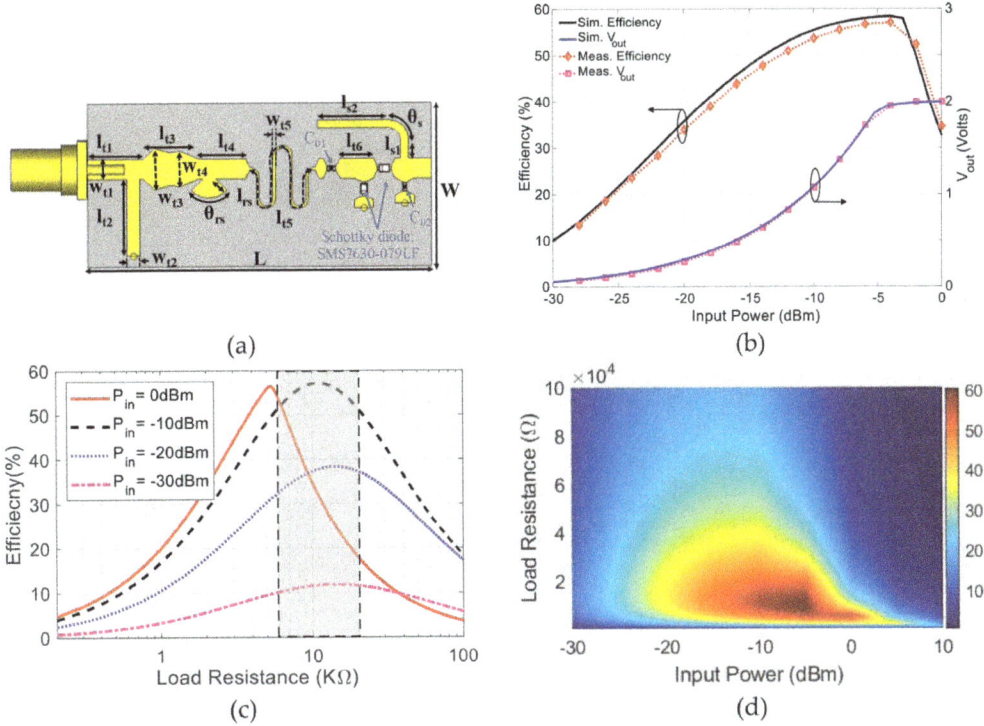

(a)

(b)

(c)

(d)

Figure 14. *Rectifier A (a) schematic with design details, (b) simulated and measured efficiency vs. input power, (c) efficiency versus termination load, and (d) efficiency as a function of input power and the load resistance. All dimensions are in mm: L= 43, l_{t1} =7, l_{t2} =9.5, l_{t3} =6.85, l_{t4} =5.67, l_{t5} =7, l_{t6} =4.14, l_{rs} =2, l_{s1} =2.2, l_{s2} =8.75, W= 20, w_{t1} = 2.4, w_{t2} = 1.6, w_{t3} = 5, w_{t4} = 4.35, w_{t5} = 0.315, θ_{rs} =60°, and θ_{s} =90°.*

Rectenna design

For most of the applications presented in the introduction, the rectifiers are not used as stand-alone components but they are part of a rectenna or a multi-stage energy harvesting system. In the literature, the term rectenna refers to the combination of an antenna cascaded with a rectifier which can convert a wirelessly received RF signal into DC voltage. This section presents a voltage doubler rectenna implementation, with a directive patch antenna suitable for targeted wireless power transfer. Another rectenna implementation with an omni-directional printed inverted F antenna (PIFA) suitable for collecting ambient RF power from random directions is presented as part of the energy harvesting discussed in the subsequent Section 4.

The implemented rectangular microstrip patch antenna has 7.6 dBi gain and 97% simulated radiation efficiency radiating effectively at 5.6 GHz. Good matching in ensured using an inset-microstrip line that has characteristic impedance 50 Ω. The directive patch antenna is cascaded with a rectifier, which is first implemented as a stand-alone device on the same Duroid material. The two standalone devices with their S-parameter measurements can be seen in **Figure 15a** and **b.** For the integration of the two components, full-wave simulations using lumped ports at the common connection point were carried out in CST Microwave Studio to take into account the effect of the radiating element on the performance of the rectifier. The diodes and the lumped capacitors were modeled using .s2p files in Circuit Design studio for co-simulations of the integrated module.

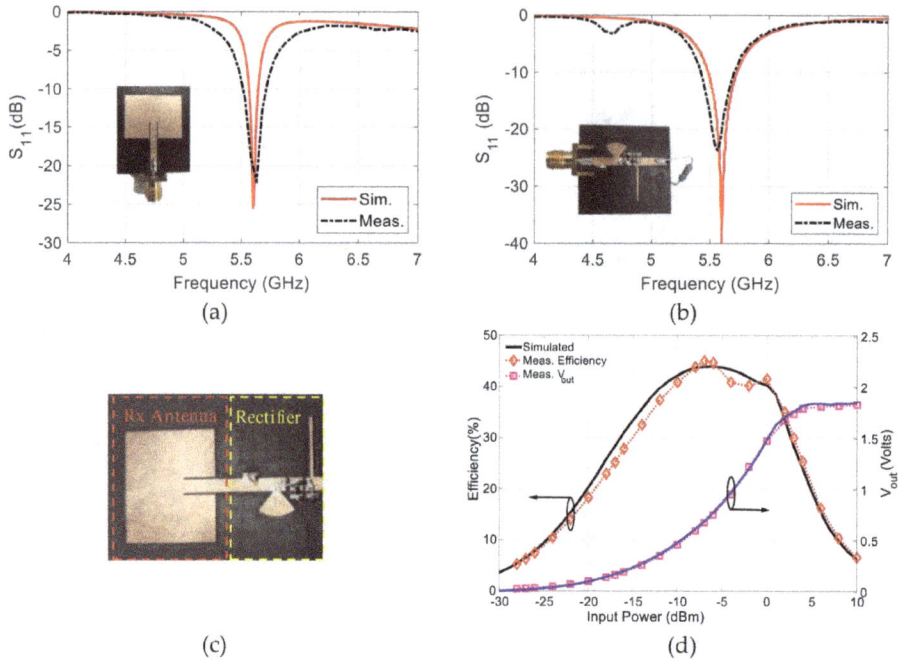

(a) (b)

(c) (d)

Figure 15. *5.6 GHz rectenna system; simulated and measured $|S_{11}|$ (a) of rectangular patch antenna, (b) 5.6 GHz rectifier, (c) 5.6 GHz rectenna on Roger RT/Duriod5880, and (d) simulated and measured efficiency and rectified voltage vs. input power [9].*

The integrated single-board directive rectenna can be seen in **Figure 15c**, and it can be used for far field wireless charging when the direction of arrival (DoA) is known and the directivity of the rectenna is aligned with it. The maximum implemented efficiency is around 43% and it occurs for -7 dBm input power.

Energy harvesting circuit

When energy harvesting (EH) or wireless power transfer (WPT) is used, the rectified DC voltage can be either temporarily stored until it reaches certain level, or it can be used directly without the need of any storage device. The most common storage devices are either rechargeable batteries or capacitors. When the rectified DC voltage is not stored, usually some kind of booster is needed in order to increase the low DC voltage level to make it suitable for the device that needs to be powered. Furthermore, the addition of subsequent stages unavoidably decreases the overall RF-to-DC efficiency. However, in many cases, the use of a booster is necessary.

Boosters can be either active which means that an external DC power source is used to bias the booster, or it can be entirely passive. The schematic of such an EH system is presented in **Figure 16**. In this section, the implementation of an energy harvesting circuit that consists of an omnidirectional PIFA antenna and a voltage doubler matched at 1.6 GHz, cascaded with an active DC-to-DC boost converter is discussed.

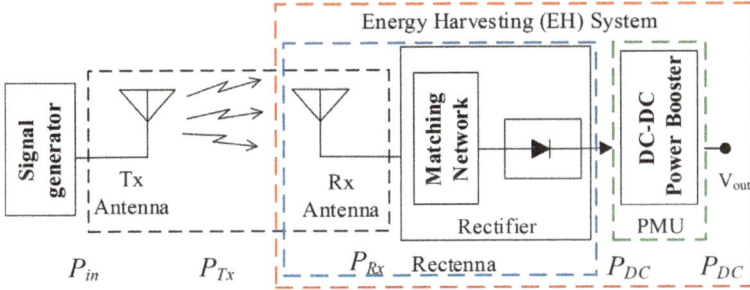

Figure 16. *Schematic diagram of RF energy harvesting system.*

The module, which is presented in **Figure 17**, was built on a Rogers 4003 material as a system on package (SoP) using a milling machine for the traces and the landing pads. The required lumped components, inductors, capacitors, Schotkky diodes, and the IC module were manually soldered on the traces. The PIFA used is presented in the inset photograph of **Figure 17a** along with the measured $|S_{11}|$, and the implemented voltage doubler presented in the inset photograph of **Figure 17b** along with its $|S_{11}|$. Considering the space limitations, the rectifier was designed in a Γ-shape, and radial stubs were used to ensure wideband matching in order to overcome the $|S_{11}|$ resonance shift in correspondence to the input power variations. The rectifier was terminated with the DC-to-DC booster built around the Texas Instruments TPS60301 charge pump IC model [21] that can be seen in **Figure 18a**. As mentioned earlier, the RF-to-DC efficiency depends on the termination load which in this case it is equal to the input impedance of the subsequent booster device. Although the input impedance of the power booster depends on its operation conditions, it is approximated to be 5.1 KΩ, and this is the assumed termination load for the rectifier design.

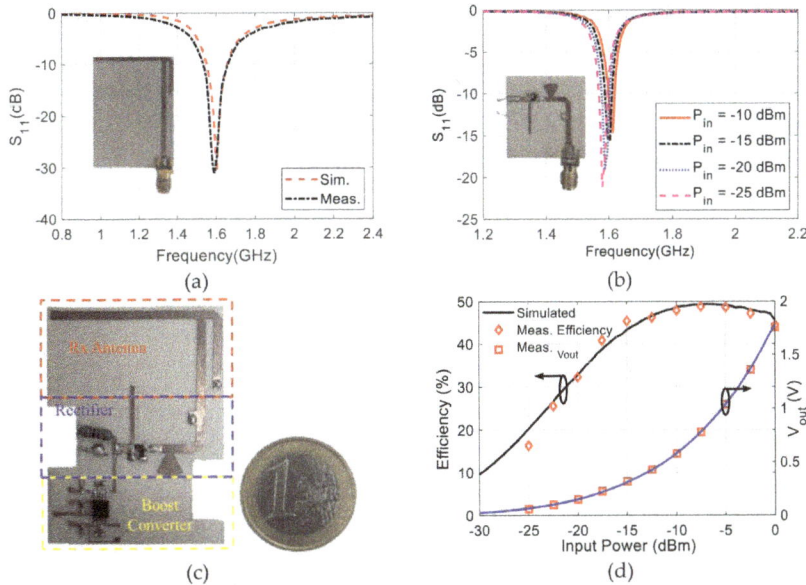

Figure 17. *1.6 GHz energy harvesting system (a) simulated and measured $|S_{11}|$ of PIFA antennae, (b) measured $|S_{11}|$ of 1.6 GHz rectifier, (c) 1.6 GHz rectenna system on Roger RO4003C, and (d) simulated and measured efficiency and rectified voltage vs. input power [10].*

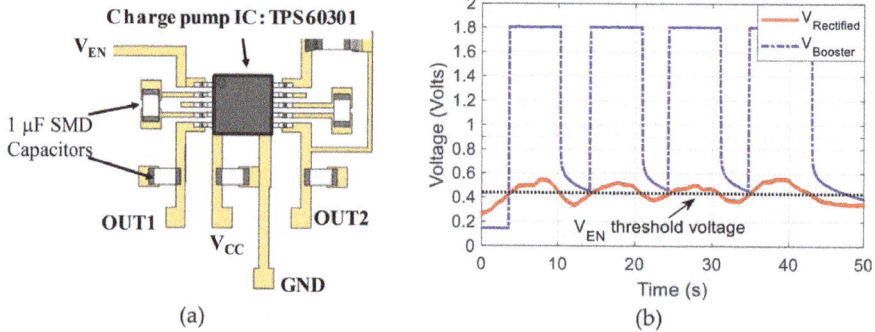

Figure 18. *Power management unit (a) schematic of DC-to-DC power booster and (b) measured voltage across DC-to-DC power booster and rectifier output with variable distance between transmitter and rectenna at 15 dBm transmitted power from the signal generator [10].*

As can be seen in the IC schematic, the enabling terminal (EN) must be connected with the output of the voltage doubler. When the rectified voltage and thus the V_{EN} voltage gets higher than 0.43 V, the DC voltage on the output terminal becomes equal to the Vcc biasing voltage, which can be anywhere between 0.9 and 1.8 V. This way the DC-to-DC booster output voltage can be elevated up to 1.8 V assuming that the rectified voltage is higher than 0.43 V. **Figure 18b** indicates how the output voltage of the booster ($V_{Booster}$) goes from 0 to 1.8 V depending on the continuous variation of the rectified voltage at the output of the voltage-doubler rectifier ($V_{Rectified}$) and whether this is larger or smaller than the minimum V_{EN} voltage, which is 0.43 V.

Conclusion

For many WPT applications, the voltage doubler is the preferred rectifier topology. This book chapter has outlined the operating principles of voltage-doubler rectifiers and has presented the most important design considerations for the implementation of such rectifying devices as standalone circuits. It has also presented examples of the use of voltage-doubler rectifiers as part of a directive rectenna system and as part of an energy harvesting system where a voltage- doubler rectifier was cascaded with an active DC-to-DC booster, demonstrating the successful implementation of wireless power transfer systems using voltage-doubler circuits as the preferred rectifier topologies.

Acknowledgements

This research was partially funded from the Cyprus RPF, under the RESTART 2016-2020 projects: SWITCH-EXCELLENCE/1216/376 and RF-META- INFRASTRUCTURES/1216/0042. The authors would like to thank Dr. Muhammad Babar Ali Abbasi from the Centre of Wireless Innovation (CWI) at Queen's University Belfast UK for his assistance and contribution.

Conflict of interest

All authors listed have contributed sufficiently to the project to be included as authors. The authors declare that there is no conflict of interest, financial or other regarding the publication of this book chapter.

Author details

Abdul Quddious1*, Marco A. Antoniades1, Photos Vryonides2 and Symeon Nikolaou2

1 Department of Electrical and Computer Engineering, University of Cyprus, Nicosia, Cyprus

2 Frederick University and Frederick Research Center, Nicosia, Cyprus

*Address all correspondence to: quddious.abdul@ucy.ac.cy

References

[1] Zhang H, Guo YX, Zhong Z, Wu W. Cooperative integration of RF energy harvesting and dedicated WPT for wireless sensor networks. IEEE Microwave and Wireless Components Letters. 2019;29(4):291-293. DOI: 10.1109/LMWC.2019.2902047

[2] Kim S, Vyas R, Bito J, Niotaki K, Collado A, Georgiadis A, et al. Ambient RF energy-harvesting technologies for self-sustainable standalone wireless sensor platforms. Proceedings of the IEEE. 2014;102(11): 1649-1666. DOI: 10.1109/JPROC.2014. 2357031

[3] Vandelle E, Vuong TP, Ardila G, Wu K, Hemour S. Harvesting ambient RF energy efficiently with optimal angular coverage. IEEE Transactions on Antennas and Propagation. 2018;67(3): 1862-1873. DOI: 10.1109/TAP.2018. 2888957

[4] Valenta CR, Durgin GD. Harvesting wireless power: Survey of energy- harvester conversion efficiency in far- field, wireless power transfer systems. IEEE Microwave Magazine. 2014;15(4): 108-120. DOI: 10.1109/MMM.2014. 2309499

[5] Mansour MM, Kanaya H. High- efficient broadband CPW RF rectifier for wireless energy harvesting. IEEE Microwave and Wireless Components Letters. 2019;29(4):288-290. DOI: 10.1109/LMWC.2019.2902461

[6] Zhang Q, Ou JH, Wu Z, Tan HZ. Novel microwave rectifier optimizing method and its application in rectenna designs. IEEE Access. 2018;6: 53557-53565. DOI: 10.1109/ACCESS. 2018.2871087

[7] Neophytou K, Antoniades MA. DC voltage boosting technique in RF wireless power transfer systems utilizing high PAPR digital modulations. IET Microwaves, Antennas and Propagation. 2019. DOI: 10.1049/iet- map.2018.5888

[8] Pan N, Belo D, Rajabi M, Schreurs D, Carvalho NB, Pollin S. Bandwidth analysis of RF-DC converters under multisine excitation. IEEE Transactions on Microwave Theory and Techniques. 2017;66(2):791-802. DOI: 10.1109/ TMTT.2017.2757473

[9] Quddious A, Abbasi MA, Tahir FA, Antoniades MA, Vryonides P, Nikolaou S. UWB antenna with dynamically reconfigurable notch-band using rectenna and active booster. IET Microwave Antennas and Propagation. 2019. DOI: 10.1049/iet-map.2018. 56642019

[10] Quddious A, Abbasi MA, Saghir A, Arain S, Antoniades MA, Polycarpou A, et al. Dynamically reconfigurable SIR filter using rectenna and active booster. IEEE Transactions on Microwave Theory and Techniques. 2019;67(4):1504-1515. DOI: 10.1109/ TMTT.2019.2891524

[11] Quddious A, Abbasi MA, Vryonides P, Nikolaou S, Antoniades MA, Manhaval B, et al. Reconfigurable notch-band UWB antenna with RF-to-DC rectifier for dynamic reconfigurability. In: 2018 IEEE International Symposium on Antennas and Propagation and USNC/ URSI National Radio Science Meeting. IEEE; 8th July; 2018. pp. 283-284. DOI: 10.1109/APUSN-CURSINRSM.2018. 8608685

[12] Quddious A, Abbasi MA, Tariq MH, Antoniades MA, Vryonides P, Nikolaou S. On the use of tunable power splitter for simultaneous wireless information and power transfer receivers. International Journal of Antennas and Propagation. 2018;2018: 12. DOI: 10.1155/2018/6183412. Article no: 6183412

[13] Pournoori N, Khan MW, Ukkonen L, Björninen T. RF energy harvesting system integrating a passive UHF RFID tag as a charge storage indicator. In: 2018 IEEE International Symposium on Antennas and Propagation and USNC/URSI National Radio Science Meeting. IEEE; 8th July; 2018. pp. 685-686. DOI: 10.1109/ APUSNCURSINRSM.2018.8608530

[14] Tentzeris MM, Nikolaou S. RFID- enabled ultrasensitive wireless sensors utilizing inkjet-printed antennas and carbon nanotubes for gas detection applications. In: 2009 IEEE International Conference on Microwaves, Communications, Antennas and Electronics Systems. IEEE; 9th November; 2009. pp. 1-5. DOI: 10.1109/ COMCAS.2009.5385940

[15] Rida A, Yang L, Reynolds T, Tan E, Nikolaou S, Tentzeris MM. Inkjet- printing UHF antenna for RFID and sensing applications on liquid crystal polymer. In: 2009 IEEE Antennas and Propagation Society International Symposium. IEEE; 1st June; 2009. pp. 1-4. DOI: 10.1109/APS.2009.5171791

[16] Rida A, Nikolaou S, Tentzeris MM. Broadband UHF RFID/sensor modules for pervasive cognition applications. In: 2009 3rd European Conference on Antennas and Propagation. IEEE; 23rd March 2009. pp. 2344-2347

[17] Abdulhadi AE, Abhari R. Multiport UHF RFID-tag antenna for enhanced energy harvesting of self-powered wireless sensors. IEEE Transactions on Industrial Informatics. 2015;12(2): 801-808. DOI: 10.1109/TII.2015. 2470538

[18] Yuan F. CMOS Circuits for Passive Wireless Microsystems. New York: Springer; 2011. DOI: 10.1007/978-1- 4419-7680-2

[19] Skyworks. Solutions. (February). Schottky Diode [online]. Available from: http://www.skyworksinc. com/ Product/511/SMS7630 [Accessed: February, 2019]

[20] Broadcom Inc. (February). Schottky Diode [online]. Available from: https:// docs.broadcom.com/ docs/hsms- 285x_equivalent_circuit [Accessed: February, 2019]

[21] Texas Instruments (February). Charge Pump IC: TPS60301. Available from: http://www.ti.com/product/ TPS60301 [Accessed: February, 2019]

8

NFC Sensors Based on Energy Harvesting for IoT Applications

Antonio Lazaro, Marti Boada, Ramon Villarino and David Girbau

Abstract

The availability of low-cost near-field communication (NFC) devices, the incorporation of NFC readers into most current mobile phones, and the inclusion of energy-harvesting (EH) capabilities in NFC chips make NFC a key technology for the development of green Internet of Things (IoT) applications. In this chapter, an overview of recent advances in the field of battery-less NFC sensors at 13.56 MHz is provided, and a comparison to other short-range RFID technologies is given. After reviewing power transfer in NFC, recommendations for the practical design of NFC-based sensor tags and NFC readers are made. A list of commercial NFC integrated circuits with energy-harvesting capabilities is also provided. A survey of recent battery-less NFC sensors developed by the group including soil moisture, water content, pH, color, and implanted NFC sensors is done.

Keywords: battery-less, near-field communication (NFC), radiofrequency identification (RFID), energy harvesting, implantable medical devices, wireless power transfer (WPT), internet of things (IoT)

Introduction

In recent years, radio identification (RFID) technology [1, 2] has grown extraordinarily. Initially, it came to replace barcodes for product identification [3]. Nowadays, there are several frequency bands and a number of applications [2]. RIFD can be classified depending on the frequency band used: low frequency (LF), high frequency (HF), ultrahigh frequency (UHF), or microwave bands [2]. Another classification refers to the power source: tags can be passive if they harvest the energy from the RF interrogating signal of the reader or active if the energy to feed the electronics is obtained from a battery. The communication in passive tags is based on backscattering communication and consists on modulating the load that modifies the radar cross section of the antenna [4, 5]. Although UHF readers are expensive (1–2 k$), the inlays designed for traceability are cheap, and the return on investment (ROI) is possible due to the large number of units involved. The read range can be increased by improving the IC sensitivity (e.g., -22 dBm for Impinj Monza R6) allowing to reach several meters (about 10–15 m). There exist also semipassive or battery-assisted tags (BAP) that use battery to feed only the electronics needed to improve the sensitivity, but the communication is based on backscattering as in the passive tags [6–8]. The cost of UHF BAP tags is often lower than active tags based on transceivers such as Bluetooth low energy (BLE), Zigbee, etc. Recently, several UHF BAP transponders have appeared in the market such as the AMS-SL900A IC [7, 9]. This chip integrates a 10-bit ADC and an internal temperature sensor. However, the sensitivity in passive mode (without battery) is lower (-6.9 dBm) than the BAP mode (-15 dBm) that allows a longer rad range and data logging [7]. Semi-passive tags are microwave frequencies, or zero power tags are recently under intense research for sensing [10, 11].

At LF and HF, the radio link is established by near-field communication (NFC) because the distance is smaller than the wavelength. Therefore, the tag is in the near-field region, and the magnetic field decays proportionally to the square of the distance. Therefore, the communication between the loop antennas of the reader and the tag is produced by inductive coupling. However, at UHF and microwave bands, the communication is based on the modulation of the far fields, and hence the electromagnetic fields decrease and become inversely proportional to the distance. As a result, the read range of these tags is higher than those based on nearfield communication [2, 4, 5].

Another recent approach to radio identification is the one known as chipless RFID [12, 13]. In these tags, the identification is done without a chip by a permanent modification of the structure that determines a specific electromagnetic signature in the frequency domain [12–15] or in the time domain [16, 17]. Initially, the main research efforts have been oriented to increase the number of bits that the electromagnetic structure can encode [13]. In frequency-coded chipless RFID, the number of bits is limited by the number of resonators with high-quality factors that can be integrated and detected in the bandwidth [13]. In time-coded chipless RFID, the number of bits is encoded in the length of the line, and it is limited by the time domain resolution of the radar that depends on its bandwidth and the losses of the lines [5]. The most important drawback of this technology is the detection of the tags. The change in the radar cross section (RCS) introduced by the chipless tag in frequency domain or in time domain is small compared with the clutter and objects in which the tag is attached. Therefore, in order to detect the tag, the background subtraction technique is often used in the literature. This technique consists of the subtraction of the background from the measurements using a previous saved measurement without the presence of the tag. In addition, the small perturbations introduced for the tag requires highly sensitive receivers; therefore, expensive vector network analyzers (VNAs) are often used as readers [18]. Another important drawback is the detection of several chipless tags in a small area due to the inexistence of an anti-collision protocol. The tags can only be identified by time gating if their spacing is higher than the spatial radar resolution [19] that is the function of the reader bandwidth. In order to mitigate these important issues, depolarizing tags [20] that reduce the clutter contribution and improve detection techniques have been developed in the literature [5, 21, 22]. In spite of these advances, the read range in frequency-coded chipless RFID is about 0.5–1 and 2–3 m for time-coded chipless tags. The inexistence of standardized readers in the market [18] and the limitations in the number of identification codes [13], in the read range, and in the cost (because it often uses high-frequency substrates, much more expensive than the substrates used in the inlay tags) are nowadays the challenges of this approach compared to UHF or other well-established RFID technologies. However, the interest of sensors based on chipless technology [23–26] is recently growing in those cases in which the robustness to harsh environments [26, 27] and the cost may be competitive with other sensor technologies. In sensor applications, the anti-collision issues are not critical in those applications in which the number of sensors is small, and few bits can be used for identification or other techniques can be employed.

Small number of resonators can be used for identification, and other resonators can be dedicated for sensing. The small number of resonators in chipless sensors simplifies its detection too and the bandwidth requirements [28].

Near-field communication technology at 13.56 MHz [29] is massively used for payment systems using NFC cards. Consequently, most smartphones include an NFC reader as it is shown in the exponential growing of number of NFC-enabled mobile devices depicted in **Figure 1** [30]. NFC employs electromagnetic induction between two loop antennas; therefore the range of these tags is constrained by the rapid drop of the magnetic field with the distance. However, enough energy can be harvested to feed low-power microcontrollers and sensors. As a consequence, the interest in the market of NFC-based sensors [31] is growing. Recently several manufacturers of NFC integrated circuits include the possibility of using energy harvesting (EH) in these integrated circuits. These components provide up to 3 mA at 2–3 V to feed low-power sensors [32]. These battery-less sensors can be interesting in numerous applications for short-range wireless reading of low-cost sensors. The use of mobiles as a reader for standardized tags enables a fast introduction of these sensors in the market. On the other hand, batteries contain toxic components that can contaminate foods and release waste pollutants to the environment [33]. Therefore, the use of battery-less sensors is a preferable choice in applications in which the devices are in contact with food or implanted in the body. In addition, in this last case, the replacement of the batteries is an important drawback.

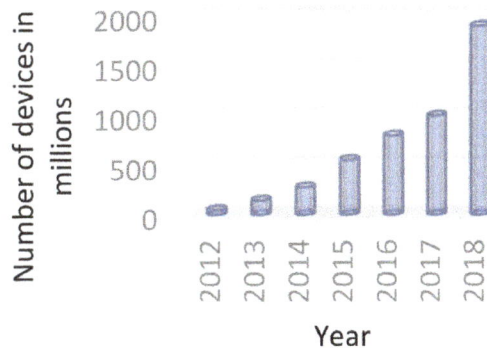

Figure 1. *Number of NFC-enabled mobile devices worldwide from 2012 to 2018 (in million units) [30].*

Nowadays, besides the traditional applications of identification, the interest of RFID technology in applications such as sensing [34] and localization [35, 36] (e.g., for robot guidance) is growing. **Table 1** summarizes the main properties of passive RFID used for sensing (chipless RFID, NFC, UHF, and RFID) and a BLE as a representative wireless technology used in modern wearable applications. The key features of RFID technologies over active wireless technologies (such as Bluetooth or Zigbee) are the cost and the short setup time due to simpler communication protocol. The absence of battery, or its long lifetime in the case of BAP tags, is another valuable feature of the RFID technology. Depending on the read range required, NFC or UHF can be used. Nowadays, chipless RFID is a promising technology for short-range sensing. However, it is not mature enough for commercial use yet. Advanced sensors require additional processing; therefore, active wireless communication technologies with integrated microcontroller are needed. For wearable applications, BLE is a good option because this technology is available in smartphones. The cost of the reader is another important point that determines the selection of the technology. While the NFC and Bluetooth are low cost because most of the smartphones are equipped with NFC and Bluetooth modules, UHF readers are expensive, and therefore this technology is used in professional or logistic users but not for personal applications. In sensing applications where the read range of the Bluetooth is limiting, other wireless communication technologies such as Wi-Fi, Zigbee networks for medium range, or GPRS, Sigfox, LoRa, or NB-IoT for long range can be considered [37].

Table 1. *Comparison of RFID technologies [32].*

Feature	Chipless RFID	NFC	UHF RFID	BLE
Communication method	Backscattering far field	Backscattering near field	Backscattering far field	Transceiver far field
Read range	Typically, <50 cm for frequency-coded 2–3 m for time-coded UWB	1–2 cm for proximity cards with energy harvesting, 0.5 m for vicinity cards	Up to 15 m with inlay tags with −22 dBm read IC sensitivity Up to 3 m UHF sensors (with −9 dBm read IC sensitivity) Up to 30 m BAP	About 10 m
Energy source	Passive	Passive or semi-passive	Passive or semi-passive	Active
Tag price	Moderate	Low	Low	High
Reader cost	High, no commercial	Low, smartphone	High 1–2 k$	Low, smartphone
Standard	No	Yes	Yes	Yes
Universal frequency regulation	No, often used UWB	Yes, ISM	No, by regions	Yes, ISM
Tag size	Large	Medium	Medium	Small
Memory capacity	<40 bits	<64 kbits	96bits EPC, typically 512 bits for users (<64kbytes)	Several Kbytes depending on the microcontroller
ID rewritable	No	Yes	Yes	Yes
Setup time	—	Less than 0.1 s	Less than 0.1 s	Approx. 6 s
Energy harvesting	No	Approx. 10 mW	Few μW	No
Tag substrate	Low-loss microwave substrates	Low cost or FR4	Low cost or FR4	FR4
Tag flexibility	Depends on the substrate	Depends on the substrate	Depends on the substrate	No
Tag robustness	High	Low (inlays)	Low (inlays)	Moderate

NFC energy harvesting

NFC communication

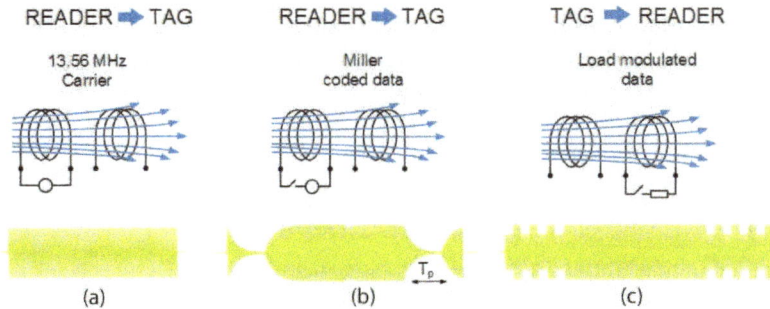

Figure 2. *Scheme of the communication between reader and tag: (a) power up the tag by the reader, (b) modulation of the carrier by the reader for transmission of the commands from the reader to the tag, (c) transmission the data from the tag to the reader by load modulation.*

NFC employs electromagnetic induction between two loop antennas. It operates within the globally available unlicensed radio frequency ISM band of 13.56 MHz on ISO/IEC 18000-3 air interface at rates ranging from 106 to 424 kbit/s [2]. The communication between reader and tag is briefly described in **Figure 2**. Firstly, the reader transmits an unmodulated carrier to activate the tag (**Figure 2a**). The receiver antenna at the tag is connected to the internal rectifier, which takes energy from the RF field to power up the tag electronics. After that, the reader sends commands by ASK modulating the carrier (**Figure 2b**), and the internal logic at the transponder demodulates the message. The tag transponder (which is assumed to be passive) responds using the passive load modulation technique (**Figure 2c**), by changing its antenna impedance [2, 29]. The passive load modulation spectrum consists of the RF carrier, two subcarriers at 12.712 and 14.408 MHz, and modulated sidebands on these two subcarrier signals (**Figure 3**). All the transmitted data are carried in the two sidebands.

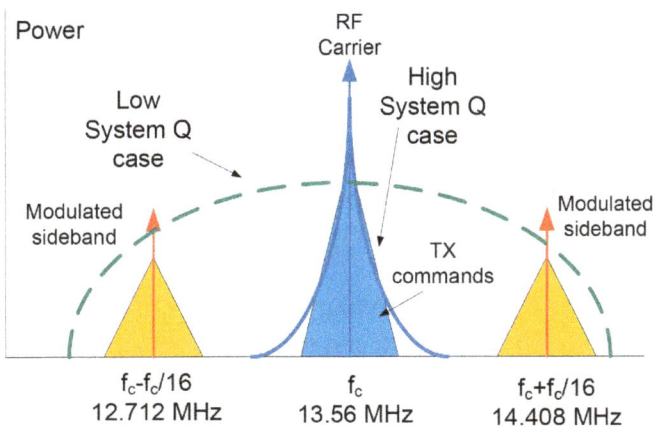

Figure 3. *A typical spectrum of an NFC system illustrating the reader command around the carrier frequency and the load modulation at the sidebands.*

An NFC tag IC is composed by two main blocks connected to the antenna ports (see **Figure 4**): the wireless power transfer unit that is responsible of harvesting the energy and powering the IC and the communication unit block that demodulates the messages and generates the clock for the transmission of data back to the reader. An RF limiter to protect the IC from a high-input voltage that could damage it, a rectifier, and a shunt regulator compose the WPT unit. The load modulator can be modeled as a shunt capacitance with the antenna. The IC impedance depends on the received power mainly due to the nonlinear behavior of the rectifier and the RF limiter. Therefore, a simplified model for the tag IC consisting of a nonlinear resistance R_{IC} in parallel with a capacitance C_{IC} that includes the parasitic elements is considered.

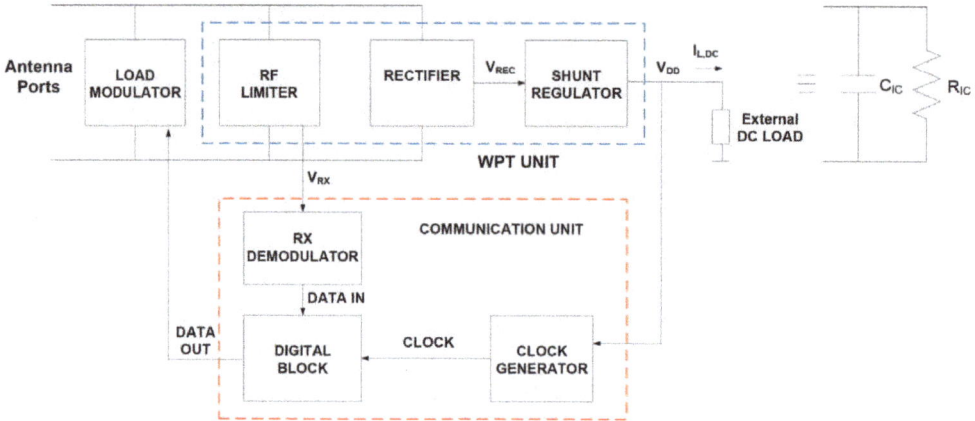

Figure 4. *Block diagram of NFC IC.*

Commercial tags IC with energy harvesting

Nowadays, the most important manufactures offer NFC IC with energyharvesting capabilities. These ICs have access to the internal output of the rectifier to feed external circuits such as microcontrollers or sensors. Some commercial NFC ICs with energy-harvesting mode are listed in **Table 2**. Most of them are compatible with standards ISO14443-3 or ISO15693 and can be connected to external microcontrollers using the I2C bus or serial-to-parallel interface (SPI). The internal EEPROM memory size to store the NFC Data Exchange Format (NDEF) message varies between 4 and 64 kbit. Although most NFC ICs are designed to be connected to a microcontroller, some models such as the MLX90129 from Melexis, the SL13 from AMS, and the SIC43x from Silicon Craft integrate an A/D for autonomous sensor acquisition without an external microcontroller, reducing the part count. In addition, the model RF430FRL152H from TI integrates a low-power microcontroller (MSP430) and a 14-bit A/D. The maximum sink current that can be drawn to power external devices connected to the EH output such as sensors or microcontrollers varies with the IC model and operation mode. Most of the models can configure different current levels depending on the range of magnetic field. The current to be provided depends on field strength, antenna size, and Q factor, but it is typically limited to the order of 5 mA with output voltages between 2 and 3 V for magnetic fields of the order of 3.5–5 A/m [32].

WPT in NFC systems and practical considerations

NFC reader

A simplified model of the RF front-end of an NFC reader transceiver is shown in **Figure 5**. The topology of the transmitter usually presents differential output, and it is connected to the loop antenna through a matching network to obtain the maximum transmitted power. The output resistance (R_{out}) is a function of the transmitter power, and the DC voltage determines the current consumption of the transmitter. The matching network is composed of a differential L matching net- work (C_s and C_p). The current that flows through the loop antenna produces the magnetic field. In order to fulfill the emission regulations, a low-pass LC filter to reduce the spurious emission at the second harmonic is included. The receiver shares the loop antenna with the transmitter; therefore, two series resistances R_x are often added to attenuate the signal, avoiding the AC voltage at the receiver input to exceed a limit value and saturating it. Sometimes, a capacitive splitter is used for that purpose.

Table 2. *Main features of some commercial NFC ICs with energy harvesting [32].*

IC model (manufacturer)	Energy harvesting Max. sink current and typical voltage	Standard	ADC	Interface bus	Memory and other comments
M24LR-E-R ST25DV-I2C (STMicro-electronics)	6 mA/3 V	ISO15693	No	I2C	4–64kBit
NT3H1101 NT3H1201 (NXP)	5 mA/2 V	ISO14443-3	No	I2C	8kBit/16kBit
NF4 (EM Microelectronics)	5 mA/3.6 V	ISO14443A	No	SPI	8kBit/32kBit/ 64KBit
GT23SC6699–1/2 (Giantec Semiconductor)	NA/3.2 V	ISO14443-3	No	I2C	8kBit/16kBit
SIC4310 SIC4340 SIC4341 (Silicon Craft)	10 mA/3.3 V	ISO14443A	No Yes	UART	220 bytes EEPROM
AS3953A (AMS AG)	5 mA/2 V	ISO14443A-4	No	SPI	1KBit
SL13 (AMS AG)	4 mA/3.4 V	ISO15693	Yes	SPI	8 kBit temperature sensor integrated
MLX90129 (Melexis)	5 mA/3 V	ISO15693	Yes	SPI	4kBit
RF430FRL152H (Texas Inst.)	NA/3 V	ISO15693	Yes	I2C/SPI	MSP430 2kB FRAM

The input impedance of the receiver is mainly capacitive, and it is modeled with the capacitance C_{in} in **Figure 5**. Although the overall impedance of the R_x and C_{in} is high, the effect of the receiver impedance must be considered in the design of the matching network. The steps to design this matching network can be found in [32]. In addition, recently, modern transceivers such as the ST25R391x incorporate an automatic antenna tuning method, enabling the matching network design in which C_p may be modified by switching an external bank of capacitors [38].

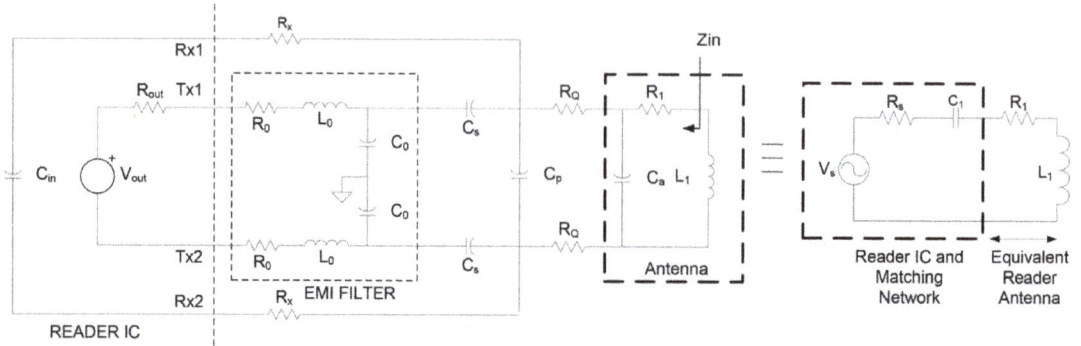

Figure 5.*Model of the reader including the matching network, EMI filter, and antenna model and the simplified equivalent circuit model (right).*

Printed loop antennas are used in NFC reader designs. There is a great variety of them depending on the reader type, but they are often designed to maximize the coupling with a payment card. Therefore, relatively large antennas between 4 and 5 cm square loops are integrated. The quality factor of these printed antennas is often high (more than 70). As the quality factor of the transmitter determines its bandwidth B ($B = f_c/Q$), the quality factor (Q_T) must be reduced to allow the transmission of the sidebands of the carrier. To this end, two series resistances R_Q are added at the input. The maximum quality factor of the transmitter is limited by the duration of the pause T_p in the modified Miller bit coding (see **Figure 2**):

$$Q_{Tmax} = fc \cdot T_p \tag{1}$$

The quality factor is limited to 40 and 128, for standards ISO 14443 and ISO 15693, respectively (35 and 100 considering design tolerances). Condition (1) represents an additional restriction for energy harvesting compared with other WPT systems for battery charging (with or without very small communication rate) where high Q coils can be used in the transmitter and receiver to increase the power efficiency.

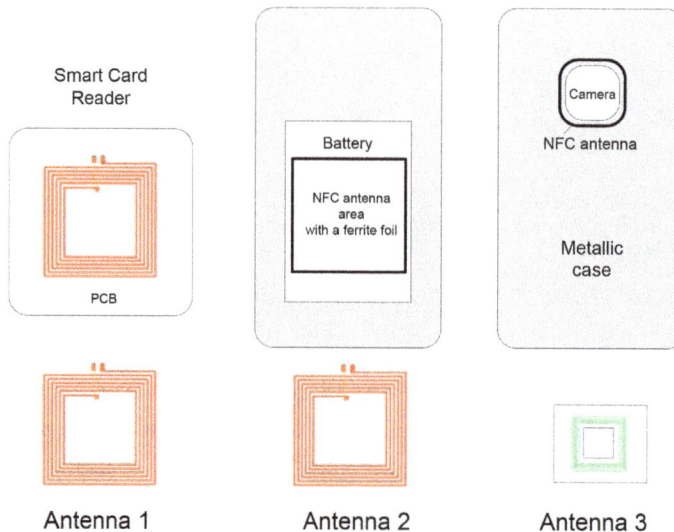

Figure 6. *Typical NFC reader antennas [41]: Antenna over a PCB (antenna 1), antenna on top of a mobile battery (antenna 2), and antenna around the camera hole (antenna 3).*

Despite the transmitter's complexity, it can be modeled using a simplified Thevenin equivalent circuit tuned at f_c = 13.56 MHz, used for circuit analysis (see **Figure 5**, right). The parasitic capacitance of the transmit antenna is included within the equivalent circuit. The relatively low-quality factor of the coils simplifies the antenna reader design. Antenna inductance varies typically between 1 and 3.5 µH. Several formulas for inductance calculation can be found depending on the shape of the printed antenna [39, 40]. In addition, the inductance and the equivalent resistance (or quality factor) can be analyzed using electromagnetic simulators such as Keysight ADS, HFSS, or CST studio, among others. The main challenge in the antenna design arises when it must be integrated into the reader case, since the proximity of metallic parts or other components (e.g., displays, batteries, or PCB boards) can modify the inductance and the parasitic capacitance. The worst case is the presence of NFC transceiver in modern smartphones, due to space limitations. **Figure 6** considers three different types of reader antennas including the typical locations of the NFC antenna into the smartphone [41]. **Table 3** compares the main parameters analyzed. Antenna 1 corresponds to a simple printed antenna over a PCB such as used on readers connected to PC through the USB port. Antenna 2 corresponds to a typical smartphone with a plastic case in which the antenna is attached to the battery pack of the mobile. A sheet of ferrite isolates the traces from the metal on the bottom plane. Antenna 3 is known as Murata solution [42], and it is based on the integration of the loop antenna around the hole of the camera.

Table 3. *Dimensions and simulated parameters of the reader antennas of* **Figure 6**.

Parameter	Antenna 1	Antenna 2	Antenna 3
Loop area (mm × mm)	50 × 50	50 × 50	25 × 25
Trace width (mm)	0.7	0.7	0.25
Spacing between traces (mm)	0.7	0.7	0.25
Number of turns	6	6	8
Substrate relative permittivity	4.7	4.7	3
Substrate thickness (mm)	0.8	0.8	0.1
Ferrite relative permeability*	—	150-j5	150-j5
Inductance (µH)	2.68	3.92	3.63
Self-resonance frequency (MHz)	77.0	37.4	79.05
Quality factor at 13.56 MHz	153	132	74

Antenna 1 and Antenna 2 have the same dimensions, but Antenna 2 has a sheet of ferrite.
100 µm thickness with 50 µm of adhesive plastic sheet of MHLL12060-200 from Laird (with μ_r = 150-j5) is considered.

The magnetic field generated or collected by a coil over or close to a metallic surface is almost parallel to this surface. Hence the magnetic flux is concentrated in the proximity of the coil [43], whereas it is zero in its center because of the cancelation of the field produced by the image currents with opposite sign. The effect is a considerable reduction in the read range and the increasing of the losses. A layer of ferrite material is placed under the coil, in order to isolate the antenna in the presence of metal. Modern sintered ferrite sheets with $Re(\mu_r)$ between 100 and 190 and $Im(\mu_r)$ losses typically of 5–10 at 13.56 MHz can be found in the market (e.g., MHLL12060-000 from Laird). As it can be appreciated in **Table 3**, the high magnetic permeability (μ_r) increases the inductance depending on the thickness of the ferrite foil. The effect of the ferrite loss produces an increasing of the antenna resistance and therefore a reduction of the antenna quality factor as it can be seen in **Table 3**. This is not an inconvenience so that it will only be necessary to readjust the matching network. Consequently, the inclusion of a ferrite layer modifies the field boundary conditions, making the magnetic field almost perpendicular, creating a scenario similar to the case of free space [32, 44].

Tag design considerations

Figure 7 shows a simplified equivalent circuit of the NFC system. The transmitter and the IC are modeled with the Thevenin model of **Figure 5** and its equivalent input impedance (**Figure 4**), respectively. The tag antenna is modeled with an inductance L_2, its equivalent resistance R_2 that takes into account the antenna losses, and its parasitic capacitance C_p. The coupling between the transmitter and receiver coils allows the transfer of energy between the transmitter and the tag. In order to achieve high efficiency, the resonance frequency of the tag must be adjusted to the operation frequency (13.56 MHz).

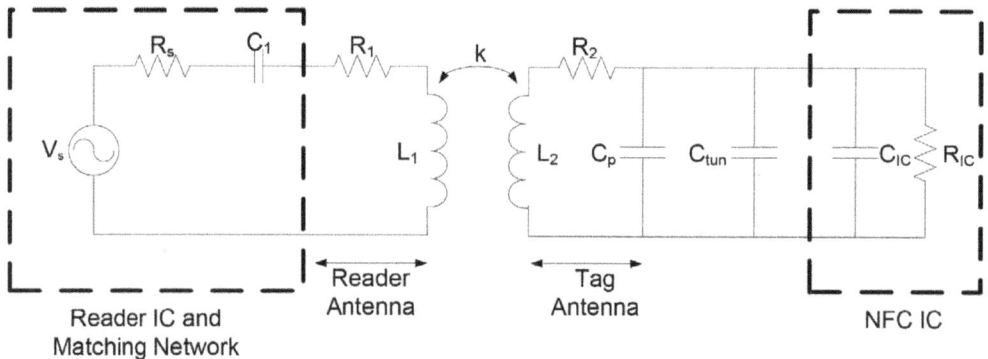

Figure 7. *Wireless power transfer model between the reader and the tag.*

The design of an NFC tag is relatively easy. The resonance frequency is obtained from the simplified equivalent circuit of **Figure 4**, and it is given by:

$$f_r \approx \frac{1}{2\pi\sqrt{L_2\left(C_{IC} + C_p + C_{tun}\right)}} \tag{2}$$

where L_2 is the inductance of the tag antenna, C_{IC} is the equivalent capacitance of the NFC IC, C_p is the parasitic capacitance of the antenna and interconnections, and C_{tun} is an additional tuning capacitance connected in parallel at the input, used to adjust the resonance frequency. C_{IC} is nonlinear and varies about 5% with the received power. The typical values of C_{IC} provided by the IC manufactures vary between 20 and 50 pF when low-input levels are applied on them.

Similar to the transmitter, the loaded quality factor of the tag (Q_{2L}) must be low enough to achieve the communication bandwidth, avoiding the attenuation of the modulation of the subcarrier in the backscattering link. Therefore, the maximum value of Q_{2L} is approximately $Q_{2L} = 8\pi \approx 25$ [29]. The loaded quality factor (Q_{2L}) is the hyperbolic average of the quality factor of the tag's antenna (Q_2) and the loaded quality factor of the IC (Q_L). Therefore, for typical values of printed loop antennas ($Q_2 > 50$), the loaded quality factor is approximately Q_L:

$$Q_{2L} = \frac{1}{\frac{1}{Q_2} + \frac{1}{Q_L}} \approx Q_L \tag{3}$$

$$Q_L = \frac{R_{IC}}{\omega L_2} \tag{4}$$

Consequently, the tag quality factor $Q_{2L} < Q_{2Lmax}$ if the following condition is fulfilled:

$$L_2 > \frac{R_{IC}}{\omega_0 Q_{2L,\,max}} \tag{5}$$

R_{IC} is highly nonlinear and decreases with the input level. R_{IC} takes typical values between 1500 and 400 in commercial ICs for the activation input level [45]. Therefore, condition (5) can be easily fulfilled for tag inductances higher than 0.7 μH, and in consequence, tag design is simple because it only is necessary to tune the resonance frequency with the capacitance C_{tun} for the desired loop antenna.

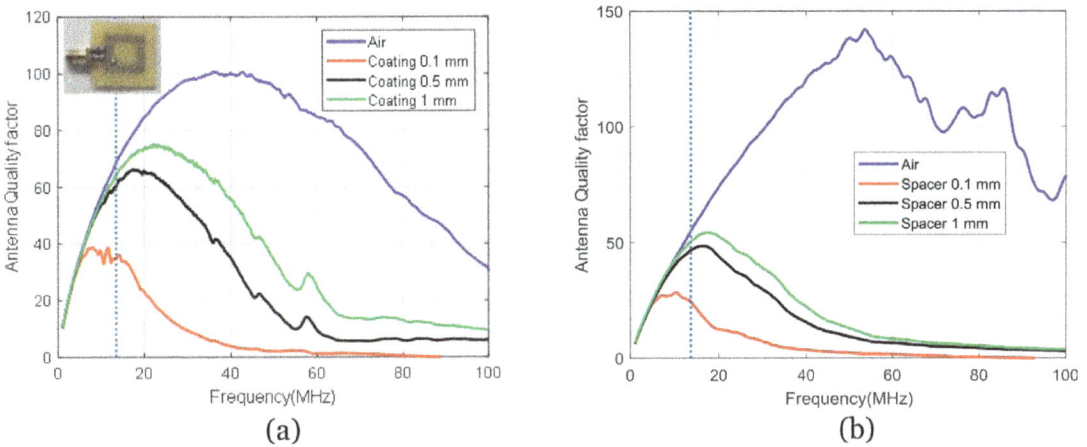

Figure 8. *(a) Measured quality factor of a coil in the air and inside of a phantom for different coating thickness, (b) measured quality factor of a thin coil over the skin as a function of the spacer thickness.*

To this end, it is important to know the environment in which the tag will be used. The presence of high-permittivity materials such as the body or, for example, on wearable or implanted applications can detune the tag. These materials are nonmagnetic; therefore, they do not modify the loop inductance, but change the parasitic capacitance, reducing the resonance frequency and increasing losses. **Figure 8a** shows the measured results of an implanted antenna in a phantom that consists of a piece of pork steak. The antenna is a 15 x 15 mm square coil consisting of six loops, printed in duplicates on each of the layers of a FR4 board whose thickness is 0.8 mm. The introduction of a low-permittivity cover (e.g., silicone, $\varepsilon_r = 3$) allows to reduce the losses of the phantom when the thickness is larger than 1 mm. Similar results are found for a coil close to the body surface (**Figure 8b**) considering a low-permittivity spacer ($\varepsilon_r = 3$) located between the coil and the body. Here the coil is a 25 x 25 mm square consisting of four loops printed on a thin flexible substrate (Ultralam 3000, thickness 100 μm).

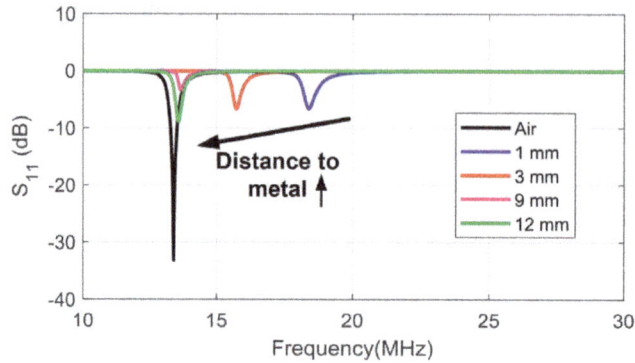

Figure 9. *Measured S_{11} of the test antenna as a function of frequency for different distances between the tag and the mobile [32].*

Another typical situation is the detuning due to the presence of nearby metal surfaces. If the NFC sensor is going to work on a metallic surface, the antenna can be isolated using a ferrite foil as in the case of the reader. But the typical situation takes place in the tags that will be read by smartphones with a metallic case. In this case, the inductance of the tag will be reduced by the presence of the metal due to the opposite image currents induced on it. The result is a detuning of the tag as can be shown in **Figure 9**. In order to check if a tag is correctly tuned, a test coil can be connected to the port 1 of a vector network analyzer. An absorption peak will be observed at the resonance frequency if the test coil is located close to the tag (but sufficiently spaced to avoid strong coupling between the coils). **Figure 10** shows the inductance of a coil as a function of the metal distance. A considerable reduction of its value is observed. Therefore, the tag must be tuned depending on the application and the desired read range of the tag.

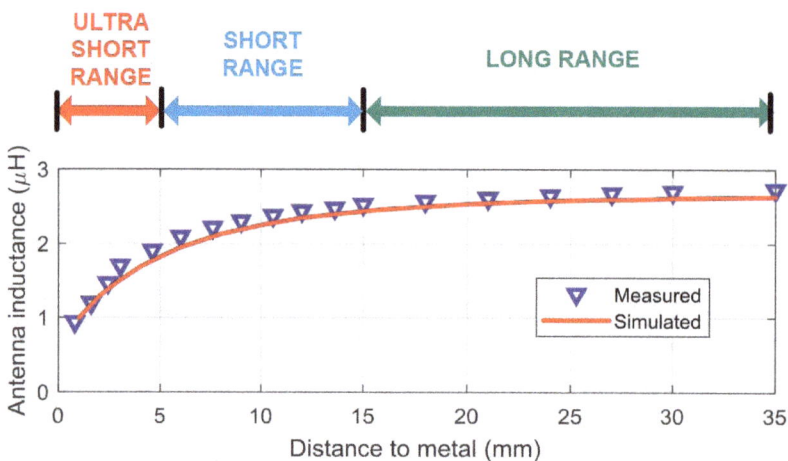

Figure 10. *Simulated and measured antenna inductance as a function of the distance to a ground plane for a 50 x 50 mm loop antenna with 0.7 mm trace width and six turns, manufactured on a 0.8-mm-thick FR4 [32].*

The area available for the antenna depends on the application. For example, **Table 4** shows the design of a tag sensor using the coil of **Figure 10** and an M24LR04E-R IC from ST for short range

(5 mm) and long range (15 mm). For long ranges (distances higher than 15 mm), the inductance and the other antenna parameters (resistance and capacitance) are nearly constant and close to the values of the free space case. The tuning capacitor (C_{tun}) that must be mounted in parallel at the input of the NFC IC is obtained from (2), and it is approximately 25 pF. However, for short read range applications in the presence of metal, the values of inductance for a distance of 5 mm must be considered. In this case, the tuning capacitance must be increased up to 36 pF. The inductance for each case is obtained from the measured impedance with the VNA, and the parasitic capacitance (C_p) is obtained from the self-resonance frequency of the antenna (without considering the tuning capacitor). If the antenna cannot be measured, e.g., due to the lack of laboratory instruments, the tuning capacitance can be obtained considering that the nominal inductance (value measured or computed in free space) decreases about 16% for short-range case. For very short ranges, the coupling coefficient k between the transmitter and tag antennas is very high (typically $k > 0.1$). Therefore, the input impedance at the transceiver increases (see **Figure** 7) due to the loading effect of the tag, and it is detuned because the matching network is designed considering that it is far from the reader and hence uncoupled to it. This is the reason why sometimes the tags are read worse when are on contact or very close to the reader than when are more spaced.

Table 4. *Example of design for short-range application (5 mm) and long-range application (15 mm).*

Parameter	Tuning for short range (5 mm)	Tuning for long range (15 mm)
Loop area (mm × mm)	50 × 50	50 × 50
Trace width (mm)	0.7	0.7
Spacing between traces (mm)	0.7	0.7
Number of turns	6	6
FR4 substrate thickness (mm)	0.8	0.8
Metal thickness (μm)	34	34
Chip capacitance C_{IC} (pF)	27.5	27.5
Inductance L_2 (μH)	2.68	3.92
Self-resonance frequency (MHz)	52.9	55.7
Antenna quality factor Q_2	87	118
Parasitic capacitance C_p (pF)	2.3	1.7
Tuning capacitance C_{tun} (pF)	36	25

WPT and read range

The communication in an NFC system can be limited by the uplink (reader to tag) when the tag does not receive enough energy to power the transponder IC or by the downlink (tag to reader) if the backscattered level received at the NFC reader is not sufficiently high to demodulate the message. For an NFC IC equipped with energy harvesting, the main limitation is the WPT in the uplink because it is necessary to provide higher energy than in a conventional NFC IC without energy harvesting. Thus, the read range in NFC tags with the EH mode activated is lower than the range for reading a previously stored data in the memory.

At this point it is interesting to take into account the factors that limit the read range of an EH NFC sensor. To this end, the wireless power transfer between the reader and tag can be analyzed considering the circuit of **Figure 7**. Analytical formulas found in the literature or a circuit simulator such as Keysight ADS can be used. In this model, $L1$ is the reader antenna inductance and R_1 is its losses. C_1 is the equivalent series capacitance that allows the circuit to resonate at 13.56 MHz. $L1$ and $R1$ can be obtained from the measurement of the antenna impedance or from electromagnetic simulations. R_s is the Thevenin output resistance, and it is chosen to achieve the maximum loaded Q factor in the transmitter to ensure the communication.

As the reader in the smartphone is designed to read both ISO 14443A and ISO 15693 standards, the Q_T is often limited to 35. Therefore, R_s is obtained from the loaded quality factor of the transmitter at 13.56 MHz ($R_s = \omega L_1/Q_T - R_1$). The inductance L_2 in series with a loss resistance R_2 is used to model the tag antenna. A key parameter is the mutual inductance M between the coils, as a function of the distance. The coupling coefficient k is obtained from the Z parameters from electromagnetic simulations or S parameters measurements performed with VNA:

$$k = \frac{M}{\sqrt{L_1 L_2}} = \frac{|\, Im\,(Z_{12})|}{\sqrt{Im\,(Z_{11}) \cdot Im\,(Z_{22})}} \tag{6}$$

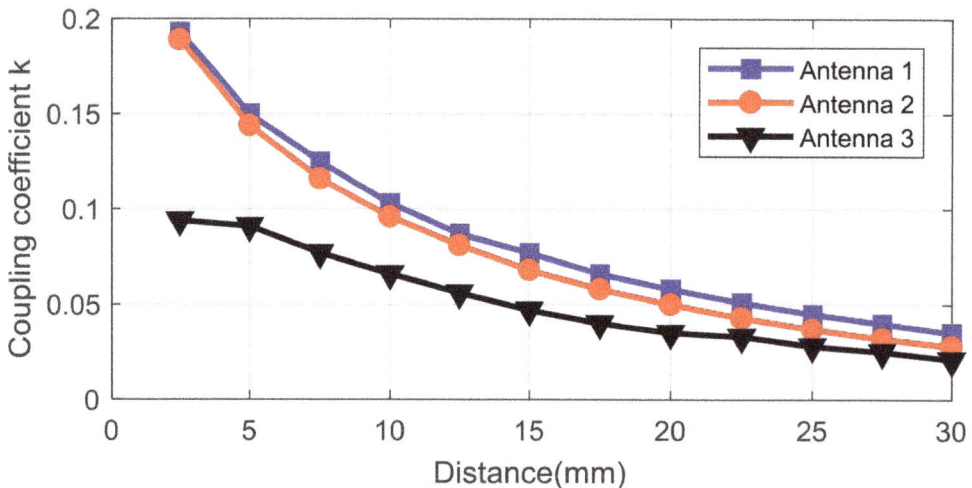

Figure 11. *Simulated coupling coefficient between a tag using antenna 1 and different reader antennas of* **Figure 6**.

Figure 11 compares the simulated coupling coefficient for each reader antenna as a function of the distance between the tag and reader antenna. Antenna 1 has been used as the tag antenna since it has a size similar to a smart card. The simulations have been done using Keysight Momentum. The coupling coefficient is obtained using (6) from the simulated Z parameters. **Figure 11** shows a similar coupling coefficient for Antennas 1 and 2 (see **Figure 6**). Therefore, the introduction of the ferrite foil effectively mitigates the effect of the metal. On the other hand, the coupling coefficient of Antenna 3 (**Figure 6**) is smaller than in the previous cases because its area is the smallest.

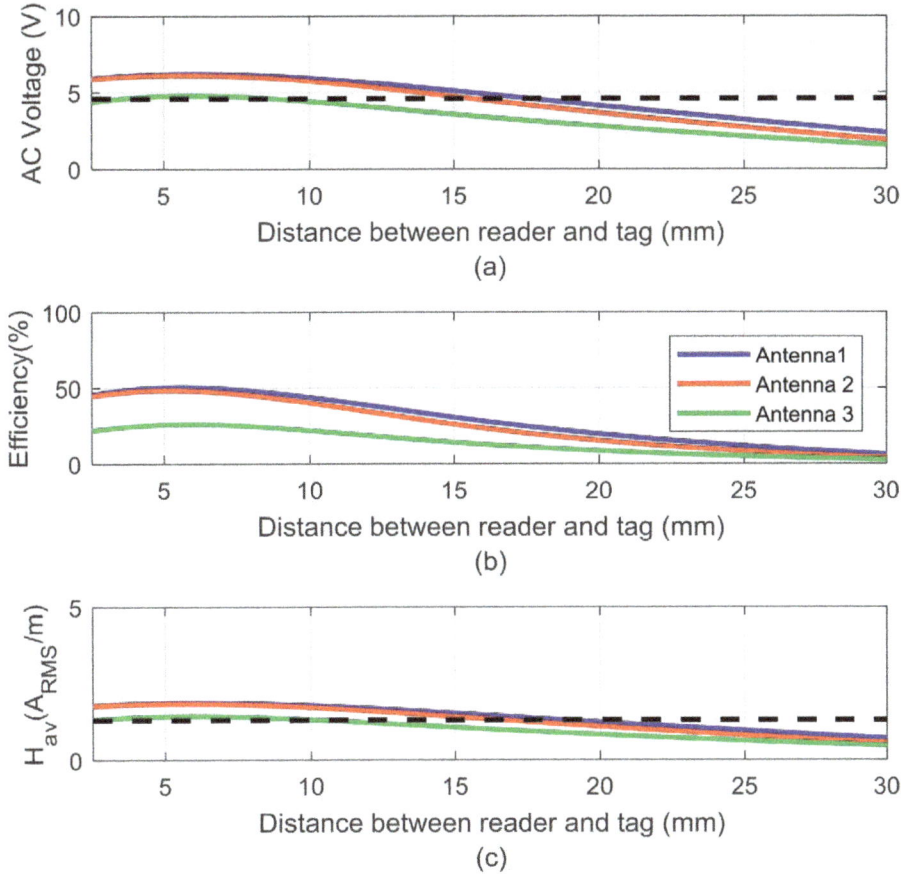

Figure 12. *(a) Simulated voltage at the input of the tag at 13.56 MHz (V_{AC}), (b) efficiency, and (c) average magnetic field H_{av} as a function of the distance for 20 dBm of transmitted power. The threshold limit is shown in dashed line.*

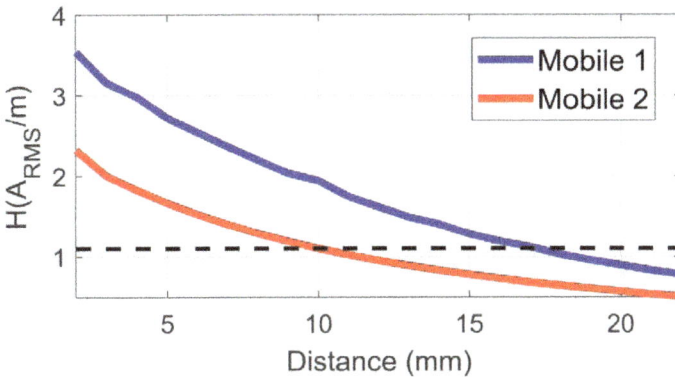

Figure 13. *Measured average magnetic field H_{av} (A_{RMS}/m) for a tag using antenna 1 as a function of the tag-to-reader distance for two mobile models.*

The circuit of **Figure 7** can be analyzed both analytically [46] or using a circuit simulator (e.g., Keysight ADS). In any case, the NFC IC is considered linear and modeled with its equivalent circuit. The values of R_{IC} and C_{IC} depend on the power. In addition, the manufacturer does not often give R_{IC}. This equivalent circuit can be measured using a VNA. However, the output level of commercial VNA does not achieve the typical level in a tag under real conditions. A modified high-power VNA with an external reflectometer and a power amplifier can be used, as it is proposed in Refs. [45, 47]. An alternative setup with an oscilloscope has been recently presented in [48]. However, these setups are not often available for the users interested in the design of NFC-based sensors. In order to estimate the maximum read range, the value of R_{IC} at that distance must be known. It can be estimated from the minimum average magnetic field, H_{min}. This is the threshold average magnetic field that ensures the RF to DC conversion at the read range and can be computed from Ref. [45]:

$$H_{min} \approx \frac{\sqrt{\left[1 - \left(\frac{f}{f_r}\right)^2\right]^2 + \frac{1}{Q_{2L}^2}}}{2\pi f \mu_0 A \cdot N} \cdot U_{min} \qquad (7)$$

where f_r is the resonance frequency of the tag, A is the loop area, N is the number of loops, and U_{min} is the minimum AC threshold voltage at the IC input required for the NFC IC operation. Therefore, H_{min} depends only on the tag parameters, and it can be considered a figure of merit of the tag. It can be experimentally found measuring the average magnetic field generated by the reader as a function of the distance between the tag and reader. A measurement procedure can be found in [32, 41]. U_{min} can be measured with an oscilloscope using a probe with low capacitance at the maximum distance where EH is activated.

Taking into account the coupling coefficients of **Figure 11**, **Figure 12** compares simulated results for the antennas considered in **Figure 6**. It is assumed that the transmitter power is 20 dBm and that EH output is activated if the AC voltage at the input of the rectifier is high enough. In these conditions, the average field is equal to H_{min}. Considering $U_{min} = 4.8$ V, $H_{min} = 1.2$ A_{RMS}/m, and $R_{IC} = 550$ Ω, a distance between 12 and 18 mm is obtained. This estimation is in agreement with measurements of received H_{av} for a tag using the same antenna than in the simulation.

Experimentally, a value of $H_{min} \approx 1.1$ A_{RMS}/m (see **Figure 13**) corresponding to an EH range of 10 and 17 mm has been obtained, depending on the mobile used.

NFC sensors

Figure 14 shows the block diagram of an NFC-based data logger assisted with a complementary (optional) energy source (e.g., a solar cell). It is composed of an antenna that receives the interrogation signal emitted from a reader, usually, a mobile equipped with an NFC reader, an NFC integrated circuit, a microcontroller, and the sensor and its conditioning electronics. **Table 2** shows that there are several commercial NFC ICs with energy-harvesting capabilities available in the market. These devices can operate in two modes: battery-assisted (semi-passive) and battery-less (passive). Battery-assisted devices are used in data logger applications where stand-alone continuous and autonomous monitoring is necessary. The NFC sensor is fed from a battery or assisted from an additional harvesting source. These devices can communicate with external microcontrollers using I2C or SPI interface.

In the passive operation mode, the integrated NFC circuit is in charge of capturing the energy from the RF signal, rectifying it and extracting energy for the electronic circuits, microcontroller, or sensors. Sensors based on NFC IC with energy harvesting may operate in both modes: in semi-passive mode until the battery is exhausted and thereafter in passive mode. Data are stored in the EEPROM of the NFC IC to be read when the user taps the sensor. It should be highlighted that for tags in battery-less mode, data can only be updated with a new sensor reading in the presence of field from the reader. Another challenge is the amount of power that can be harvested from the RF signal. According to market research summarized in **Table 2**, commercial ICs can provide up to 15 mW. Despite this figure, battery-less sensors based on NFC have a wide range of applications. The most important key is that a specific reader is not required because the own smartphone equipped with NFC can be used as reader, and the data can be uploaded to cloud services using the Internet connectivity of the mobile. This fact is probably the important factor for the diffusion of this technology as can be derived from the large number of batteryless sensors based on NFC that are recently reported in the literature. Next section will review some examples.

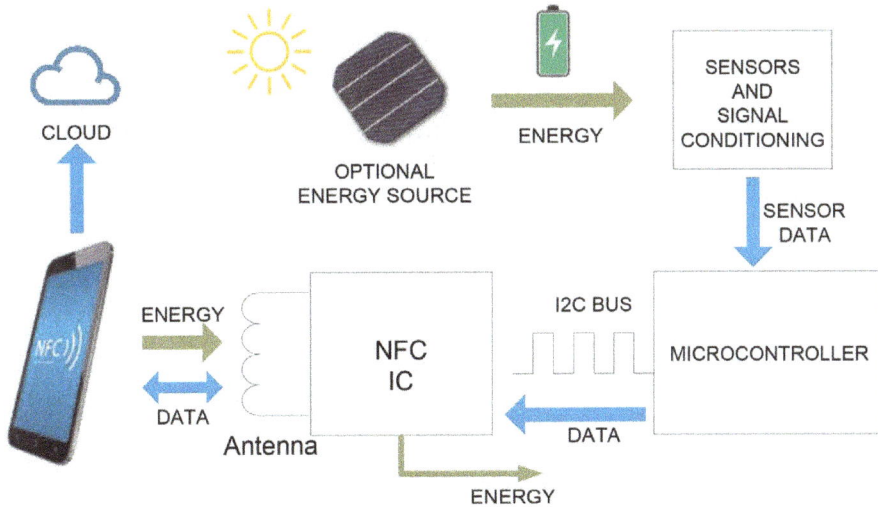

Figure 14. *Block diagram of a green NFC-based sensor system.*

Survey of NFC-based sensors

Several temperature sensors based on NFC have been proposed in the literature for different applications. In the pioneering works, based on NFC-WISP [49, 50] due to the inexistence of commercial NFC IC with energy harvesting, the rectification was performed with an external full-wave rectifier with discrete diodes. NFC-WISP communicated using the ISO-14443 protocol implemented in a low-power microcontroller (TI MSP430). As an option, the data could be shown in an E-ink display.

The monitoring of environment temperature in a cold chain or body temperature without using batteries that contain toxic products has aroused great interest. Several temperature sensors based on commercial IC such as MLX90129 [51] or RF430FRL152H [52, 53] have been reported in the literature for these applications. Other physical magnitudes can be measured with NFC sensors. For example, [54] reports a wireless tire pressure sensor based on a custom ASIC compatible with ISO14443 protocol.

A soil moisture battery-less NFC sensor for low-cost irrigation control at home, in a greenhouse, or at a garden center has been presented in [55]. The tag is based on an M24LR04E-R from ST connected to a low-cost microcontroller (ATtiny85 from Atmel). Volumetric water content of soil is obtained from capacitance measurement based on a low-power timer 555 working as an oscillator and a diode detector whose output is measured by the ADC of the microcontroller. Additionally, the temperature is measured using an I2C temperature sensor (LM75A), while air humidity is detected by reading the analog output from the HIH-5030 humidity sensor from Honeywell. The external circuitry requires less than 1 mA at 3 V to operate. **Figure 15** shows an image of the soil volumetric water content sensor inserted in a flower pot, the mobile app that shows the information in the screen and a comparison between the measured and the real volumetric water content (soil moisture) [55].

A second example is a smart diaper [56]. The tag is based on the M24LR04E-R from ST and an ATtiny85 from Atmel. The microcontroller senses the capacitance between two electrodes based on the discharge time of an RC circuit. The capacitance changes as a function of the urine level, as it is shown in **Figure 16**. Chemical sensors inserted in NFC-based tags have also been proposed. The presence of gas sensors based on functionalized CWNT that change the resonance frequency of an NFC has also been proposed in [57].

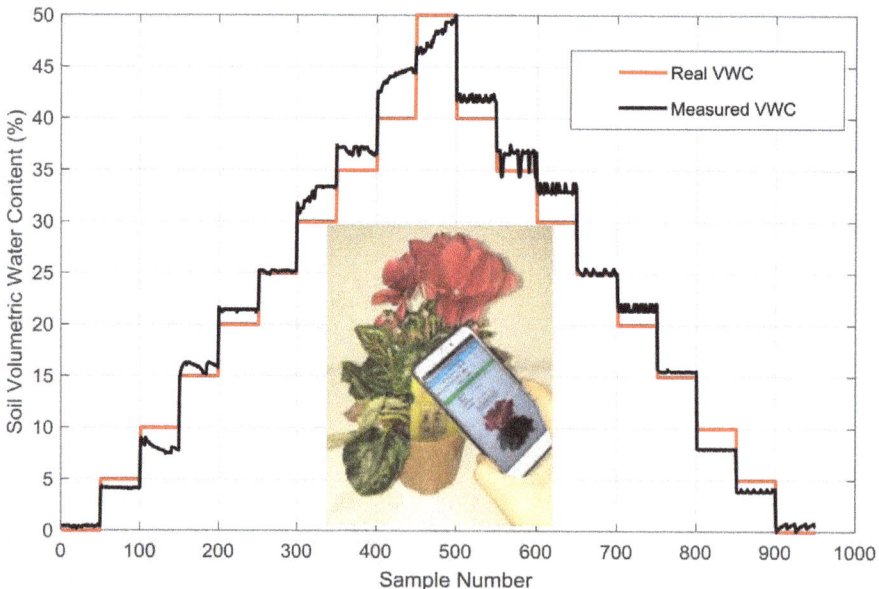

Figure 15. *Comparison of the volumetric water content (VWC) obtained with the sensor and real [55].*

Another advantage of NFC technology over other types of RFID working at UHF frequencies is that the NFC antennas can be designed to be compatible with the body. The effect of the body on NFC antennas is lower than at UHF band, where a considerable reduction of antenna efficiency and therefore the read range is experienced. Therefore, although the body losses reduce the quality factor of the coil and can detune the resonance frequency, the introduction of a small spacer can mitigate these effects, as shown in the previous section. Several wearable NFC-based sensors can be found in the literature that exploits this fact. In addition, the short read range inherent in the NFC technology provides a degree of privacy and security over undesired access to sensible information by nearby third parties [58, 59]. Prototypes of patches to perform measurements of body temperature based on commercial NFC IC can be found in [52, 53]. Another example is a prototype for monitoring the degree of hydration based on Melexis MLX90129 [60]. This sensor measures the concentration of NaCl in sweat, reading the surface temperature and sensing the potential difference between two electrodes. A noninvasive, NFC pH sensing system for monitoring wound healing and identifying the possibility of early-stage infection is reported in [61]. A low-power CMOS ISFET array for pH sensing is inductively powered using SIC4310 from Silicon Craft NFC IC in [62].

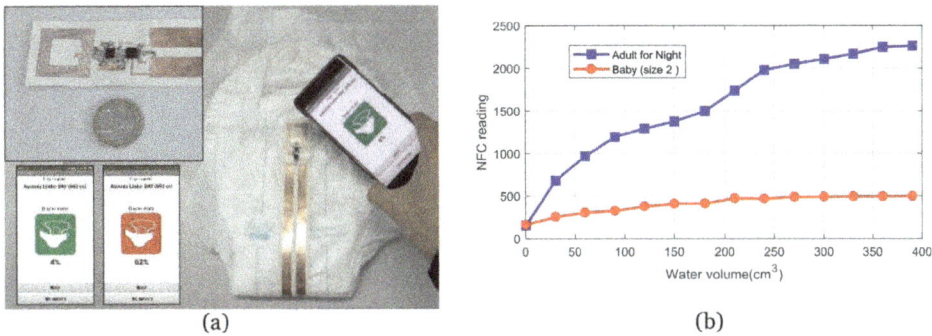

(a)   (b)

Figure 16. *(a) Photograph of the layout of the tag and the screen of the mobile app. (b) NFC reading as a function of the level of saline water for two diapers (adult for night and day and baby size 2).*

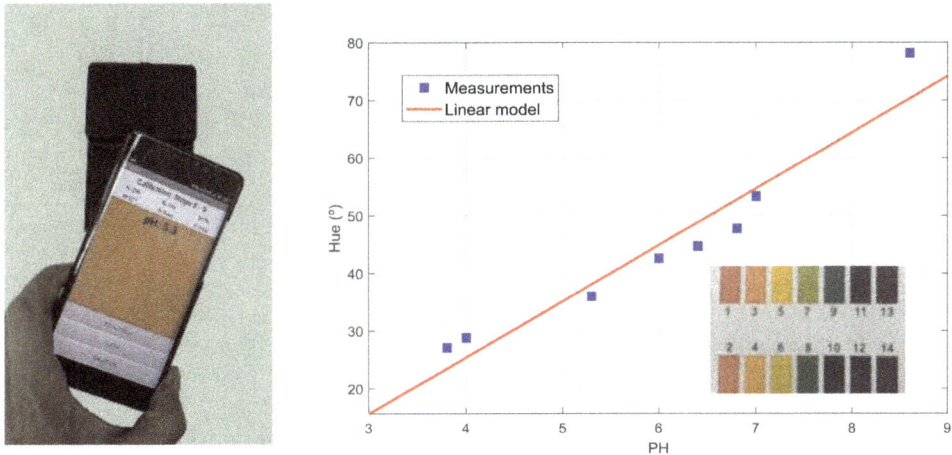

Figure 17. *Photograph of an NFC PH sensor from the measurement of the color of a strip with the android application and relation between the measured hue and the pH [41].*

The authors have demonstrated that colorimetric measurements can be done with battery-less NFC. The advantage over other methods based on cameras is that the illumination conditions are similar, and therefore the measurements have higher repeatability and accuracy. A prototype of NFC colorimeter with a low-cost color light-to-digital TCS3472 converter from TAOS is integrated with an ST M24LR04E-R NFC IC and a microcontroller (see **Figure 17**). It is used to measure the change in color of pH tires [41]. The same method can be used for other sensors based on the measurement of the color changes due to another sensitive chemical component (e.g., urine tests). **Figure 17** shows an image of the prototype and a comparison of the measured hue component as a function of the pH, showing that a simple linear model can be employed for pH calibration. The same NFC colorimeter is adapted to the design of a color-based classification system for grading the ripeness of fruit in [63] (**Figure 18**).

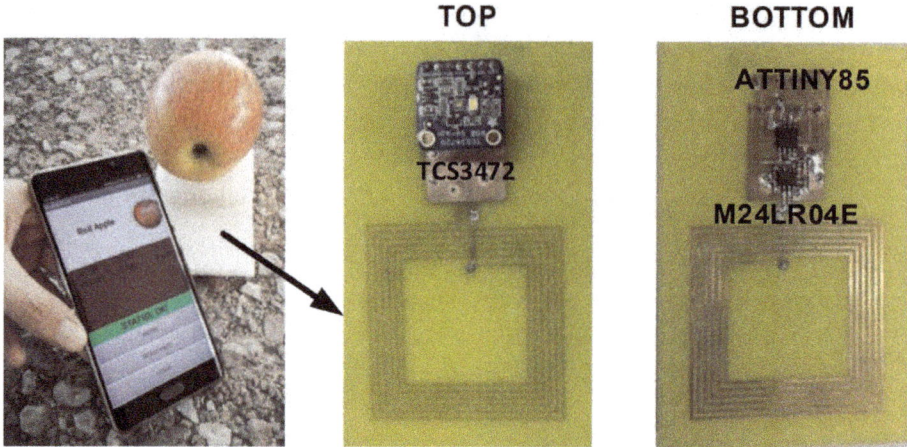

Figure 18. *Photograph of the NFC tag within the 3D printed enclosure in real application to detect apple ripeness and top/ bottom photographs of the prototype.*

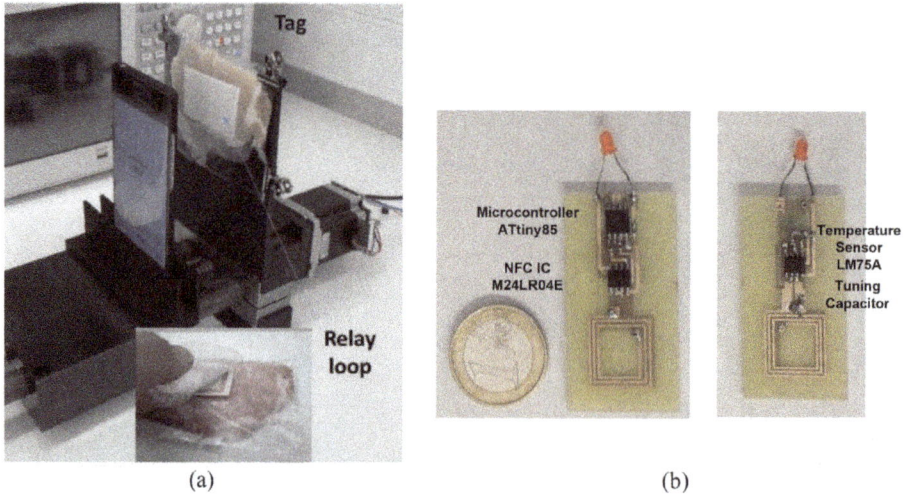

Figure 19. *(a) Setup used for the measurement of the average magnetic field, (b) photograph of a designed tag for testing: Top view (left) and bottom view (right).*

Finally, another application of NFC sensors is implantable medical devices (IMDs). Long-term implantable continuous glucose monitoring based on custom ASIC has been reported in [64, 65]. Wireless communication with IMDs is fundamental for monitoring and the reconfiguration of these devices to reduce the surgical operations [58]. Due to the size of the implanted coil, the coupling coefficient with the antennas integrated into the mobile is small, resulting in a short read range. One method to improve the readability with a modern smartphone and to increase the energy-harvesting range is including a relay coil on the body surface integrated on a patch.

Figure 19a shows the setup for measuring the average magnetic field and a prototype of implanted NFC sensor integrating a LED, a temperature sensor, and a microcontroller as proof of concept (**Figure 19b**).

Figure 20 shows the measurements with a commercial mobile of the average magnetic field as a function of the distance of an implanted coil of 15 x 15 mm within a phantom of pork steak for different depths of the implanted tag. Two systems are compared: the conventional two-coil system and a three-coil system where a relay resonant coil is on a patch over the skin. Using the last system, implanted NFC sensors can be powered by conventional smartphones at a depth up to 16 mm and read at distance between 1 and 2 cm from the skin.

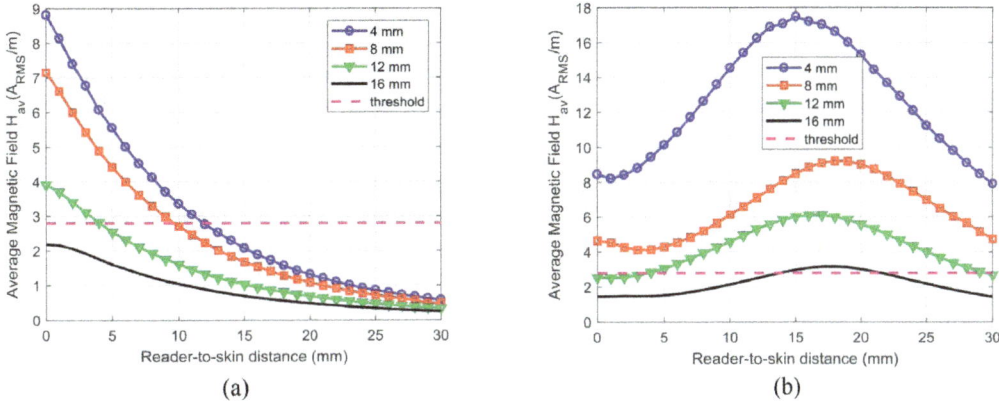

(a)   (b)

Figure 20. *Measured average magnetic field as a function of the mobile to skin distance for an implanted tag in a phantom at different depths between 4 and 16 mm for the two-coil system (a) and the three-coil system with a resonant relay loop on the skin (b). The threshold H_{min} is shown in dashed line.*

Conclusion

The availability in the market of low-cost near-field communication devices with energy-harvesting capabilities allows to feed small sensors enabling the development of low-cost battery-less sensors. One important advantage over other RFID is that additional readers are not required since most of the modern smartphones incorporate NFC readers. The limited read range for a selected number of applications is an advantage rather than a drawback because it ensures the privacy and improves the security under undesired access to sensible information by nearby third parties. In this chapter, an overview of recent advances in the field of battery-less NFC sensors at 13.56 MHz is provided, and it also briefly compares these sensors to other short-range RFID technologies. After reviewing power transfer in NFC, recommendations are made for the practical design of NFC-based sensor tags and NFC readers. A list of commercial NFC integrated circuits with energy-harvesting capabilities is also provided. A survey of recent battery-less NFC- based sensors for different applications has been done showing that conventional low-power sensors can be integrated within NFC tags for the new generation of IoT devices.

Acknowledgements

This work was supported by grant BES-2016-077291 and Spanish Government Project RTI2018-096019-B-C31.

Author details

Antonio Lazaro*, Marti Boada, Ramon Villarino and David Girbau Department of Electronic, Electric and Automatic Control Engineering, Rovira i Virgili University, Tarragona, Spain

*Address all correspondence to: antonioramon.lazaro@urv.cat

References

[1] Landt J. The history of RFID. IEEE Potentials. 2005;24(4):8-11. DOI: 10.1109/MP.2005.1549751

[2] Finkenzeller K. Rfid Handbook: Fundamentals and Applications in Contactless Smart Cards, Radio Frequency Identification and Near-Field Communication. 3rd ed. Chichester: Wiley-Backwell; 2010. ISBN: 978–0470695067

[3] Want R. An introduction to RFID technology. IEEE Pervasive Computing. 2006;5(1):25-33. DOI: 10.1109/MPRV.2006.2

[4] Rao KVS, Nikitin PV, Lam SF. Antenna design for UHF RFID tags: A review and a practical application. IEEE Transactions on Antennas and Propagation. 2005;53(12):3870-3876. DOI: 10.1109/TAP.2005.859919

[5] Nikitin PV, KVS R. Performance limitations of passive UHF RFID systems. In: 2006 IEEE Antennas and Propagation Society International Symposium. Albuquerque: IEEE; 2006. pp. 1011-1014. DOI: 10.1109/APS.2006.1710704

[6] Sample AP, Yeager DJ, Powledge PS, Mamishev AV, Smith JR. Design of an RFID-based battery-free programmable sensing platform. IEEE Transactions on Instrumentation and Measurement. 2008;57(11):2608-2615. DOI: 10.1109/TIM.2008.925019

[7] Nappi S, Mazzaracchio V, Fiore L, Arduini F, Marrocco G. Flexible pH sensor for wireless monitoring of the human skin from the Medimun distances. In: 2019 IEEE International Conference on Flexible and Printable Sensors and Systems (FLEPS). Glasgow: IEEE; 2019. pp. 1-3. DOI: 10.1109/ FLEPS.2019.8792291

[8] Ramos A, Lazaro A, Girbau D. Semi- passive time-domain UWB RFID system. IEEE Transactions on Microwave Theory and Techniques. 2013;61(4):1700-1708. DOI: 10.1109/TMTT.2013.2247056

[9] Wang B, Law M-K, Yi J, Tsui C-Y, Bermak A. A -12.3 dBm UHF passive RFID sense tag for grid thermal monitoring. IEEE Transactions on Industrial Electronics. 2019;66(11):8811-8820. DOI: 10.1109/TIE.2019.2891447

[10] Lorenzo J, Lazaro A, Villarino R, Girbau D. Modulated frequency selective surfaces for wearable RFID and sensor applications. IEEE Transactions on Antennas and Propagation. 2016; 64(10):4447-4456. DOI: 10.1109/TAP.2016.2596798

[11] Memon ML, Saxena N, Roy A, Singh S, Shin DR. Ambient backscatter communications to energize IoT devices. IETE Technical Review. 2019: 1-15. DOI: 10.1080/02564602.2019.1592717

[12] Preradovic S, Balbin I, Karmakar NC, Swiegers GF. Multiresonator-based chipless RFID system for low-cost item tracking. IEEE Transactions on Microwave Theory and Techniques. 2009;57(5):1411-1419. DOI: 10.1109/TMTT.2009.2017323

[13] Preradovic S, Karmakar NC. Design of fully printable planar chipless RFID transponder with 35-bit data capacity. In: 2009 European Microwave Conference (EuMC). Rome: IEEE; 2009. pp. 013-016. DOI: 10.23919/ EUMC.2009.5296182

[14] Vena A, Perret E, Tedjini S. Chipless RFID tag using hybrid coding technique. IEEE Transactions on Microwave Theory and Techniques. 2011;59(12): 3356-3364. DOI: 10.1109/TMTT.2011. 2171001

[15] Costa F, Genovesi S, Monorchio A. A chipless RFID based on multiresonant high-impedance surfaces. IEEE Transactions on Microwave Theory and Techniques. 2013;61(1):146-153. DOI: 10.1109/TMTT.2012.2227777

[16] Hu S, Law C, Dou W. A balloon- shaped monopole antenna for passive UWB-RFID tag applications. IEEE Antennas and Wireless Propagation Letters. 2008;7:366-368. DOI: 10.1109/ LAWP.2008.928462

[17] Lazaro A, Ramos A, Girbau D, Villarino R. Chipless UWB RFID tag detection using continuous wavelet transform. IEEE Antennas and Wireless Propagation Letters. 2011;10:520-523. DOI: 10.1109/LAWP.2011.2157299

[18] Garbati M, Perret E, Siragusa R, Halope C. Ultrawideband chipless RFID: Reader technology from SFCW to IR- UWB. IEEE Microwave Magazine. 2019; 20(6):74-88. DOI: 10.1109/MMM.2019. 2904408

[19] Rezaiesarlak R, Manteghi M. A space–time–frequency anticollision algorithm for identifying chipless RFID tags. IEEE Transactions on Antennas and Propagation. 2014;62(3): 1425-1432. DOI: 10.1109/TAP.2013. 2295393

[20] Vena A, Perret E, Tedjini S. High- capacity chipless RFID tag insensitive to the polarization. IEEE Transactions on Antennas and Propagation. 2012; 60(10):4509-4515. DOI: 10.1109/TAP. 2012.2207347

[21] Rezaiesarlak R, Manteghi M. Chipless RFID: Design Procedure and Detection Techniques. Switzerland: Springer; 2014. ISBN: 978-3319101682

[22] Ramos A, Perret E, Rance O, Tedjini S, Lazaro A, Girbau D. Temporal separation detection for chipless depolarizing frequency-coded RFID. IEEE Transactions on Microwave Theory and Techniques. 2016;64(7): 2326-2337. DOI: 10.1109/TMTT.2016. 2568180

[23] Shrestha S, Balachandran M, Agarwal M, Phoha VV, Varahramyan K. A chipless RFID sensor system for cyber centric monitoring applications. IEEE Transactions on Microwave Theory and Techniques. 2009;57(5):1303-1309. DOI: 10.1109/ TMTT.2009.2017298

[24] Girbau D, Ramos A, Lazaro A, Rima S, Villarino R. Passive wireless temperature sensor based on time-coded UWB chipless RFID tags. IEEE Transactions on Antennas and Pro agation. 2012;60(11):3623-3632. DOI: 10.1109/TMTT.2012.2213838

[25] Feng Y, Xie L, Chen Q, Zheng L-R. Low-cost printed chipless RFID humidity sensor tag for intelligent packaging. IEEE Sensors Journal. 2015; 15(6):3201-3208. DOI: 10.1109/JSEN. 2014.2385154

[26] Dey S, Saha JK, Karmakar NC. Smart sensing: Chipless RFID solutions for the internet of everything. IEEE Microwave Magazine. 2015;16(10): 26-39. DOI: 10.1109/MMM.2015. 2465711

[27] Lazaro A, Villarino R, Costa F, Genovesi S, Gentile A, Buoncristiani L, et al. Chipless dielectric constant sensor for structural health testing. IEEE Sensors Journal. 2018;18(13):5576-5585. DOI: 10.1109/ JSEN.2018.2839689

[28] El Matbouly H, Tedjini S, Zannas K, Duroc Y. Chipless sensing system compliant with the standard radio frequency regulations. IEEE Journal of Radio Frequency Identification. 2019; 3(2):83-90. DOI: 10.1109/JRFID.2019. 2909092

[29] Paret D. Design Constraints for NFC Devices. London: Wiley-ISTE; 2016. ISBN: 978-1-84821-884-0

[30] NFC-enabled mobile devices worldwide 2012–2018 | Statistic. Available from: https://www.statista.com/statistics/461494/nfc-enabled- mobile-devices-worldwide/ [Acccessed: August 19, 2019]

[31] Want R. Enabling ubiquitous sensing with RFID. Computer (Long Beach Calif). 2004;37(4):84-86. DOI: 10.1109/MC.2004.1297315

[32] Lazaro A, Villarino R, Girbau D, Lazaro A, Villarino R, Girbau D. A survey of NFC sensors based on energy harvesting for IoT applications. Sensors. 2018;18(11):3746. DOI: 10.3390/ s18113746

[33] Wang X, Gaustad G, Babbitt CW, Bailey C, Ganter MJ, Landi BJ. Economic and environmental characterization of an evolving Li-ion battery waste stream. Journal of Environmental Management. 2014;135:126-134. DOI: 10.1016/j. jenvman.2014.01.021

[34] Marrocco G. Pervasive electromagnetics: Sensing paradigms by passive RFID technology. IEEE Wireless Communications. 2010;17(6): 10-17. DOI: 10.1109/ MWC.2010. 5675773

[35] Hahnel D, Burgard W, Fox D, Fishkin K, Philipose M. Mapping and localization with RFID technology. In: IEEE International Conference on Robotics and Automation, 2004 Proceedings ICRA '04. New Orleans: IEEE; 2004. pp. 1015-1020. DOI: 10.1109/RO-BOT.2004.1307283

[36] Yang P, Wu W, Moniri M, Chibelushi CC. SLAM algorithm for 2D object trajectory tracking based on RFID passive tags. In: 2008 IEEE International Conference on RFID. Las Vegas: IEEE; 2008. pp. 165-172. DOI: 10.1109/ RFID.2008.4519349

[37] Lauridsen M, Nguyen H, Vejlgaard B, Kovacs IZ, Mogensen P, Sorensen M. Coverage comparison of GPRS, NB-IoT, LoRa, and SigFox in a 7800 km^2 area. In: 2017 IEEE 85th Vehicular Technology Conference (VTC Spring). Sydney: IEEE; 2017. pp. 1-5. DOI: 10.1109/VTCSpring.2017.8108182

[38] AN4914 Application note Automatic Antenna Tuning (AAT) of ST25R3911B/ST25R391x devices Introduction. 2017 Available from: www.st.com [Acccessed: August 19, 2019]

[39] Lee Y. AN710 Antenna Circuit Design for RFID Applications. Microchip Tech. Available from: http://ww1.microchip.com/downloads/en/ appnotes/00710c.pdf [Accessed: August 19, 2019]

[40] Kim DH, Park YJ. Calculation of the inductance and AC resistance of planar rectangular coils. IEE Electronics Letters. 2016;52(15):1321-1323. DOI: 10.1049/el.2016.0696

[41] Boada M, Lazaro A, Villarino R, Girbau D. Battery-less NFC sensor for pH monitoring. IEEE Access. 2019;7: 33226-33239. DOI: 10.1109/ ACCESS.2019.2904109

[42] Liu Q, Zhou J, Chen D. NFC Antenna Assembly. US9941572B2. 2016. Available from: https://patents.google. com/patent/US9941572B2/en

[43] Zhu J-Q, Ban Y-L, Sim C-Y-D, Wu G. NFC antenna with nonuniform meandering line and partial coverage ferrite sheet for metal cover smartphone applications. IEEE Transactions on Antennas and Propagation. 2017;65(6): 2827-2835. DOI: 10.1109/TAP.2017. 2690532

[44] Gebhart M, Neubauer R, Stark M, Warnez D. Design of 13.56 MHz smartcard stickers with ferrite for payment and authentication. In: 2011 Third International Workshop on Near Field Communication. IEEE; 2011. pp. 59-64. DOI: 10.1109/ NFC.2011.14

[45] Gebhart M, Bruckbaue J, Gossar M. Chip impedance characterization for contactless proximity personal cards. In: 2010 7th International Symposium on Communication Systems, Networks & Digital Signal Processing (CSNDSP 2010), Newcastle upon Tyne. IEEE; 2010. pp. 826-830. Available from: https://ieeexplore.ieee.org/abstract/doc ument/5580309/

[46] Agbinya JI. Wireless Power Transfer. 2nd ed. Denmark: River Publishers; 2015. ISBN: 978-87- 93237-62-9

[47] Gvozdenovic N, Mayer LW, Mecklenbräuker CF, Measurement of harmonic distortions and impedance of HF RFID chips. In: The 8th European Conference on Antennas and Propagation (EuCAP 2014); The Hague; 2014. pp. 2940-2944. DOI: 10.1109/EuCAP.2014.6902444/

[48] Couraud B, Deleruyelle T, Kussener E, Vauche R. Real-time impedance characterization method for RFID-type backscatter communication devices. IEEE Transactions on Instrumentation and Measurement. 2018;67(8):288-295. DOI: 10.1109/TIM.2017.2769224

[49] Zhao Y, Smith JR, Sample A. NFC- WISP: A sensing and computationally enhanced near-field RFID

platform. In: 2015 IEEE International Conference on RFID (RFID); San Diego: IEEE; 2015. pp. 174-181. DOI: 10.1109/RFID.2015. 7113089

[50] de Oliveira Filho JI, do Prado Villarroel Zurita ME. Development of NFC TAG for temperature sensing of premature newborns in neonatal incubators. In: 2017 2nd International Symposium on Instrumentation Systems, Circuits and Transducers (INSCIT). Fortaleza: IEEE; 2017. pp. 1-4. DOI: 10.1109/INSCIT.2017. 8103519

[51] Lorite GS, Selkälä T, Sipola T, Palenzuela J, Jubete E, Viñuales A, et al. Novel, smart and RFID assisted critical temperature indicator for supply chain monitoring. Journal of Food Engineering. 2017;193(1):20-28. DOI: 10.1016/j.jfoodeng.2016.06.016

[52] Vicente JM, Avila-Navarro E, Juan CG, Garcia N, Sabater-Navarro JM. Design of a wearable bio-patch for monitoring patient's temperature. In: 2016 38th Annual International Conference of the IEEE Engineering in Medicine and Biology Society (EMBC); Orlando: IEEE; 2016. pp. 4792-4795. DOI: 10.1109/EMBC.2016.7591799/

[53] Jeong H, Ha T, Kuang I, Shen L, Dai Z, Sun N, et al. NFC-enabled, tattoo-like stretchable biosensor manufactured by "cut-and-paste" method. In: 2017 39th Annual International Conference of the IEEE Engineering in Medicine and Biology Society (EMBC); Seogwipo: IEEE; 2017. pp. 4094-4097. DOI: 10.1109/EMBC.2017.8037756

[54] Kollegger C, Greiner P, Steffan C, Wiessflecker M, Froehlich H, Kautzsch T, et al. A system-on-chip NFC bicycle tire pressure measurement system. In: 2017 IEEE 60th International Midwest Symposium on Circuits and Systems (MWSCAS). Boston: IEEE; 2017. pp. 60-63. DOI: 10.1109/ MWSCAS.2017.8052860

[55] Boada M, Lazaro A, Villarino R, Girbau D. Battery-less soil moisture measurement system based on a NFC device with energy harvesting capability. IEEE Sensors Journal. 2018; 18(13):5541-5549. DOI: 10.1109/ JSEN.2018.2837388

[56] Lazaro A, Boada M, Villarino R, Girbau D. Battery-less smart diaper based on NFC technology. IEEE Sensors Journal. 2019:1-1. DOI: 10.1109/ JSEN.2019.2933289

[57] Azzarelli JM, Mirica KA, Ravnsbæk JB, Swager TM. Wireless gas detection with a smartphone via rf communication. Proceedings of the National Academy of Sciences. 2014; 111(51):18162-18166. DOI: 10.1073/ pnas.1415403111

[58] Halperin D, Heydt-Benjamin TS, Ransford B, Clark SS, Defend B, Morgan W, et al. Pacemakers and implantable cardiac defibrillators: Software radio attacks and zero-power defenses. In: 2008 IEEE Symposium on Security and Privacy (sp 2008). Oakland: IEEE; 2008. pp. 129-142. DOI: 10.1109/SP.2008.31/

[59] Gollakota S, Hassanieh H, Ransford B, Katabi D, Fu K, Gollakota S, et al. They can hear your he

artbeats. In: Proceedings of the ACM SIGCOMM 2011 Conference on SIGCOMM - SIGCOMM '11. New York, USA: ACM Press; 2011. pp. 2-13. DOI: 10.1145/2043164.2018438

[60] Rose DP, Ratterman M, Griffin DK, Linlin Hou, Kelley-Loughnane N, Naik RK, et al. System-level design of an RFID sweat electrolyte sensor patch. In: 2014 36th Annual International Conference of the IEEE Engineering in Medicine and Biology Society. Chicago: IEEE; 2014. pp. 4038-4041. DOI: 10.1109/EMBC.2014.6944510

[61] Rahimi R, Brener U, Ochoa M, Ziaie B. Flexible and transparent pH monitoring system with NFC communication for wound monitoring applications. In: Proceedings of the IEEE International Conference on Micro Electro Mechanical Systems (MEMS). Las Vegas: IEEE; 2017. pp. 125-128. DOI: 10.1109/MEMSYS.2017.7863356

[62] Douthwaite M, Georgiou P. Live demonstration: An NFC based batteryless CMOS ISFET array for real-time pH measurements of bio-fluids. In: 2017 IEEE Sensors. Glasgow: IEEE; 2017. pp. 1-1. DOI: 10.1109/ICSENS.2017. 8234027

[63] Lazaro A, Boada M, Villarino R, Girbau D, Lazaro A, Boada M, et al. Color measurement and analysis of fruit with a battery-less NFC sensor. Sensors. 2019;19(7):1741. DOI: 10.3390/ s19071741

[64] Anabtawi N, Freeman S, Ferzli R. A fully implantable, NFC enabled, continuous interstitial glucose monitor. In: 2016 IEEE-EMBS International Conference on Biomedical and Health Informatics (BHI). Las Vegas: IEEE; 2016. pp. 612-615. DOI: 10.1109/BHI.2016.7455973

[65] DeHennis A, Getzlaff S, Grice D, Mailand M. An NFC-enabled CMOS IC for a wireless fully implantable glucose sensor. IEEE Journal of Biomedical and Health Informatics. 2016;20(1):18-28. DOI: 10.1109/JBHI.2015.2475236

9

Dual-Band Resonator Designs for Near-field Wireless Energy Transfer Applications

Lai Ly Pon, Mohamed Himdi, Sharul Kamal Abdul Rahim and Chee Yen Leow

Abstract

Dual-band near-field wireless energy transfer (WET) designs outweigh single-band system with regard to either concurrent energy and data transfer or multiple wireless charging standard functionalities. There are two major approaches in resonator designs, namely, multi-coil and single-coil. This chapter presents a review on design constraints for each design approach and rectification techniques available in counteracting impediments of dual-band near-field WET systems.

Challenges pertinent to link design are discussed primarily followed by methods implemented to mitigate detrimental impact on performance metrics. Front-end dual-band resonator design methods are accentuated in this chapter in lieu of end- to-end WET system. This is envisioned to offer insights for designers contemplating on design modes or developing ways to facilitate a boost in rectification options currently available.

Keywords: near-field, dual-band, wireless energy transfer, single-coil, multi-coil

Introduction

The concept of electrical energy propagation with physical interconnection eradication spearheaded by Nikola Tesla has been adapted into diverse low- and high-power wireless energy transfer (WET) applications ranging from wireless charging for consumer electronics to electric vehicles. The massive landscape of consumer electronics has expanded toward healthcare, wellness, and fitness apart from the indispensable smartphones that have become such an integral part of one's daily life. Enhancement on quality of life reaches beyond patients who are dependent on implanted medical devices with healthcare monitoring devices for elderly and disabled communities. Keeping track with one's physical location, activity, and fitness regime is empowered by smart wearables [1]. However, data accumulated from monitoring of heart pacing or metabolic activities require appropriate transmission for further analysis or backup and storage. As such, simultaneous power and data transfer without any physical interconnections such as cables turn out to be an alternative option made possible with the forthcoming fifth generation

(5G) technology. Ubiquitous integration of hardware compatibility to support wireless charging for all smart devices is foreseeable.

Provision of more than one frequency band is enabled by resonator which is intended for several functions either concurrent power and data transfer or concurrent wireless charging at multiple standards. At present, there are two principal standards for wireless charging of consumer electronics specifically smartphones. The Wireless Power Consortium, otherwise known as Qi, is one of the leading wireless charging standards operating in low-frequency (LF) band, 110–205 kHz [2]. AirFuel Alliance is the merger between Power Matters Alliance (PMA) and the Alliance for Wireless Power (A4WP, also known as Rezence) [3]. A4WP employs magnetic resonance coupling technique operating at 6.78 MHz ± 15 kHz [4], while both Qi and PMA engage in inductive charging technique. The range of PMA's operating frequency is from 277 to 357 kHz [5]. Nonetheless, unlicensed industrial, scientific and medical (ISM) radio bands of up to 13.56 MHz are commonly selected to be the operating frequency for both inductive and magnetic resonant techniques in the current research community [6]. This chapter commences with challenges in relation to dual-band near-field WET design and the corresponding key performance metrics followed by design approaches and rectification techniques currently employed to alleviate the adverse effects on performance metrics with the intention of providing insights for designers in deciding and further enhancing current rectification options available in the design of dual-band near-field resonators. This framework, however, solely pertains to front-end resonator designs excluding end-to-end scope of near-field WET systems.

Impediments of dual-band near-field wireless energy transfer resonator designs

For a single-band near-field WET resonator design, designers delve into achieving maximum power transfer efficiency (PTE) between a pair of coupled resonators by designing the highest possible quality factor (Q-factor). However, there is an apparent complexity for designing resonators operating in more than one frequency band. PTE for either one of the frequency bands tends to surpass its counterpart. As such, concurrent capitalization on PTE for both frequency bands, f_1 and f_2, remains as one of the ultimate challenges.

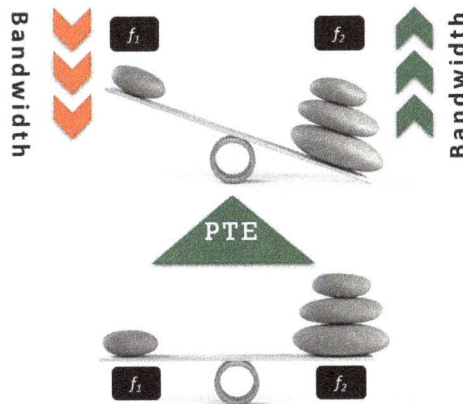

Figure 1. *Transfer efficiency equilibrium for dual-band near-field coupled resonators.*

Another concern is the inversely proportionate relation between Q-factor and bandwidth. Increasing PTE of the affected frequency band is feasible by developing resonators with enhanced

Q-factor. This comes at the expense of higher bandwidth which is pivotal especially for resonators aiming for concomitant power and data transfer functionalities as portrayed in **Figure 1**. PTE for f_1 is higher with improved Q-factor than f_2 but falls behind in terms of bandwidth since it is constrained by Q-factor. On the other hand, redesigning higher Q-factor resonator in contemplation of attaining improved PTE at f_2 unfortunately leads to bandwidth degradation. As such, there is always a dilemma between achieving PTE equilibrium for f_1 and f_2 and maximum PTE and bandwidth for each frequency bands.

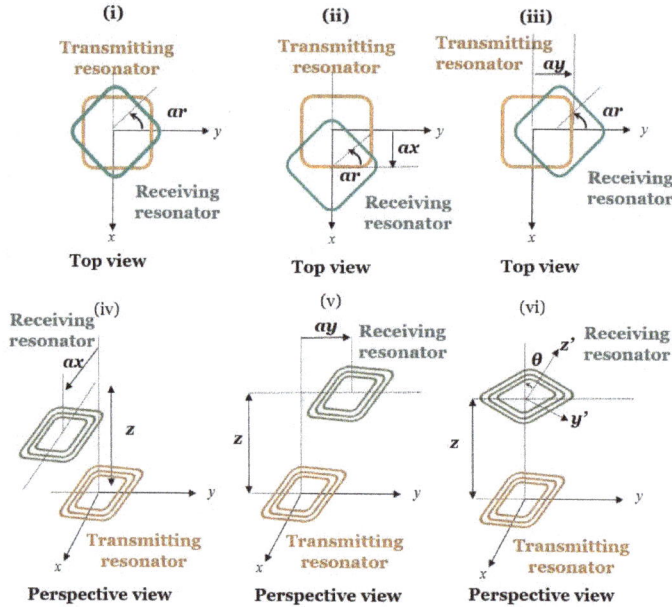

Figure 2. *Types of displacements between near-field coupled resonators: (i) planar; (ii) concurrent planar and lateral (x-axis); (iii) concurrent planar and lateral (y-axis); (iv) lateral (x-axis), (v) lateral (y-axis); (vi) angular.*

Imperfect positioning of loop resonator is yet another impairment of WET system which impacts its performance metrics specifically coupling coefficient, k, and PTE [7–9]. Strict adherence toward perfect alignment between a pair of loop resonator in assuring maximum transfer efficiency is seemingly impossible in practical sense because of misalignment be it planar, lateral, or angular frequently supervened [10]. Referring to **Figure 2**, planar displacement refers to the angle of rotation ar when both centers are axially aligned. Separated at a fixed axial distance, z, the center of receiving loop resonators is shifted by a distance, ax, known as horizontal lateral displacement, while ay is referred as vertical lateral displacement. The occurrence of simultaneous planar and the respective lateral displacements are visualized by ar with either ax or ay offsets. Angular displacement occurs when the plane of receiving loop resonator is being tilted by an angle θ [11]. Similarly, the consequences of displacement should be taken into consideration in the design of dual-band coupled resonators.

Performance metrics

Figure-of-merit (FOM) consideration yields a comparative insight for a diverse WET system design [12]. Derivation of FOM originates from analytical expression of link transfer efficiency (PTE) based on circuit theory (CT) [13] and reflected load theory (RLT) [14] as shown in Eq.

(1) and Eq. (2). FOM is reliant on coupling coefficient, k, and Q-factor parameters which are derived from Eq. (3) and Eq. (4). Proportional relationship is shared between coupling coefficient, k, and mutual inductance, M, with an inverse correlation shared between the latter and product of square root of transmitting and receiving self-inductances, $L_{1,2}$. Q-factor, $Q_{1,2}$, is proportional to angular frequency, ω, and self-inductances but inversely proportional to transmitting and receiving resistances, $R_{1,2}$:

$$PTE = FOM[\{1 + (1 + FOM)^{0.5}\}]^{-1} \tag{1}$$

$$FOM = k^2 Q_1 Q_2 \tag{2}$$

$$k = M(L_1 L_2)^{-0.5} \tag{3}$$

$$Q_{1,2} = \omega L_{1,2}(R_{1,2}) \tag{4}$$

Performance metrics that are commonly used are power transfer efficiency, Q-factor, and coupling coefficient. As discussed in Section 2, PTE and Q-factor share a positive correlation which is further validated from Eq. (1) and Eq. (2). Nevertheless, optimum coupling coefficient is often desirable in maximizing PTE. Since k is dependent on M and M is negatively correlated to transfer distance, z, a cautionary reminder is to ensure optimum transfer distance between transmitting and receiving resonators. This is validated with subsequent Equations [15] that reveal the correlation between M, z, and k [16]. M_{ij} indicates partial mutual inductance between each two turns on a pair of loops with turn radii, r_i and r_j. μ and $n_{1,2}$ represent conductor permeability and number of loops' turns, while ρ is the factor which is dependent on loop profile. K and E refer to complete elliptic integrals of the first and second kind:

Table 1. *Figure-of-merits (FOM).*

FOM	Concerned parameters	Dual-band	Significance
$k^2 Q_1 Q_2$	Coupling coefficient, Q-factor transmitting, receiving resonator [12]	No	Independent optimization of k and Q parameters
$PTE^2 P_L(V_s)^{-2}$	Transfer efficiency, load power, and voltage source [17]	No	Take into account of transfer efficiency and power delivered to load
$PTEz^3(A_{RX})^{-1.5}$	Transfer efficiency, transfer distance, and area of receiver [18, 19]	No	Equitable assessment between various transfer distance and geometry designs
$PTEz(d_o)^{-1}$	Transfer efficiency, transfer distance, and outermost side length [20]	No	Take into account of transfer distance and outermost side lengths
$PTEz(A)^{-0.5}$	Transfer efficiency, transfer distance, and area [21]	No	Equitable assessment between various transfer distance and geometry designs
$PTEz/$ $(d_{o\_tx} \times d_{o\_rx})^{0.5}$	Transfer efficiency, transfer distance, and outermost side lengths of transmitter and receiver [22]	No	Equitable assessment between various transfer distance and asymmetrical pair of resonators
$PTEz(A)^{-0.5}$	Transfer efficiency, transfer distance, and area [23, 24]	Yes	Equitable assessment between various transfer distance and geometry designs

$$M_{ij} = 2\mu\left(r_i r_j\right)^{0.5}\rho^{-1}\left[\left(1 - 0.5\rho^2\right)K(\rho) - E(\rho)\right] \tag{5}$$

$$\rho = 2\left(r_i r_j\right)^{0.5}\left[\left(r_i + r_j\right)^2 + z^2\right]^{-0.5} \tag{6}$$

$$M = \rho\sum_{i=1}^{n_1}\sum_{j=1}^{n_2}M_{ij}\left(r_i, r_j, z\right) \tag{7}$$

Other FOMs used as performance quantification are summarized in **Table 1**.

Frequency selection

There are accompanying implications with the choice of operating frequency on WET links. With higher frequency, the manifestation of eddy currents results in a nonuniformly distributed current density in conductive trace. The probability of current flowing toward the surface of a conductor known as skin effect amplifies alternate current resistance followed by power and heat dissipations [25]. Proximity effect is yet another dilemma of nonuniform current distribution due to current crowding on the inner loop edge caused by nearby conductive traces or layers [26]. On the contrary, current density is uniformly distributed at lower frequency in which eddy current effects are negligible. As such, feasibility of achieving higher power transfer efficiency is established with lower operating frequency at kHz than higher operating frequency at MHz at the expense of transfer range extension and reduced displacement sensitivity [27].

Table 2. *Summary of frequency selections for dual-band near-field WET resonator.*

References	f_1	f_2	Ratio, f_2/f_1	Variance, ΔPTE_{f1_f2} (%)
[30]	100 kHz	13.56 MHz	136	<10
[29]	100 kHz	6.78 MHz	68	3
[31, 32]	200 kHz	6.78 MHz	34	7.4, <8
[33]	300 MHz	900 MHz	3	8
[24]	300 MHz	675 MHz	2	7
[34–36]	6.78 MHz	13.56 MHz	2	15, 11.93, 1.68
[37]	90.3 MHz	138.8 MHz	2	1
[23]	300 MHz	700 MHz	2	1
[38]	470 MHz	730 MHz	2	11.4

For dual-band WET links, the ratio between frequencies selected is equally important since it is proportional to the transfer efficiency variance between f_1 and f_2 [28]. Additionally, selecting a larger ratio between frequencies with both power and data being transmitted at significantly lower and higher operating frequency bands turns out to be one of the isolation options available in curbing interference [23]. **Table 2** summarizes frequency selections opted for dual-band resonator designs. For example, variance of PTE between kHz and MHz frequencies is larger than MHz frequencies with reduced ratio of f_2 to f_1. However, research work demonstrated in [29] reports on reduced PTE variance with higher frequency ratio by implementing other rectification techniques which will be discussed in subsequent sections. In general, it is observed that there is higher probability of concurrently capitalizing PTE for both frequencies by designing resonators with lowest possible ratio of f_2 to f_1.

Design approaches

Table 3. *Comparisons of design approaches for dual-band near-field WET resonator.*

Factors	Single-coil	Multi-coil
Cross-coupling	Reduced	More
Geometrical area for resonators	Reduced	More
Concurrent transfer efficiency capitalization for both frequencies	Possible	Increased possibility
Dual-band support configuration	• Dual-band TX and dual-band RX resonators • Dual-band TX and single-band RX resonators	• Dual-band TX and dual-band RX resonators • Dual-band TX and single-band RX resonators

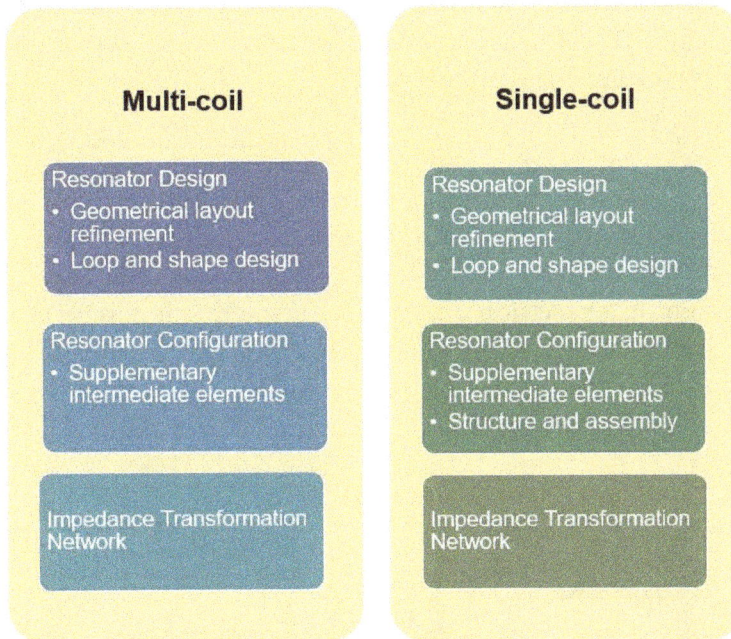

Figure 3. *Summary of rectification techniques for dual-band near-field WET resonator.*

Multi-coil [31, 32, 39–42] and single-coil [34, 43, 44] are two predominant strategies in designing dual-band near-field coupled resonators for WET links. Each design approach comes with their respective strengths and weaknesses which deserved appropriate assessment as tabulated in **Table 3**. Single-coil approach is preferred owing to minimum cross-coupling as compared to multi-coil [34]. Even though designing two separate coils in multi-coil approach necessitate more geometrical area, there is a greater potential in exploiting on highest possible transfer efficiency at two different frequencies in a single transmitter as it allows independent selection of inductance and quality factor [31].

Apart from deciding on the type of design approach to engage with, dual-band support configuration ought to be determined as well. Applicable configurations [28] involve designing either identical support mode as in dual-band transmitter (TX) and dual-band receiver (RX) or nonidentical support mode as in dual-band transmitter and single-band receivers. Bearing in mind that there is no superior design approach, rectification techniques employed address the respective shortcomings. Referring to **Figure 3**, rectification techniques for dual-band resonator designs which will be discussed in the following subsections encompass aspects relating to resonator design, resonator configuration, and impedance transformation network.

Rectification techniques

Resonator design

There is a penchant in designing printed spiral coils which encompassed conductive trace etched on dielectric substrates over conventional solenoid coils owing to lightweight and compactness features apart from having privilege of freedom in geometrical optimization [45] and conformity capability on malleable substrates [46] as compared to its predecessor. Equivalent series resistance (ESR) and substrate's dielectric losses are commonly contemplated to boost Q-factor of loop resonator designed [47] since the performance metric indicator, kQ, product can be derived from impedance matrix components given by [48]:

$$kQ = |Z_{21}|ESR^{-1} \tag{8}$$

where ESR $= [(R_{11}R_{22}) - (R_{12}R_{21})]^{0.5}$ for two-port system. Despite the fact that higher inductance could be acquired by increasing number of turns, ESR is not exempted. Direct current (DC) and alternate current (AC) resistances are components of ESR for a loop resonator. Hence, it is obligatory to exercise caution in geometrical layout alteration specifically number of turns, n, conductor width, w_c, and spacing between conductor trace, s_c. **Figure 4** illustrates comparison between a conventional layout of spiral resonator with uniform distribution of w_c and s_c and nonuniform layout refinement between each loop turns.

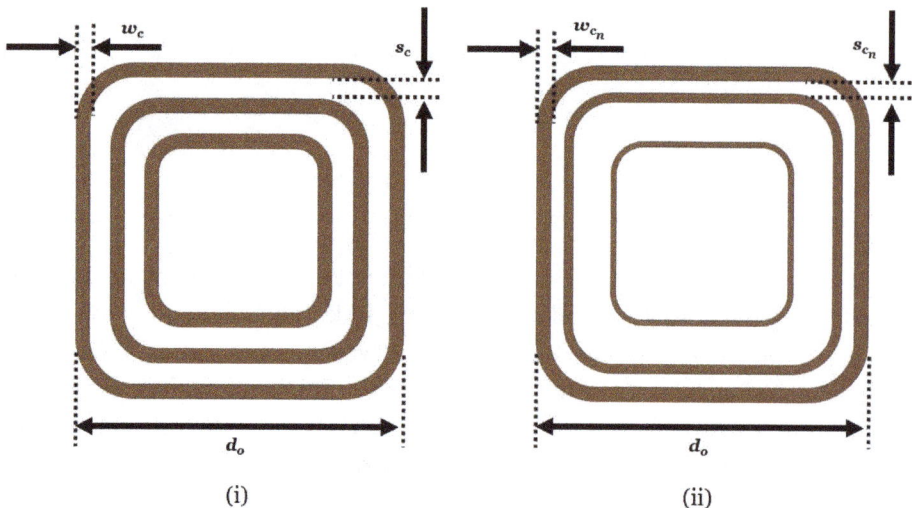

(i)　　　　　　　　　　　　　　　　　　　　　　　(ii)

Figure 4. *Geometrical layout refinement: (i) uniform; (ii) nonuniform.*

By designing inconstant conductor width distributions and spatial distributions, Q-factor enhancement is attained in double-layer printed spiral resonator [49], printed circular resonator [50], and stacked multilayer printed spiral resonator [51]. This is made possible as losses induced by eddy current are hindered with designs that involve gradual decrement of conductive trace widths from the outermost loop to the innermost loop [52] unlike conventional constant width for all loops. It is worth keeping in mind that designing larger constant conductor width for all turns and smaller constant spacing between adjacent conductor trace will only reinforce the total resistance [53]. However, ensuring ratio between conductor width and spacing to be relatively small aids in the reduction of proximity effects [54].

As such, geometrical layout refinement validated in performance enhancement for single-band resonator designs [55–58] can be adopted in resonators designed to operate at more than one frequency band. In [36], multi-coil approach with meticulous geometrical manipulation yields a nonuniform conductor widths and spacing between conductor trace of spiral resonators, collocating in a single-layer substrate. Combined with independent impedance transformation network, which will be covered in ensuing subsection, this leads to minimum transfer efficiency variation between both frequencies.

The prevailing shapes for loop resonator design are circular and square. Never- theless, there is an upsurge research trend in venturing into other types of resonators with diverse and irregular shapes such as defected ground structure (DGS). DGS resonator designs offer yet another miniaturization technique without forgoing WET performance especially high Q-factor such as circular DGS [23], interlaced DGS [24], C-shaped DGS [33], and bow-tie DGS [38]. However, the lowest frequency for all these designs hitherto is 300 MHz with ratio between frequency bands of not more than 3.

Resonator configuration

Transfer range deficiency associated with conventional two elements of single- band WET system architecture steers toward the adoption of supplementary elements such as the inclusion of repeaters or relay resonators in between source and load as depicted in **Figure 5**. Hence, sequence of four-coil strongly coupled magnetic resonance (SCMR) configuration introduced in [59], namely, driver, primary, secondary, and load coils, demonstrates extended transfer distance of greater than threefold compared to the diameter of primary resonator. Generally, identical designs can be observed from the driver and load resonators aside from primary and secondary resonators for conventional SCMR configuration. Incorporating repeater as a third element rather than four-element configuration substantiates viability of transfer range extension in single-band WET system [60].

Likewise, incorporating intermediate elements such as repeater in a single-coil approach is a remedy to the predicament in achieving simultaneous high-energy transfer for dual-band as demonstrated in [28, 61]. It is worth noting that additional repeater entails more space allocation apart from frequency splitting manifestation of either one of the frequency bands owing to strong coupling. Moreover, designers ought to take into account that transfer efficiency is greatly affected by parasitic resistance originating from repeater.

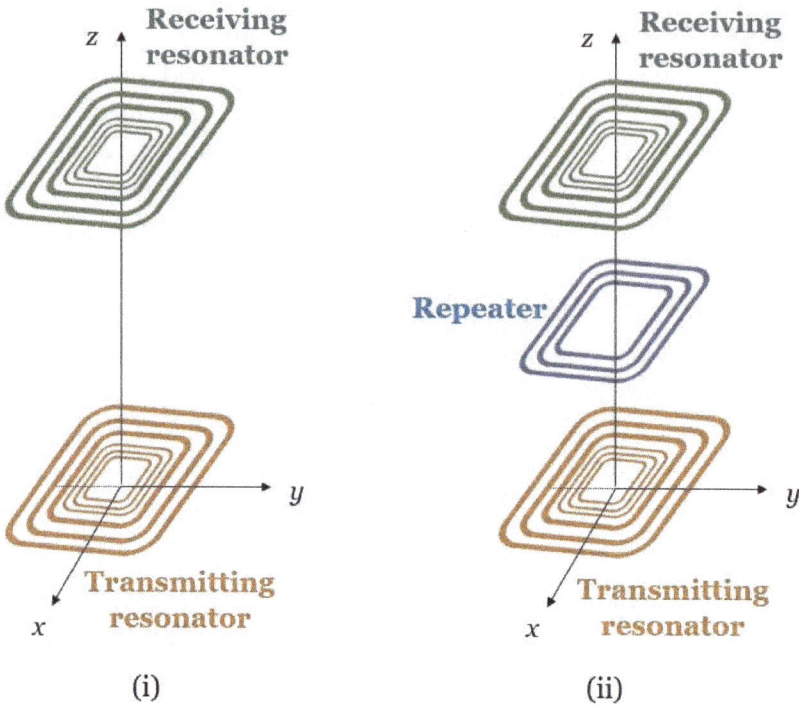

Figure 5. *Resonator configurations: (i) without intermediate element; (ii) with intermediate element (repeater).*

Composition and construction of resonator is a complementary technique to geometrical layout modification as well as loop and shape design. In multi-coil approach, each resonator operating at f_1 and f_2 can be designed either sharing similar axis or surface as illustrated in **Figure 6** (i) and (ii). The latter alleviates cross-coupling, but larger geometrical area is deemed necessary [39]. Nevertheless, this can be suppressed by designing a pair of asymmetrical coaxial resonators as shown in **Figure 6** (iii) whereby inductance for transmitting resonator is higher than receiving resonator. It is also inferred that optimizing power transfer foremost reduces interference between power carrier and data telemetry since improved coupling that leads to higher transfer power efficiency counterbalances interference between power and data. Further miniaturization effort is described in [24] by designing a superimposed dual-band DGS instead of coplanar or coaxial configurations.

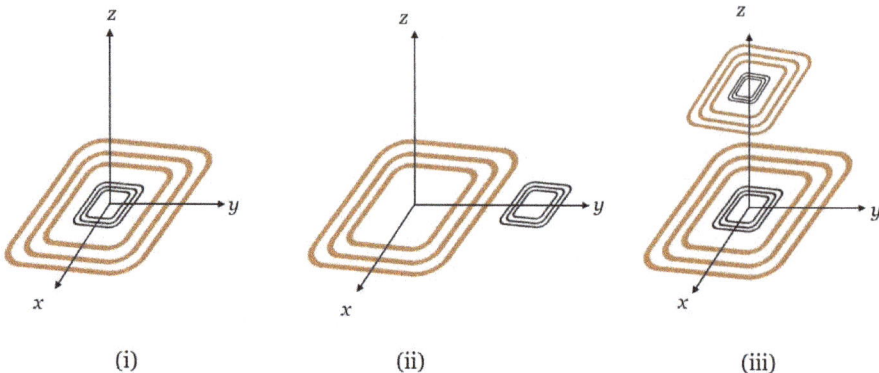

Figure 6. *Resonator configurations: (i) coaxial; (ii) coplanar; (iii) asymmetrical configuration.*

Furthermore, isolation techniques such as antiparallel loop structure [42] and frequency selective loop [40] structure functioning as band-pass filter have proven effective in rectifying interference caused by multi-coil mode. Band-stop filter in [31] filters' undesired parasitic eddy current from higher frequency induced across lower frequency coil path. The design is then revised in [32] with minimization of large spacing between low and high frequency coils and discrete inductor size along with total number of inverters. With careful combination of impedance values and coil-winding track, magnitude and phase of higher frequency voltage are fine-tuned to attain nil summation of total voltage across lower frequency path. As for displacement countermeasure, constructing more than one loop in an array structure reinforces tolerance toward detrimental consequences caused by imperfect orientations between transmitting and receiving resonators [62]. Robustness toward lateral displacement with wider coverage area is validated in [31] by incorporating asymmetrical configuration technique between a pair of coupled resonators.

Impedance transformation network

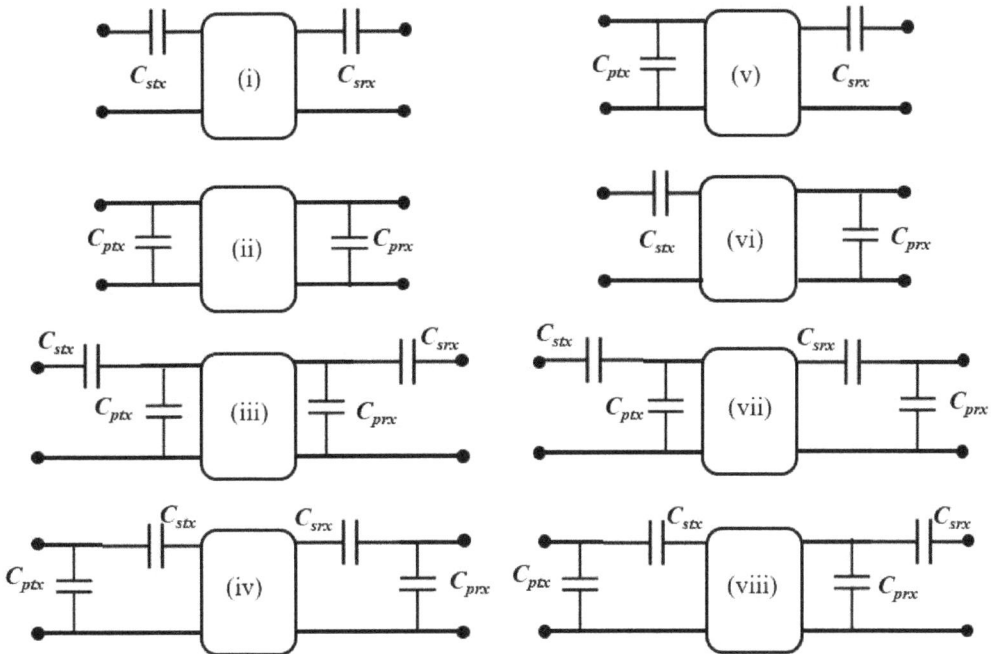

Figure 7. Symmetrical compensation network topologies: (i) series-series (S-S); (ii) parallel-parallel (P–P); (iii) series parallel-series parallel (SP-SP); (iv) parallel series-parallel series (PS-PS). Asymmetrical compensation network topologies: (v) parallel-series (P-S); (vi) series-parallel (S-P); (vii) series parallel-parallel series (SP-PS); (viii) parallel series-series parallel (PS-SP).

Impedance transformation network functions as alleviative measures of mutual inductance disparities triggered by spatial distance or load fluctuations between near-field coupled resonators. Resonance tuning and impedance matching, also referred to as compensation network, are commonly applied as front-end resonator design before AC-DC rectification. Implementing appropriate reactive compensation is necessary toward realizing maximum power transfer efficiency at preferred resonance frequency. **Figure 7** illustrates capacitive compensation topologies which can be generally categorized into symmetrical and asymmetrical compensation network (CN) for single-band near-field WET system. Single capacitive compensation commonly employed each at transmitting and receiving resonator sides encompasses series-series (S-S), parallel-parallel (P–P), series-parallel (S-P), and parallel-series (P-S). Double capacitive CN connected either in series or parallel is also known as L-match network. WET system implemented with complex conjugate matching yields other combination of CN topologies apart from these topologies such as T-match network [19] as well as hybrid T-match network of inductor and capacitors [63].

Comprehensive analysis on impedance matching is discussed in [34], which proposes single-coil mode supporting two resonant frequencies at 6.78 and 13.56 MHz thanks to two diverse impedance matching networks comprising of LC branch in series and LC tank in parallel. Unfortunately, this method is not applicable when variance between two target frequencies is considerably large particularly if one of the target frequencies originates from kHz band [31]. A way to overcome limitation of vast frequency separation is through specific coil arrangement by positioning inner lower band coil and outer higher band coil [31].

As for multi-coil approach, designers do not have to concern much on the total target frequency separation since impedance matching can be executed independently. In [42], a high and a low Q impedance matching are performed separately at two different matching distances specifically closer distance for higher transfer efficiency aimed at wireless charging. Conversely, the low Q impedance match is completed at further distance as a trade-off for higher bandwidth intended for data transfer applications.

Eq. (9) is employed in [36] before performing simultaneous conjugate matching using symmetrical double capacitive CN in order to acquire PTE equilibrium between both frequencies selected. Since the design is based on multi-coil mode, compromise should be made in selecting optimum separation distance between outermost side length of loop operating at f_1 and f_2. The relationship between optimal transfer distance, z_{op_fn}, and outermost side length of loop, $d_{o\_f1,f2}$, at maximum excited magnetic field derived in [58] for a square spiral resonator yields:

$$z_{op\_fn} = 0.3931 d_{o\_fn} \tag{9}$$

On the other hand, simultaneous high-energy transfer is validated in [23] for both frequency bands using a compact two circular DGS resonators with independent coupling whereby series capacitance is implemented together with single stub for matching method. Correspondingly, comparable findings are also reported in [37] by adopting single-coil approach and single CN with dual-mode printed inductor at a minimized geometrical area requirement. In contrast, DGS resonator designs executed with impedance transformation network which consists of series capacitive lumped element and single stub matching in [24, 38] are unable to acquire minimum PTE variance between frequencies, but these designs excel in terms of maximizing FOM.

Simulation of dual-band resonator design with multiple rectification techniques

By applying single-coil approach, **Figure 8** depicts an example of dual-band printed spiral resonator simulated with full-wave electromagnetic simulator, CST Microwave Studio separated at transfer distance of 15 mm. Three rectification techniques are implemented specifically resonator design, resonator configuration, and impedance transformation network. Since dimension constrictions are of paramount concern in near-field WET system, the overall dimension of transmitting resonator is limited to 75 mm by 82.5 mm. Other parameter properties are detailed in **Table 4** after optimization of parametric studies via geometrical layout tuning.

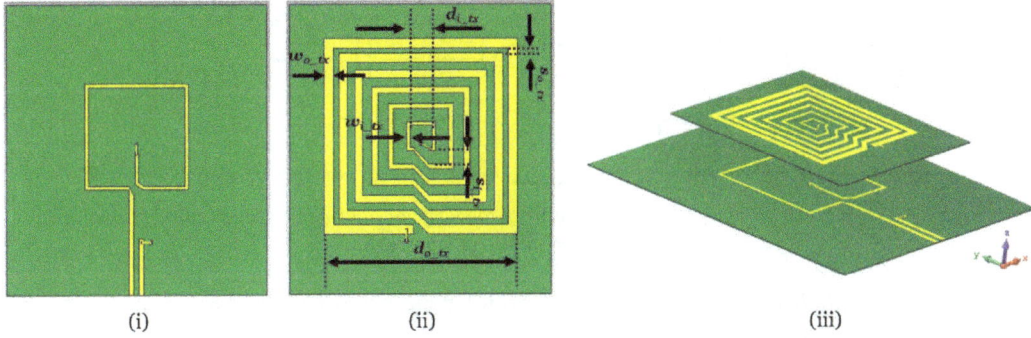

(i)　　　　　　　　(ii)　　　　　　　　(iii)

Figure 8. *Dual-band printed spiral resonator design adopting single-coil approach: (i) transmitter's top layer; (ii) transmitter's bottom layer; (iii) 3D model of transmitting and receiving resonators.*

Table 4. *Parameter properties for dual-band printed spiral resonator design adopting single-coil approach.*

Parameters	Value (mm)
Substrate thickness, T_s	0.4
Conductor thickness, t_c	0.035
Overall dimension (width × length)	75 × 82.5
Outermost side length of transmitting resonator, d_{o_tx}	55.8
Innermost side length of transmitting resonator, d_{i_tx}	6.45
Outermost conductor's width of transmitting resonator, w_{o_tx}	2.85
Innermost conductor's width of transmitting resonator, w_{i_tx}	0.98
Outermost conductor's spacing of transmitting resonator, s_{o_tx}	1.5
Innermost conductor's spacing of transmitting resonator, s_{i_tx}	3.75
Size ratio of transmitting to receiving resonators, Tx:Rx	1.5:1

Combined with asymmetrical configuration technique, receiving resonator is designed at two-thirds of transmitting resonator's size. Symmetrical hybrid compensation network topology which comprises of inductor and capacitors is incorporated in order to achieve simultaneous conjugate matching at 6.78 and 13.56 MHz as shown in **Figure 9**. LC tank is added with series-parallel capacitive compensation topology for single-coil approach. Values of optimized lumped elements are detailed in **Table 5**. **Figure 10** shows simulated S_{11} and S_{21} plots. Despite attaining two distinct resonance frequencies at the intended f_1 and f_2, feasibility in supporting data transmission at higher frequency is affected as the corresponding -3 dB fractional bandwidth is unfortunately low at approximately 5%. Nevertheless, it is observed that the highest simulated PTE is realized at lower frequency, while minimum variation of PTEs is preserved as depicted in **Figure 11**. Positioned at perfect alignment, PTE for each frequency is 90.11 and 80.56% at 6.78 and 13.56 MHz, respectively. Variation of PTEs at 9.55% is satisfactory considering ratio between frequencies is significantly small at only two for single-coil approach.

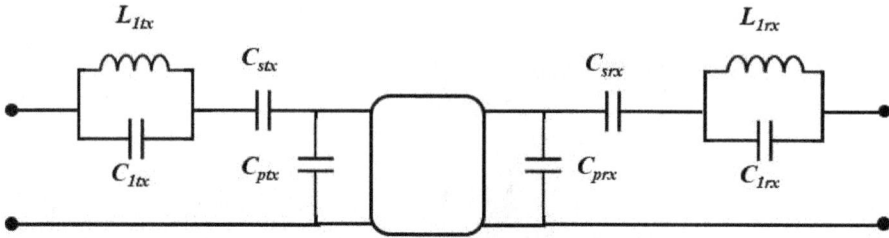

Figure 9. *Symmetrical hybrid compensation network topology for dual-band printed spiral resonator design adopting single-coil approach.*

Table 5. *Symmetrical hybrid compensation properties for dual-band printed spiral resonator design adopting single-coil approach.*

Parameters	Value
L_{1tx}	4.15 µH
C_{1tx}	8.5 pF
C_{stx}	94.65 pF
C_{ptx}	184.69 pF
L_{1rx}	2.25 µH
C_{1rx}	4.32 pF
C_{srx}	77.58 pF
C_{prx}	269.10 pF

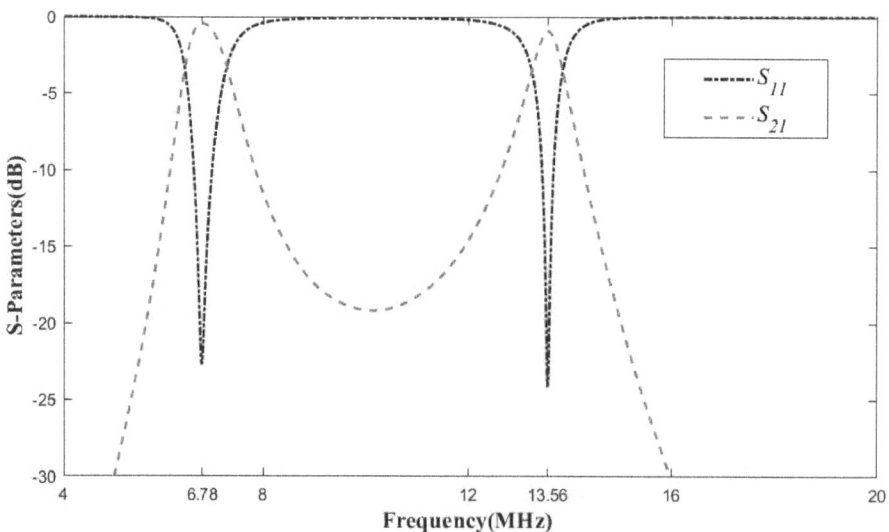

Figure 10. *Simulated S-parameter plots of dual-band printed spiral resonator design adopting single-coil approach.*

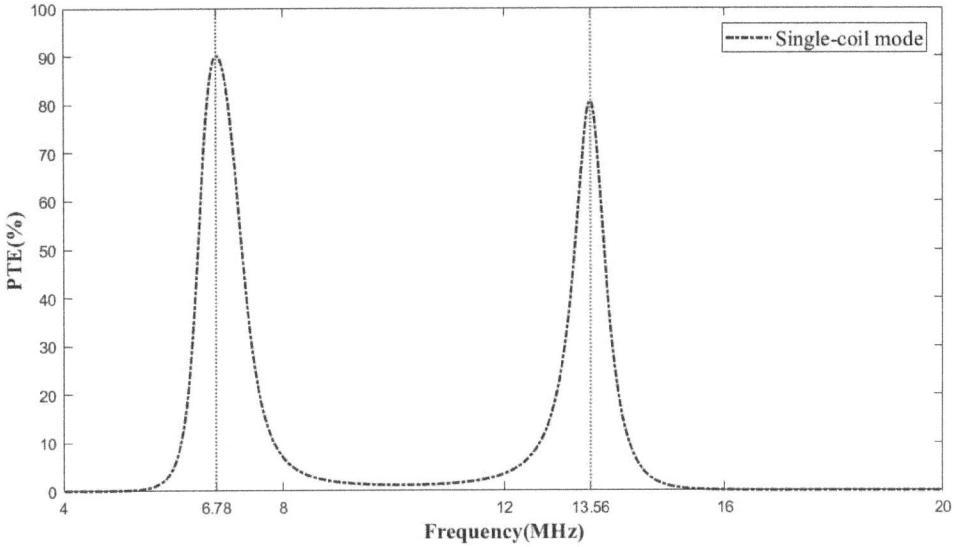

Figure 11. *Simulated power transfer efficiency plot of dual-band printed spiral resonator design adopting single-coil approach.*

Conclusions

Constraints of dual-band near-field wireless energy transfer links are presented. Front-end mitigation techniques in sustaining acceptable performance metrics for dual-band WET system are reviewed. Accomplishment of satisfactory transfer efficiency at a relevant axial distance regardless of resonance frequencies preferred is imperative besides having resilience against coupling variations and displacement offset from ideal orientation. Deciding on the design approaches conveyed above is narrowed down to either meeting specifications of transfer efficiency proportionality between selected frequencies or sufficient energy and data transfer capability. As such, precautionary steps should be undertaken in determining technique adoption in an effort to strike a balance between performance goals, design complexity, and physical and budgetary restraints. All the aforementioned techniques discussed in this chapter serve as a benchmark or recommended framework for designers' discretion.

Acknowledgements

This work is jointly supported by CREST under Grant 4B151 and the Institute of Electronics and Telecommunication of Rennes (IETR).

Conflict of interest

The authors declare no conflict of interest.

Notes/Thanks/Other declarations

Thanks to Kristina Tijan for invitation to contribute this chapter to the book "Wireless Energy Transfer Technology".

Appendixes and nomenclature

Symbol	Parameter
$f_{1,2}$	frequency 1, 2
k	coupling coefficient
ar	planar displacement
ax	horizontal lateral displacement
ay	vertical lateral displacement
θ	angular displacement
z	transfer distance
z_{op_fn}	optimal transfer distance
$Q_{1,2}$	Q-factor transmitting or receiving resonator
M	mutual inductance
$L_{1,2}$	inductance of transmitting or receiving resonator
P_L	load power
V_s	voltage source
A_{RX}	area, area of receiving resonator
$d_{o\_tx,0\_rx}$	outermost side lengths of transmitting or receiving resonator
d_{i_tx,i_rx}	innermost side length of transmitting or receiving resonator
n	turns
w_c	conductor's width
w_{o_tx}	outermost conductor's width of transmitting resonator
w_{i_tx}	innermost conductor's width of transmitting resonator
s_c	conductor's spacing
s_{o_tx}	outermost conductor's spacing of transmitting resonator
s_{i_tx}	innermost conductor's spacing of transmitting resonator
T_s	substrate thickness
t_c	conductor thickness
$L_{1tx,1rx}$	simulated inductor at transmitting or receiving resonator
$C_{1tx,1rx}$	simulated capacitor at transmitting or receiving resonator
$C_{stx,srx}$	simulated series capacitor at transmitting or receiving resonator
$C_{ptx,prx}$	simulated parallel capacitor at transmitting or receiving resonator

Author details

Lai Ly Pon[1], Mohamed Himdi[2*], Sharul Kamal Abdul Rahim[1] and Chee Yen Leow[1]

1 Wireless Communication Center, School of Electrical Engineering, Faculty of Engineering, Universiti Teknologi Malaysia, Johor Bahru, Malaysia

2 Institute of Electronics and Telecommunication of Rennes, University of Rennes 1, Rennes, France

*Address all correspondence to: mohamed.himdi@univ-rennes1.fr

References

[1] Piwek L, Ellis DA, Andrews S, Joinson A. The rise of consumer health wearables: Promises and barriers. PLoS Medicine. 2016;13:1-9. DOI: 10.1371/journal.pmed.1001953

[2] Wireless Power Consortium. The Qi Wireless Power Transfer System Power Class 0 Specification Parts 1 and 2: Interface Definitions; 2016

[3] Merger Information. Retrieved from Airfuel website: http://www.airfuel.org/ news/merger-information, (n.d.)

[4] Technical Working Committees (TWC 1+2) of the Alliance for Wireless Power (A4WP), A4WP Wireless Power Transfer System Baseline System Specification (BSS) v 1.2.1, A4WP Stand; 2014

[5] Chandrasekar Rao TS, Geetha K. Categories, standards and recent trends in wireless power transfer: A survey. Indian Journal of Science and Technology.2016;9:11. DOI: 10.17485/ijst/2016/v9i20/91041

[6] Trivedi Y. Wireless power transfer: "Look ma, no hands, no wires! IEEE Communications Magazine. 2016;54:8.DOI: 10.1109/MCOM.2016.7514157

[7] Hui SY. Planar wireless charging technology for portable electronic products and qi. Proceedings of the IEEE. 2013;101:1290-1301. DOI: 10.1109/JPROC.2013.2246531

[8] Liu D, Hu H, Georgakopoulos SV. Misalignment sensitivity of strongly coupled wireless power transfer systems. IEEE Transactions on Power Electronics. 2017;32:5509-5519. DOI: 10.1109/TPEL.2016.2605698

[9] Song M, Belov P, Kapitanova P. Wireless power transfer inspired by the modern trends in electromagnetics. Applied Physics Reviews. 2017;4: 021102. DOI: 10.1063/1.4981396

[10] Joy ER, Dalal A, Kumar P. Accurate computation of mutual inductance of two air core square coils with lateral and angular misalignments in a flat planar surface. IEEE Transactions on Magnetics. 2014;50:1-9. DOI: 10.1109/ TMAG.2013.2279130

[11] Flynn BW, Fotopoulou K. Rectifying loose coils: Wireless power transfer in loosely coupled inductive links with lateral and angular misalignment. IEEE Microwave Magazine. 2013;14:48-54. DOI: 10.1109/ MMM.2012.2234634

[12] Sample AP, Meyer DA, Smith JR. Analysis, experimental results, and range adaptation of magnetically coupled resonators for wireless power transfer. IEEE Transactions on Industrial Electronics. 2011;58:544-554. DOI: 10.1109/TIE.2010.2046002

[13] Dionigi M, Mongiardo M, Perfetti R. Rigorous network and full-wave electromagnetic Modeling of wireless power transfer links. IEEE Transactions on Microwave Theory and Techniques. 2015;63:65-75. DOI: 10.1109/ TMTT.2014.2376555

[14] Kiani M, Ghovanloo M. The circuit theory behind coupled-mode magnetic resonance-based wireless power transmission. IEEE Transactions on Circuits and Systems I: Regular Papers. 2012;59:2065-2074. DOI: 10.1109/ TCSI.2011.2180446

[15] Jow U, Ghovanloo M. Design and optimization of printed spiral coils for efficient transcutaneous inductive power transmission. IEEE Transactions on Biomedical Circuits and Systems. 2007;1:193-202. DOI: 10.1109/TBCAS. 2007.913130

[16] Schormans M, Valente V, Demosthenous A. Frequency splitting analysis and compensation method for inductive wireless powering of implantable biosensors. Sensors (Switzerland). 2016;16:1229. DOI: 10.3390/s16081229

[17] Kiani M, Ghovanloo M. A figure-of- merit for designing high-performance inductive power transmission links. IEEE Transactions on Industrial Electronics. 2013;60:5292-5305. DOI: 10.1109/TIE.2012.2227914

[18] Zargham M, Gulak PG. Fully integrated on-chip coil in 0.13 μm CMOS for wireless power transfer through biological media. IEEE Transactions on Biomedical Circuits and Systems. 2015;9:259-271. DOI: 10.1109/ TBCAS.2014.2328318

[19] Yang C, Chang C, Lee S, Chang S, Chiou L. Efficient four-coil wireless power transfer for deep brain stimulation. IEEE Transactions on Microwave Theory and Techniques. 2017;65:2496-2507. DOI: 10.1109/ TMTT.2017.2658560

[20] Tahar F, Saad R, Barakat A, Pokharel RK. 1.06 FoM and compact wireless power transfer system using rectangular defected ground structure resonators. IEEE Microwave and Wireless Components Letters. 2017;27: 1025-1027. DOI: 10.1109/LMWC.2017. 2750032

[21] Hekal S, Abdel-Rahman AB, Jia H, Allam A, Barakat A, Pokharel RK. A Novel technique for compact size wireless power transfer applications using defected ground structures. IEEE Transactions on Microwave Theory and Techniques. 2017;65:591-599. DOI: 10.1109/TMTT.2016.2618919

[22] Liu Z, Chen Z, Li J, Guo Y, Xu B. A planar L-shape transmitter for wireless power transfer system. IEEE Antennas and Wireless Propagation Letters. 2016; 1225:960-963. DOI: 10.1109/LAWP. 2016.2615112

[23] Tahar F, Barakat A, Saad R, Yoshitomi K, Pokharel RK. Dual-band defected ground structures wireless power transfer system with independent external and inter- resonator coupling. IEEE Transactions on Circuits and Systems II: Express Briefs. 2017;64:1372-1376. DOI: 10.1109/ TCSII.2017.2740401

[24] Sharaf R, Abdel-Rahman AB, El- Hameed ASA, Barakat A, Hekal S, Allam A. A new compact dual-band wireless power transfer system using interlaced resonators. IEEE Microwave and Wireless Components Letters. 2019; 49:498-500. DOI: 10.1109/ LMWC.2019. 2917747

[25] Bahl I. Lumped Elements for RF and Microwave Circuits. Artech House; 2003

[26] Kuhn WB, Ibrahim NM. Analysis of current crowding effects in multiturn spiral inductors. IEEE Transactions on Microwave Theory and Techniques. 2001;49:31-38. DOI: 10.1109/22.899959

[27] Liu M, Chen M. Dual-band wireless power transfer with reactance steering network and reconfigurable receivers. IEEE Transactions on Power Electronics. 2019;8993:1-1. DOI: 10.1109/TPEL.2019.2913991

[28] Kung M-L, Lin K-H. Dual-band coil module with repeaters for diverse wireless power transfer applications. IEEE Transactions on Microwave Theory and Techniques. 2018;66: 332-345. DOI: 10.1109/ TMTT.2017. 2711010

[29] Riehl PS, Satyamoorthy A, Akram H, Yen Y-C, Yang J-C, Juan B, et al. Wireless power Systems for Mobile Devices Supporting Inductive and Resonant Operating Modes. IEEE Transactions on Microwave Theory and Techniques. 2015;63:780-790. DOI: 10.1109/ TMTT.2015.2398413

[30] Liu M, Chen M. Dual-band multi- receiver wireless power transfer: Architecture, topology, and control. In: 2019 IEEE Appl. Power Electron. Conf. Expo. New York: IEEE; 2019. pp. 851-859. DOI: 10.1109/ APEC.2019. 8721837

[31] Ahn D, Mercier PP. Wireless power transfer with concurrent 200-kHz and 6.78-MHz operation in a single- transmitter device. IEEE Transactions on Power Electronics. 2016;31: 5018-5029

[32] Ahn D, Kim S, Kim S-W, Moon J, Cho I. Wireless power transmitter and receiver supporting 200-kHz and 6.78- MHz dual-band operation without magnetic field canceling. IEEE Transactions on Power Electronics. 2017;32:7068-7082. DOI: 10.1109/ TPEL.2016.2629494

[33] Atallah HA, Huseein R, Abdel- Rahman AB. Novel and compact design of capacitively loaded C-shaped DGS resonators for dual band wireless power transfer (DB-WPT) systems. AEÜ - International Journal of Electronics and Communications. 2019;100:95-105. DOI: 10.1016/j.aeue.2018.12.016

[34] Kung ML, Lin KH. Enhanced analysis and design method of dual- band coil module for near-field wireless power transfer systems. IEEE Transactions on Microwave Theory and Techniques. 2015;63:821-832. DOI: 10.1109/TMTT.2015.2398415

[35] Ding K, Yu Y, Lin H. A novel dual- band scheme for magnetic resonant wireless power transfer. Progress In Electromagnetics Research Letters. 2018;80:53-59. DOI: 10.258/ PIERL18082201

[36] Pon LL, Kamal S, Rahim A, Leow CY, Chua TH. Non-radiative wireless energy transfer with single layer dual-band printed spiral resonator. Bull. Electr. Eng. Informatics. 2019;8:744-752. DOI: 10.11591/eei. v8i3.1593

[37] Barakat A, Yoshitomi K, Pokharel RK. Design and implementation of dual-mode inductors for dual-band wireless power transfer systems. IEEE Transactions on Circuits and Systems II: Express Briefs. 2018;66: 1287-1291. DOI: 10.1109/TCSII.2018. 2883671

[38] Saad MR, Tahar F, Chalise S, Barakat A, Yoshitomi K, Pokharel RK. High FOM dual band wireless power transfer using bow-tie defected ground structure resonators. In: 2018 IEEE Wirel. Power Transf. Conf. New York: IEEE; 2018. pp. 1-4. DOI: 10.1109/ WPT.2018. 8639134

[39] Wang G, Wang P, Tang Y, Liu W. Analysis of dual band power and data telemetry for biomedical implants. IEEE Transactions on Biomedical Circuits and Systems. 2012;6:208-215. DOI: 10.1109/ TB-CAS.2011.2171958

[40] Kwon D, Yeo TD, Oh KS, Yu JW, Lee WS. Dual resonance frequency selective loop of near-field wireless charging and communications systems for portable device. IEEE Microwave and Wireless Components Letters. 2015; 25:624-626. DOI: 10.1109/LMWC.2015. 2451352

[41] Chung M, Chien Y-L, Cho L, Hsu P, Yang C-F. A dual-mode antenna for wireless charging and near field communication. In: 2015 IEEE Int. Symp. Antennas Propag. Usn. Natl. Radio Sci. Meet. 2015. pp. 1288-1289. DOI: 10.1109/APS.2015.7305033

[42] Lee W-S, Kim D-Z, Yu J-W. Multi- functional high-isolation dual antenna for controllable wireless charging and NFC communication. Electronics Letters. 2014;50:912-913. DOI: 10.1049/ el.2014.0858

[43] Dionigi M, Mongiardo M. A novel resonator for simultaneous wireless power transfer and near field magnetic communications. In: IEEE MTT-S Int. Microw. Symp. Dig. New York: IEEE; 2012. pp. 8-10. DOI: 10.1109/ MWSYM.2012.6259383

[44] Han M, Kim J-M, Sohn H. Dual- mode wireless power transfer module for smartphone application. In: 2015 IEEE Int. Symp. Antennas Propag. Usn. Natl. Radio Sci. Meet. New York: IEEE; 2015. pp. 111-112. DOI: 10.1109/ APS.2015.7304441

[45] Fotopoulou K, Flynn BW. Wireless power transfer in loosely coupled links: Coil misalignment model. IEEE Transactions on Magnetics. 2011;47: 416-430. DOI: 10.1109/TMAG.2010. 2093534

[46] Heo E, Choi K-Y, Kim J, Park J-H, Lee H. A wearable textile antenna for wireless power transfer by magnetic resonance. Textile Research Journal. 2018;88:913-921. DOI: 10.1177/ 0040517517690626

[47] Eteng AA, Rahim SKA, Leow CY. Wireless nonradiative energy transfer: Antenna performance enhancement techniques. IEEE Antennas and Propagation Magazine. 2015;57:16-22. DOI: 10.1109/ MAP.2015.2437281

[48] Ohira T. The kQ product as viewed by an analog circuit engineer. IEEE Circuits and Systems Magazine. 2017;17: 27-32. DOI: 10.1109/MCAS.2016. 2642698

[49] Mehri S, Ammari AC, Ben Hadj J, Slama LL. Performance characterization of variable width square coils for inductive link wireless power transfer. In: 2016 18th Mediterr. Electrotech. Conf. New York: IEEE; 2016. pp. 1-6. DOI: 10.1109/MELCON.2016.7495447

[50] Kim J, Lee K, Kwon H, Cha C. New structure for high Q-factor printed antenna in wireless power trans-

mission. In: IEEE EUROCON 2017 -17th Int. Conf. Smart Technol. New York: IEEE; 2017. pp. 474-478. DOI: 10.1109/ EUROCON.2017.8011155

[51] Mehri S, Ammari AC, Slama JBH, Sawan M. Design optimization of multiple-layer PSCs with minimal losses for efficient and robust inductive wireless power transfer. IEEE Access. 2018;6:31924-31934. DOI: 10.1109/ ACCESS.2018.2831785

[52] Hsu H-M. Improving the quality factor of a broadened spiral inductor with arithmetic-progression step width. Microwave and Optical Technology Letters. 2005;45:118-120. DOI: 10.1002/ mop.20741

[53] Klaric Felic G, Ng D, Skafidas E. Investigation of frequency-dependent effects in inductive coils for implantable electronics. IEEE Transactions on Magnetics. 2013;49:1353-1360. DOI: 10.1109/TMAG.2012.2231091

[54] Mehri S, Ammari AC, Slama J, Sawan M. Minimizing printed spiral coil losses for inductive link wireless power transfer. In: 2016 IEEE Wirel. Power Transf. Conf. New York: IEEE; 2016. pp. 1-4. DOI: 10.1109/ WPT.2016. 7498860

[55] Zierhofer CM, Hochmair ES. Geometric approach for coupling enhancement of magnetically coupled coils. IEEE Transactions on Biomedical Engineering. 1996;43:708-714. DOI: 10.1109/10.503178

[56] Sharma A, Zuazola IJG, Gupta A, Perallos A, Batchelor JC. Non-uniformly distributed-turns coil antenna for enhanced H-field in HF-RFID. IEEE Transactions on Antennas and Propagation. 2013;61:4900-4907. DOI: 10.1109/TAP.2013.2275244

[57] Sharma A, Zuazola IJG, Gupta A, Perallos A, Batchelor JC. Enhanced H- field in HF RFID systems by optimizing the loop spacing of antenna coils. Microwave and Optical Technology Letters. 2013;55:944-948. DOI: 10.1002/mop.27406

[58] Eteng AA, Abdul Rahim SK, Leow CY, Chew BW, Vandenbosch GAE. Two-stage design method for enhanced inductive energy transmission with Q-constrained Planar Square loops. PLoS One. 2016;11:e0148808. DOI: 10.1371/journal.pone.0148808

[59] Kurs A, Karalis A, Robert M, Joannopoulos JD, Fisher P, Soljacic M. Wireless power transfer via strongly coupled magnetic resonances. Science (80-). 2007;317:83-86. DOI: 10.1126/science.1143254

[60] Ahn D, Hong S. A study on magnetic field repeater in wireless power transfer. IEEE Transactions on Industrial Electronics. 2013;60:360-371.DOI: 10.1109/ TIE.2012.2188254

[61] Kung M, Lin K. A 6.78 MHz and 13.56 MHz dual-band coil module with a repeater for wireless power transfer systems. In: 2016 IEEE Int. Symp. Antennas Propag. New York: IEEE; 2016. pp. 157-158. DOI: 10.1109/ APS.2016.7695787

[62] Lee W-S, Park S, Lee J-H, Tentzeris MM. Longitudinally misalignment-insensitive dual-band wireless power and data transfer systems for a position detection of fast- moving vehicles. IEEE; Transactions on Antennas and Propagation. 2019;67: 5614-5622. DOI: 10.1109/TAP.2019. 2916697

[63] Zhu G, Gao D, Wang S, Chen S. Misalignment tolerance improvement in wireless power transfer using LCC compensation topology. In: 2017 IEEE PELS Work. Emerg. Technol. Wirel. Power Transf. New York: IEEE; 2017. pp. 1-7. DOI: 10.1109/WoW.2017. 7959411

10

Wireless Power Transfer for Miniature Implantable Biomedical Devices

Qi Xu, Tianfeng Wang, Shitong Mao, Wenyan Jia, Zhi-Hong Mao and Mingui Sun

Abstract

Miniature implantable electronic devices play increasing roles in modern medicine. In order to implement these devices successfully, the wireless power transfer (WPT) technology is often utilized because it provides an alternative to the battery as the energy source; reduces the size of implant substantially; allows the implant to be placed in a restricted space within the body; reduces both medical cost and chances of complications; and eliminates repeated surgeries for battery replacements. In this work, we present our recent studies on WPT for miniature implants. First, a new implantable coil with a double helix winding is developed which adapts to tubularly shaped organs within the human body, such as blood vessels and nerves. This coil can be made in the planar form and then wrapped around the tubular organ, greatly simplifying the surgical procedure for device implantation. Second, in order to support a variety of experiments (e.g., drug evaluation) using a rodent animal model, we present a special WPT transceiver system with a relatively large power transmitter and a miniature implantable power receiver. We present a multi-coil design that allows steady power transfer from the floor of an animal cage to the bodies of a group of free-moving laboratory rodents.

Keywords: implantable device, wireless power transfer, animal experiment, power transmitter, power receiver, antenna, double-helix coil, blood vessel, nervous system, resonance, even magnetic field, miniaturization

Introduction

Diagnosis and treatment of human diseases using an implantable electronic device represent a new trend in modern medicine. While the developments of biosensors, bioelectric stimulators and drug release mechanisms are important in the designs of medical implants, these developments are application specific. Therefore, they cannot be studied in a unified fashion. On the other hand, essentially all implantable devices require a common component: a power supply, which is usually a battery. Recent advances in wireless power transfer (WPT) provide an alternative method to power implantable electronic devices [1–3]. The WPT technology not only eliminates the needs of repeated surgical replacements of a depleted battery within the human body, but also reduces the size of the implant, simplifies the implantation procedure, and enables the device to be placed in restricted anatomic locations prohibitive to large implants.

Due to the importance of WPT in the next-generation medical implants, there have been extensive studies over recent years [4–7]. One of the major limiting factors in battery-less implants is the low power output at the receiving end due to the weak coupling of the wireless power link. Many existing WPT components in biomedical implants operate in the low-MHz frequency range, e.g., those utilizing the widely accepted 13.56 MHz industrial, scientific, and medical (ISM) band.

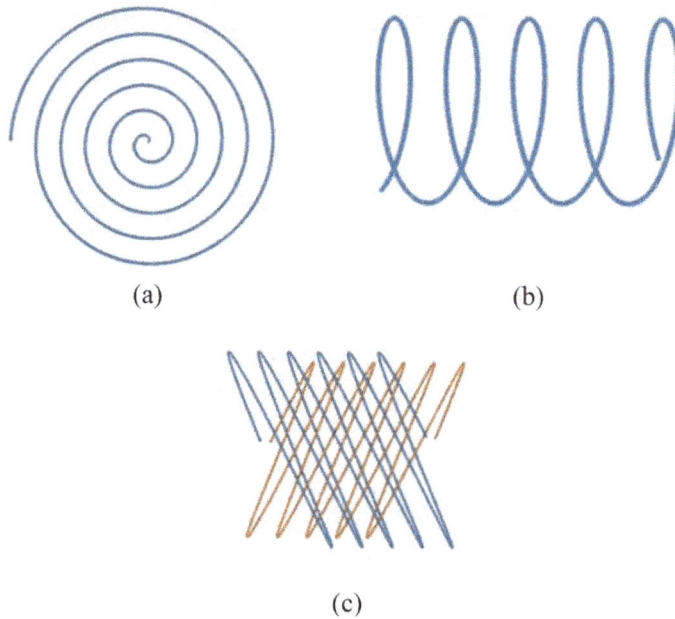

(a)   (b)

(c)

Figure 1. *Power receiver coils within the body: (a) planar spiral coil, (b) solenoidal coil, and (c) DH coil.*

Although the WPT component in this frequency range is easy to design and robust, a relatively large receiver antenna is required, which limits its application to implantable devices in millimeter scales [8]. In recent years, it has been demonstrated that this challenge can be overcome by increasing the operating frequency and producing spatially focused regions within the biological tissue [9]. This approach effectively makes the WPT component smaller; however, the design of the power receiver coils that both operate in a high frequency and adapt to anatomical features of biological organs or tissues has not be well studied. In essentially all magnetic resonance based WPT systems reported, the power receiver within the implant utilizes a spiral coil in either a planar or a solenoidal form, as shown in **Figure 1a** and b, respectively [4–6]. In general, the planar spiral coil (PSC) [4, 5, 7] has been used in implants having a relatively larger surface, such as certain cardiac pacemaker [10]. In other cases, medical implants are often designed in a cylindrical or capsular shape such as the Bion® microstimulator [11]. For these cases, the use of a solenoidal coil is more common [12–14]. Despite successful designs exist, these two forms of coils cannot meet the requirements for all medical implants. For example, the human body has many tubularshaped organs, such as nerves, lymphatic channels and blood vessels. An implant that wraps around such a biological structure to perform sensing and therapeutic functions is often desirable. In these cases, serious problems are encountered, with rare exceptions, because the structure is not allowed to be cut, and it is difficult to cut and rejoin a solenoidal coil during surgery. In addition, the optimal orientation of the implant may not comply with the orientation of the power transmitter coil outside the human body. In order to solve these problems, we will present a new form of coil, called double-helix (DH) coil (**Figure 1c**), to be applied to tubular organs within the human body.

Implantable devices are used not only for diagnosis and treatment of human diseases, but also for developments of new drugs and therapeutic mechanisms (e.g., electric stimulation). In the early stage of these developments, an animal model (e.g., the rodent model) is often utilized to study both treatment efficacies and side effects. In these studies, specially designed microsensors are often implanted within the body of a laboratory rodent to measure certain variables of interest [9]. Frequently, animal behaviors are also monitored by videotaping and other means [15]. This approach often encounters a significant problem of lacking a suitable power supply because the use of either a battery or a wire connection to the implant inside the body seriously interferes with animal's mobility and behavior. In this case, the application of the WPT technology is essential because it allows much reduced weight and size of the system being carried by, or implanted within, the rodent [16]. In order to provide the animal with a sufficient space for free movements, a special WPT system with a large stationary transmitter (in which the coil is embedded under the floor of an animal cage) and a miniature receiver (implanted within or carried by the animal) is required. In order for the WPT system to perform properly regardless of the animal's location within the cage, the transmitter must produce an even radio-frequency (RF) magnetic field throughout the floor of the animal cage. As a result, the wirelessly delivered energy is relatively even everywhere over the entire floor. This chapter studies this problem and presents a seven-coil design with several desirable properties, including the theoretical optimality and ease of implementation.

This chapter is organized as follows. In Section 2, we describe the DH coil that can be applied to tubular organs within the human body. The coupling factor and power transfer efficiency (PTE) were analyzed. To further evaluate the performance of the DH coil, both simulations and experiments were conducted and presented. In Section 3, we present a power mat consisting of an array of planar transmitter coils. This mat produces a nearly even magnetic field distribution over the entire animal cage floor. For clarity, we present our evaluation, simulation and/ or experimental results at the end of each methodological section. Finally, we conclude this chapter in Section 4.

Double-helix coil for wrap-around implants

The human body contains networks of tubular organs, such as nerves, lymphatic channels and blood vessels [17]. In order to monitor the functions or pathologic states of these organs (e.g., clogging of a certain major blood vessel) or provide therapeutic functions (e.g., stimulating a peripheral nerve), a wirelessly powered miniature implant wrapped around a tubular or rod-like biological structure is highly desirable. Although an ordinary solenoidal coil can support this wrap-around implant, an intact tubular organ usually cannot be cut and rejoined to allow a solenoidal coil to be threaded into the desired implanting position. Alternatively, one may wind the coil wire around the tubular organs manually during surgery.

This method is practically unacceptable due to the restricted time of surgery and difficulties in quality control manual winding. Another method is to wrap the tubular organ by a coil that has been cut longitudinally. To reform an intact coil, a surgeon needs to reconnect the wires by soldering or using special connectors. This approach is also unrealistic due to the high risk of infection involved and the possible failure of the connectors.

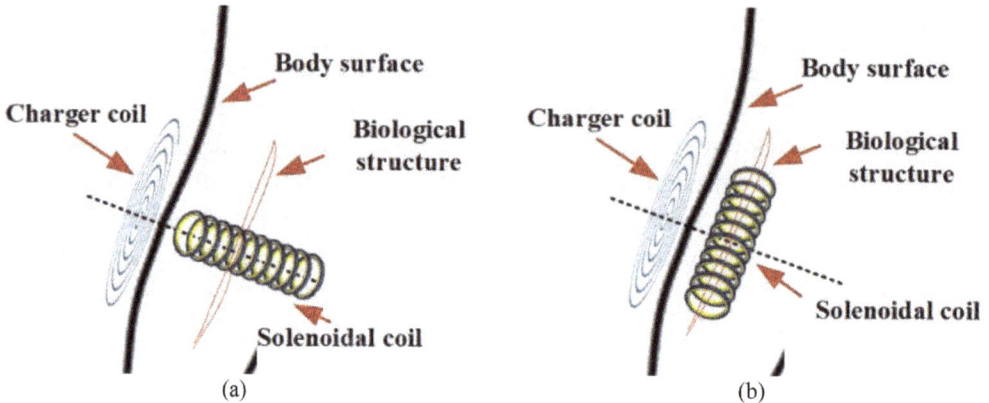

Figure 2. *The implanted solenoid coil (a) perpendicular and (b) parallel to a planar transmitter (shown in blue).*

In addition to these practical difficulties, the usage of a solenoidal coil for wrap- around implants suffers from another implementational problem. In most cases, it is desirable to use a flexible PSC for the power transmitter because this planar coil can be easily integrated with a garment, a blanket or a bedding sheet, providing a high convenience for unobtrusive recharging of the implant. In order to achieve the maximum coupling between the transmission and reception coils, the solenoidal coil of the implant is expected to be oriented perpendicularly to the body surface (**Figure 2a**). This requirement is highly problematic because, as indicated by human anatomy [17], many tubular organs within the body are oriented parallel to the body surface, which provides the worst orientation to the PSC for power transmission because of the weakest magnetic coupling (**Figure 2b**).

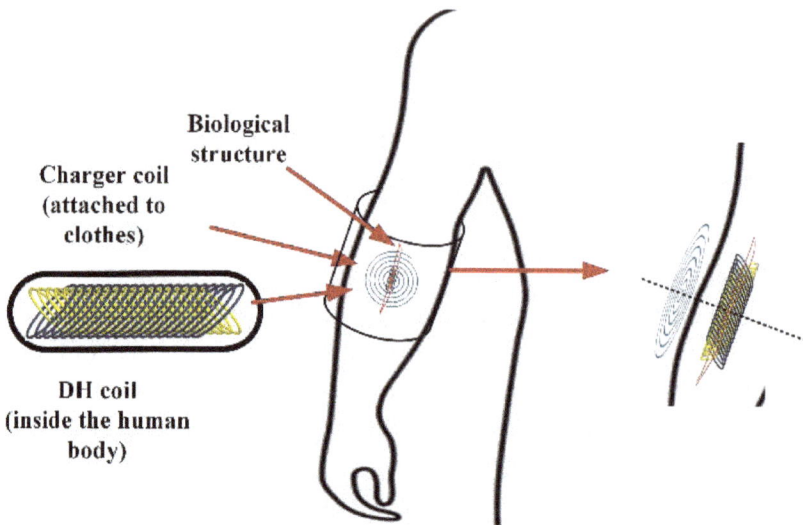

Figure 3. *A DH coil is wrapped around a biological structure (red curve) to serve as a power receiver.*

To address these significant problems, we have developed an air-core DH coil for tubular implants [18, 19]. The new coil can be printed on a flat flexible printed circuit board (PCB) and installed on a tubular biostructure during surgery. As shown in **Figure 3**, when the DH coil is wrapped around a tubular organ parallel to the skin surface, the optimal coupling can be achieved with a PSC integrated with a garment or a bedding sheet.

Structural design of DH coil

Figure 4a shows a set of parallel sinusoidal wires printed on both sides of a flexible PCB. This PCB is made of polyimide film which is characterized by high strength, low RF energy loss, small thickness, and high flexibility. The wires on two sides are connected to each other in series via a column of through holes (one of them is shown), forming closed loops. The current paths are denoted by the black arrows in **Figure 4a**. In addition to the DH coil, sensors, actuators, microprocessor and electronic elements (not shown) can be installed on the same flexible PCB. During surgery, the hermetically sealed PCB (using a biocompatible polymer material) is wrapped around the tubular structure at the position of interest forming a double helix winding along with all electronic components, as shown in **Figure 4b**. It can be observed that, after the tubular structure is formed, the closed loops on the two sides of the PCB form opposite tilt angles but maintain the clockwise/counter- clockwise current direction.

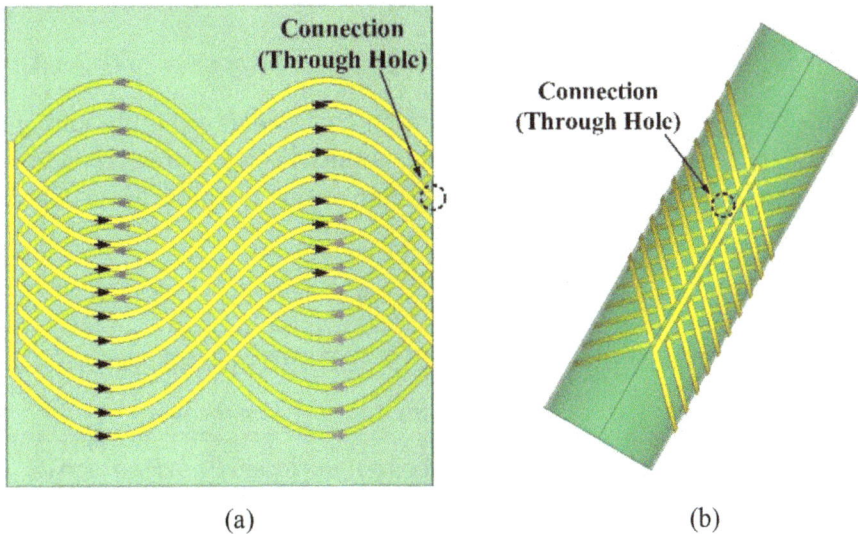

(a)   (b)

Figure 4. *(a) Schematic representation of the DH coil and current directions indicated by arrows. Conductors are printed on both sides of the flexible PCB; (b) the PCB is wrapped to form a DH coil.*

The two-layer structure in **Figure 4** can be extended to a multiple-layer structure using a PCB with more than two layers. Similarly, the 45° tilted angle in each layer can be modified to adapt to specific tubular organ orientation for optimal wireless power delivery.

Inductance and mutual inductance of DH coil

As shown in **Figure 5**, for simplicity, the DH coil is separated into several cells and each cell consists of an inner loop (Loop 1) and an outer loop (Loop 2).

By superposition, the inductance of the DH coil is equal to the sum of the cell inductance and the mutual inductance between each cell (**Figure 5a**), namely:

$$L_{DH} = \sum_{\substack{i=1 \\ }}^{n} \sum_{\substack{j=1 \\ j \neq i}}^{n} M_{i,j} + \sum_{i=1}^{n} L_i = \sum_{\substack{i=1 \\ }}^{n} \sum_{\substack{j=1 \\ j \neq i}}^{n} M_{i,j} + nL_{cell}. \tag{1}$$

where $M_{i,j}$ denotes the mutual inductance between cell i and cell j, and Lcell denotes the inductance of a single cell.

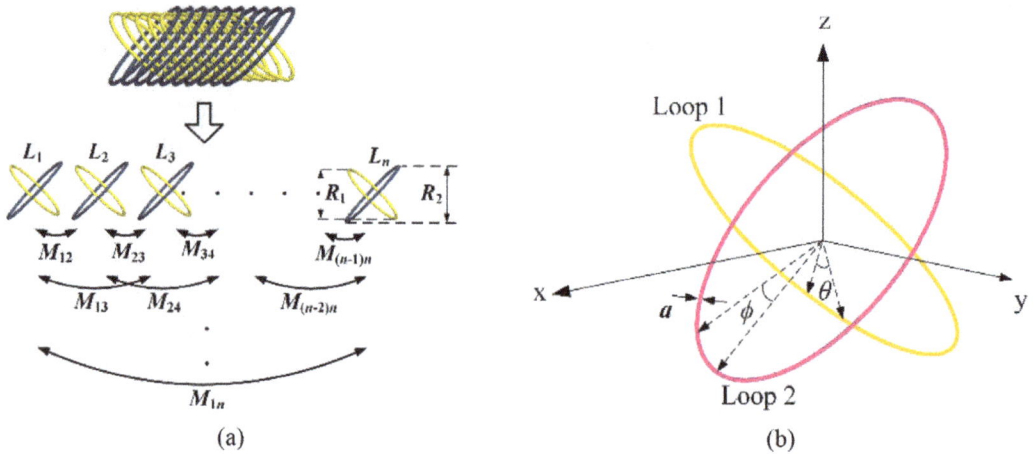

Figure 5. *(a) The DH coil can be separated into n cells, having mutual inductance with each other; (b) cell model.*

According to the cell model (**Figure 5b**), the inductance of a single cell is the sum of the inductances of Loops 1 and 2 and their mutual inductance. Since the loops are perpendicular to each other, the mutual inductance between the loops is zero so that the inductance of a single cell is given by

$$L_{cell} = L_1 + L_2 + 2M_{12} = L_1 + L_2 \tag{2}$$

where L_1 and L_2 are the inductances of Loops 1 and 2, respectively, given by

$$\begin{cases} L_1 = \dfrac{\mu_0}{4\pi} \int_0^{2\pi} \int_0^{2\pi} \dfrac{R_1(R_1 - a)(2\sin\theta\sin\phi + \cos\theta\cos\phi)}{\sqrt{2(R_1\cos\theta - (R_1 - a)\cos\phi)^2 + (R_1\sin\theta - (R_1 - a)\sin\phi)^2}} d\theta d\phi \\[4mm] L_2 = \dfrac{\mu_0}{4\pi} \int_0^{2\pi} \int_0^{2\pi} \dfrac{R_2(R_2 - a)(2\sin\theta\sin\phi + \cos\theta\cos\theta')}{\sqrt{2(R_2\cos\theta - (R_2 - a)\cos\phi)^2 + (R_2\sin\theta - (R_2 - a)\sin\phi)^2}} d\theta d\phi \end{cases} \tag{3}$$

where R_1 and R_2 are the projected radii of Loops 1 and 2 along the z axis (**Figure 5a**).

Figure 6 shows a model containing two cells. The mutual inductance between the two cells is calculated by

$$M_{i,j} = M_{i-1,\,j-1} + M_{i-1,\,j-2} + M_{i-2,\,j-1} + M_{i-2,\,j-2}$$

$$= M_{i-1,\,j-1} + 2M_{i-1,\,j-2} + M_{i-2,\,j-2} \tag{4}$$

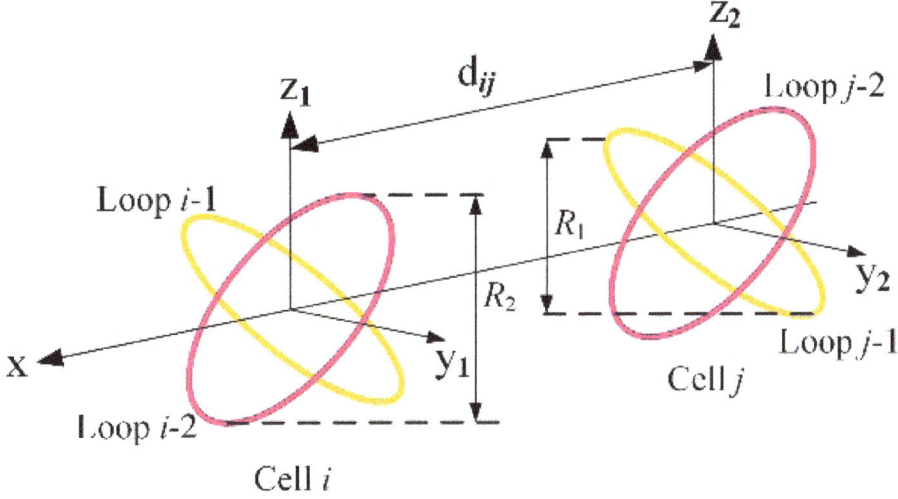

Figure 6. *Modeling of the mutual inductance between two cells.*

where,

$$
\begin{cases}
M_{i-1,j-2} = \dfrac{\mu_0}{4\pi} \displaystyle\int_{2\pi}^{0}\int_{0}^{2\pi} \dfrac{R_1 R_2 \cos\theta \cos\phi}{\sqrt{\left(R_1\cos\theta - R_2\cos\phi + d_{ij}\right)^2 + R_1^2 + R_2^2 + 2R_1 R_2 \cos(\theta+\phi)}} d\theta d\phi \\[4mm]
M_{i-1,j-1} = \dfrac{\mu_0}{4\pi} \displaystyle\int_{0}^{2\pi}\int_{0}^{2\pi} \dfrac{R_1^2(2\sin\theta\sin\phi + \cos\theta\cos\phi)}{\sqrt{\left(R_1\cos\theta - R_1\cos\phi + d_{ij}\right)^2 + 2R_1^2 - 2R_1^2\cos(\theta-\phi)}} d\theta d\phi \\[4mm]
M_{i-2,j-2} = \dfrac{\mu_0}{4\pi} \displaystyle\int_{0}^{2\pi}\int_{0}^{2\pi} \dfrac{R_2^2(2\sin\theta\sin\phi + \cos\theta\cos\phi)}{\sqrt{\left(R_2\cos\theta - R_2\cos\phi + d_{ij}\right)^2 + 2R_2^2 - 2R_2^2\cos(\theta-\phi)}} d\theta d\phi
\end{cases} \quad (5)
$$

Therefore, the total inductance of the DH is derived based on the superpositions of the calculated inductances by Eq. (2).

As shown in **Figure 7a**, the mutual inductance between the DH coil and the transmitter (Tx) can also be regarded as the sum of all individual mutual inductances between Tx and each loop in the DH coil, namely,

$$
M_{Total} = \sum_{i=1}^{n} M_{Ti-1} + \sum_{i=1}^{n} M_{Ti-2} = \sum_{i=1}^{n}(M_{Ti-1} + M_{Ti-2}) = \sum_{i=1}^{n} M_{Ti}. \quad (16)
$$

where M_{Ti-1} and M_{Ti-2} are the mutual inductances between Tx and the ith inner and outer loops, respectively, M_{Ti} is the mutual inductance between Tx and the ith cell, and n denotes the cell number. For simplicity, Tx is modeled as a circular coil with only one turn as shown in **Figure 7b**.

Then, M_{Ti-1} and M_{Ti-2} are given by

$$
\begin{cases}
M_{Ti-1} = \dfrac{\mu_0 m}{4\pi} \displaystyle\int_{0}^{2\pi}\int_{0}^{2\pi} \dfrac{R_1 R_{TX} \cos(\theta-\gamma)}{\sqrt{(R_1\cos\theta - R_{TX}\cos\gamma)^2 + (-R_1\sin\theta - R_{TX}\sin\gamma)^2 + (R_1\cos\theta - d)^2}} d\theta d\gamma \\[4mm]
M_{Ti-2} = \dfrac{\mu_0 m}{4\pi} \displaystyle\int_{0}^{2\pi}\int_{0}^{2\pi} \dfrac{R_2 R_{TX} \cos(\phi-\gamma)}{\sqrt{(R_2\cos\phi - R_{TX}\cos\gamma)^2 + (-R_2\sin\phi - R_{TX}\sin\gamma)^2 + (R_2\cos\phi + d)^2}} d\phi d\gamma
\end{cases} \quad (17)
$$

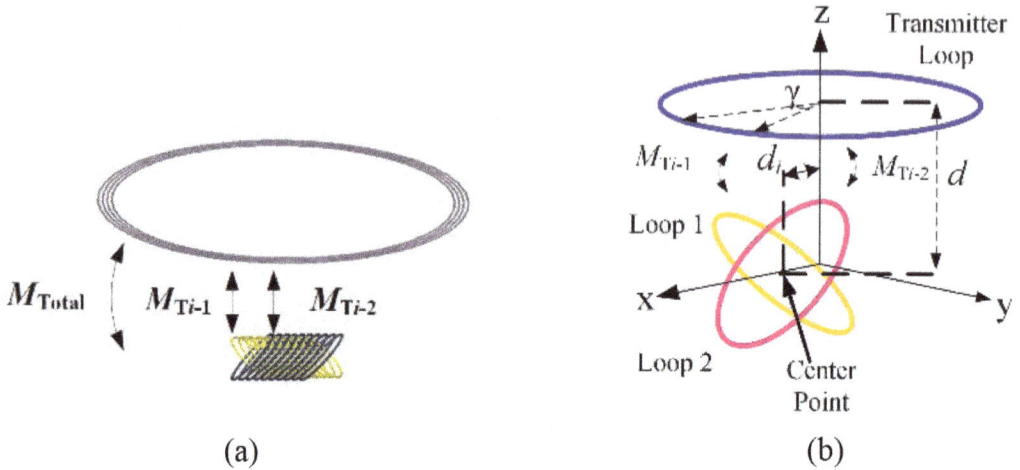

Figure 7. *(a) The total mutual inductance between Tx and the DH coil is calculated by the superpositions of the mutual inductances between Tx and all cells; (b) Modeling of the mutual inductance between Tx and one cell.*

where R_{TX} is the radius of Tx, μ_o is the magnetic permeability of free space, and d is the vertical distance between the cell and Tx. Accordingly, we have

$$M_{Ti} = M_{Ti-1} + M_{Ti-2};$$ (8)

The coupling factor k between Tx and Rx is given by

$$k = \frac{M_{Total}}{\sqrt{L_{DH}L_{TX}}}$$ (9)

where

$$L_{TX} = \mu_0 R_{TX}\left(\ln\left(\frac{R_{TX}}{a_{TX}}\right) - 1.75\right).$$ (10)

Coupling factor simulations

According to the magnetic resonant WPT theory, the coupling factor k largely influences system performance [20]. In order to evaluate the coupling factor in the system with the misalignment between the transmitter and receiver coils, the proposed DH coil was compared to a conventional double-layer solenoid by simulation. The transmitter was modeled as a PSC with the outer and inner radii being 30 and 15 mm, respectively. For the DH coil and conventional solenoid, R_1 and R_2 were 5 and 6 mm, respectively. The variation of k was investigated with respect to the lateral and angular misalignments.

Figure 8 shows the simulation models with lateral or angular misalignments. X_o and Y_o in **Figure 8a** and b indicate the displacements along the x-direction and y-direction, respectively. α and β indicate the rotating angles around the x-axis and y-axis, respectively. The vertical distance d in **Figure 8c** and d is 20 mm in all cases. The simulation was used to calculate the coupling factor in different scenarios. The results are shown in **Figure 9**.

It is seen that offers the maximum coupling factor k of the conventional solenoid much smaller than that of the DH coil, comparing **Figure 9a** and b. Additionally, the coupling factor k of the

DH coil is larger than that of the conventional solenoid in most of the measurement range. **Figure 9c** showed that k is essentially invariable as the conventional solenoid rotates around the x-axis. However, the DH coil has an optimal angle as which the largest k is achieved. With β decreases, the central axes of both the DH coil and conventional solenoid change from being perpendicular to being parallel with the plane of Tx as shown in **Figure 9d**. Within such a process, the coupling factor of the solenoid decreases, while the coupling factor increases for the DH coil. In addition, when β is 0, the coupling factor of the DH coil is much larger than that of the conventional solenoid. Accordingly, the orthogonal-coil structure enhances the mutual inductance. This phenomenon verifies the superiority of the DH coil over the traditional solenoid.

Experimental results

We constructed several prototypes of DH coils with variable turns, gaps and widths. Then, the coil with the largest quality factor was chosen and tuned to a resonant frequency of 5.2 MHz using capacitors. This DH coil is presented in **Figure 10**.

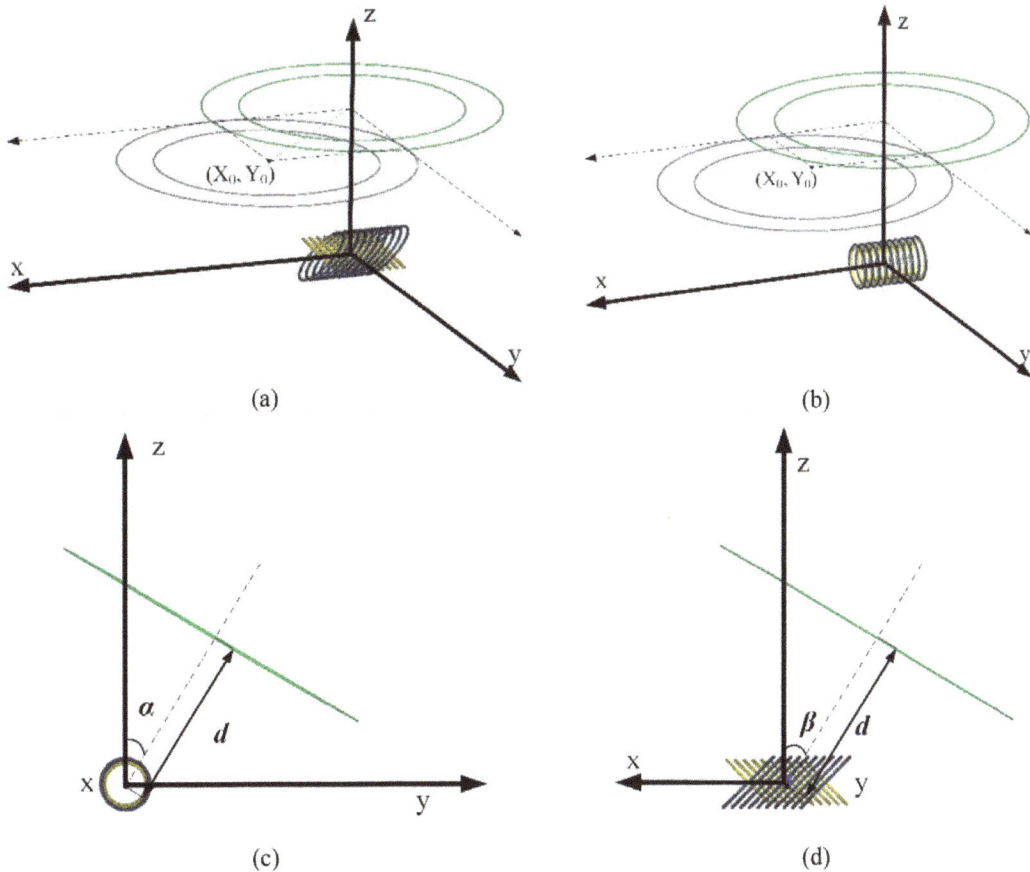

Figure 8. *Simulation models of (a) DH coil and (b) solenoid coil for lateral misalignment, and for angular shifting around (c) the x-axis and (d) the y-axis, respectively.*

To study WPT performances in different misalignment scenarios, the PTE was chosen as an evaluation index. The PTE was measured based on the scatter parameters measured by a network analyzer. PTE measurements were also performed for both lateral and angular misalignments. The results shown in **Figure 11** indicate that the WPT system achieves the maximum PTE at the resonant frequency. However, the PTE decreases as the misalignment increases, similar to the variation of the coupling factor.

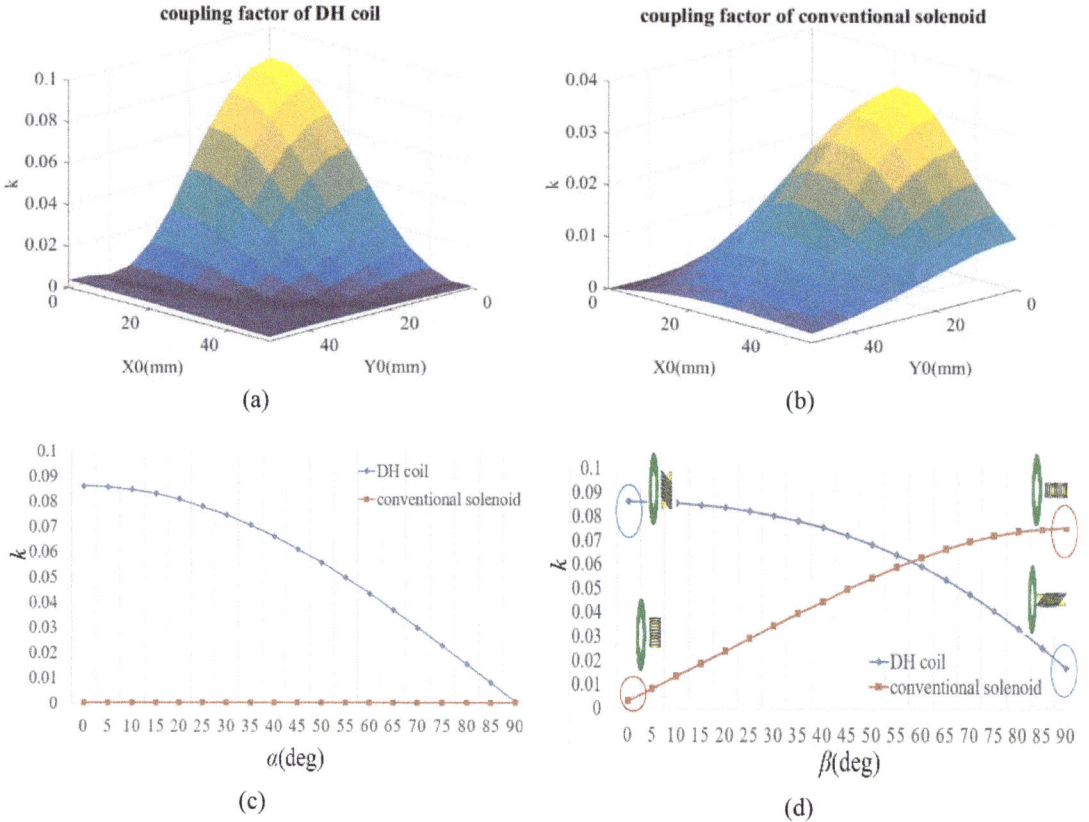

Figure 9. *The coupling factor as a function of the relative position between the receiver and transmitter coils in different cases: (a) DH coil with lateral misalignment, (b) traditional coil with lateral misalignment, (c) the angular shifting around the x-axis, and (d) the angular shifting around the y-axis.*

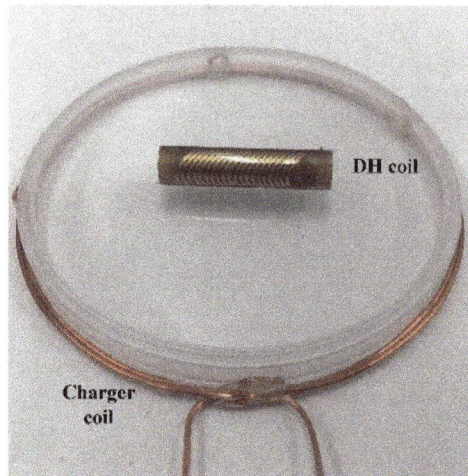

Figure 10. *Experimental setups for efficiency measurement.*

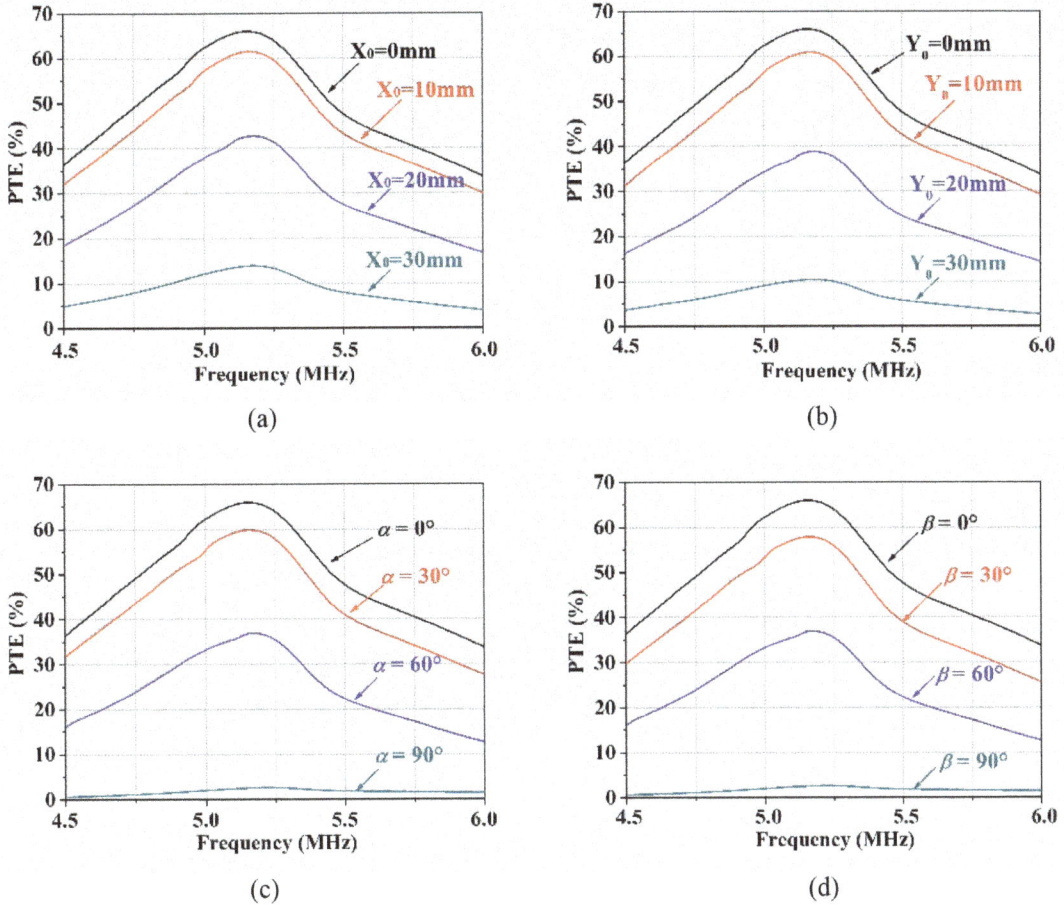

Figure 11. *PTE measurements vs. frequency with variable misalignments: (a) lateral misalignment in the x-direction, (a) lateral misalignment in the y-direction, (c) angular misalignment around x-axis, and (d) angular misalignment around y-axis.*

It can be observed that the PTE is strongly dependent on the inductive coupling, which was discussed in the previous section. Therefore, the proposed DH coil offers more efficient power delivery than the traditional solenoid.

Mat-based wireless power transfer to moving targets

Neural stimulation and recording provide emerging prosthetic and treatment options for spinal cord injury, stroke, sensory dysfunction, and other neurological diseases and disorders. Neural recording from awake animals with observable behavior has greatly enhanced our understanding of central and peripheral nervous systems. Although there has been substantial studies on miniaturized, implantable electronic circuits that record neural data and stimulate neuronal networks in freely moving laboratory animals, the mobility of the animal subject is often limited, and the experimental results obtained under restricted conditions may not reflect the full repertoire of brain activity corresponding to their natural behaviors [16]. There are similar problems in the study of new drugs which often requires monitoring a number of variables from the inside of the animal body and observation of their mobility and behaviors.

Traditionally, magnetic induction was used for WPT using a similar form to a transformer [21]. This form of the magnetic induction method is highly efficient (>90%) in the near-field range, but much less efficient as the transmission distance increases. In 2007, an efficient mid-range WPT via strongly coupled magnetic resonance was reported [22]. This system consists of four coils (**Figure 12**), namely, driver, primary (or transmitter), secondary (or receiver), and load coils. Inductive coupling is used between the driver and primary coils as well as between the secondary and load coils. The primary and secondary coils with the same resonant frequencies tend to exchange energy efficiently. This mechanism is valuable in the application to medical implants because biological tissues are generally non- resonant at the operating frequency in the RF range.

Because our special WPT system involves moving targets (animals, e.g., rodents), the vertical component of the magnetic field generated by the transmitter is required to be distributed as even as possible over the entire area of interest (e.g., floor of the rodent cage). When this condition is satisfied, the device carried by or implanted within each rodent can receive steady power at any location of the floor. Previously, we designed a WPT system in which multiple circular spiral coils were printed on hexagonal PCBs [23]. These PCBs were then tiled hexagonally forming a "power mat" shown in **Figure 13**. Note that the use of hexagons in the pack of coils is not an arbitrary choice, rather it has been proven that this design will leave the smallest gap between circular resonator coils [24]. The power mat is able to deliver wireless power to the implants and "carry-on" devices to multiple rodents which move freely on the floor above the mat.

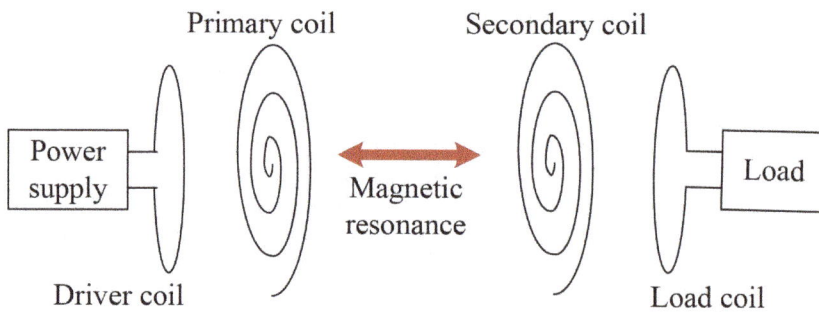

Figure 12. *A resonance based WPT system including four coils, namely driver, primary, secondary, and load coils.*

Figure 13. *WPT system in which the floor of the animal cage is located over a hexagonally packed power mat.*

It has been found from computer simulation that the packing of a circle by identical disks demonstrates interesting patterns [25]. Denser packings consist of specific numbers of disks. The first several numbers are: 1, 7, 19, 37, 61, and 91. Although the packing density increases as more disks are packed within the enclosing circle, the density is upper-bounded by $\pi^2/12 \approx 0.822$. Although, in general, using more disks produces a higher density, the implementation complexity increases as the number of disks increases. Additionally, it is difficult to connect the RF signal to numerous transmitter (Tx) coils, and the cost involved is high. On the other hand, we have previously reported that the RF signal can be easily connected to a pair of open concentric rings to power seven Tx coils simultaneously [23].

Therefore, if more than one coil is used in the Tx, the seven-coil design is often the best choice for most practical applications to power free-position devices although its packing density is not the highest.

Theoretical analysis of mat-based WPT system

As shown in **Figure 14**, the mat-based WPT system enables magnetically coupled resonance between an array of transmitter coils and a single receiver coil.

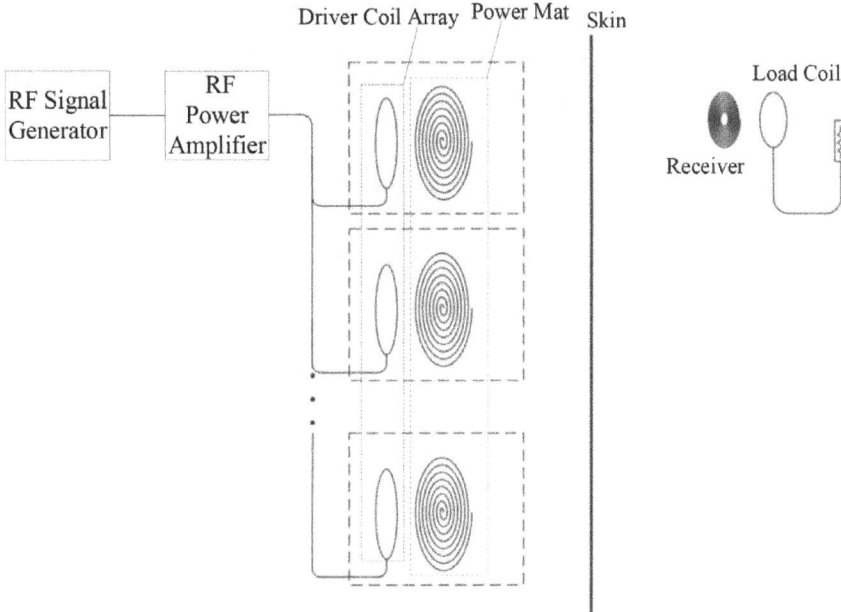

Figure 14. *Mat-based WPT system design.*

An adjustable RF oscillator produces a sinusoidal signal, which is amplified by a power amplifier. The output of the amplifier is connected to an array of driver coils which are inductively coupled with primary coils [23]. At the power receiving site (a laboratory animal), the receiver coil is inductively coupled with a load coil to supply the power to the electronic components within the implant.

In the coupled mode theory (CMT), the first eigen-mode is used to analyze a resonant system. The approximation by the first eigen-mode is quite accurate under the condition of a strong coupling in the WPT system. In practical applications, it is not possible or necessary to directly work on an arbitrary large number of resonators. Rather, for a large hexagonally packed transmitter (HPT) mat (**Figure 15**), every one of the resonators can be treated as in the middle of the mat, except for those ones at the edge, and fortunately the edge effect can be solved simply by making the mat larger than the animal cage floor. For this reason, we may simplify analysis by examining the seven-resonator case.

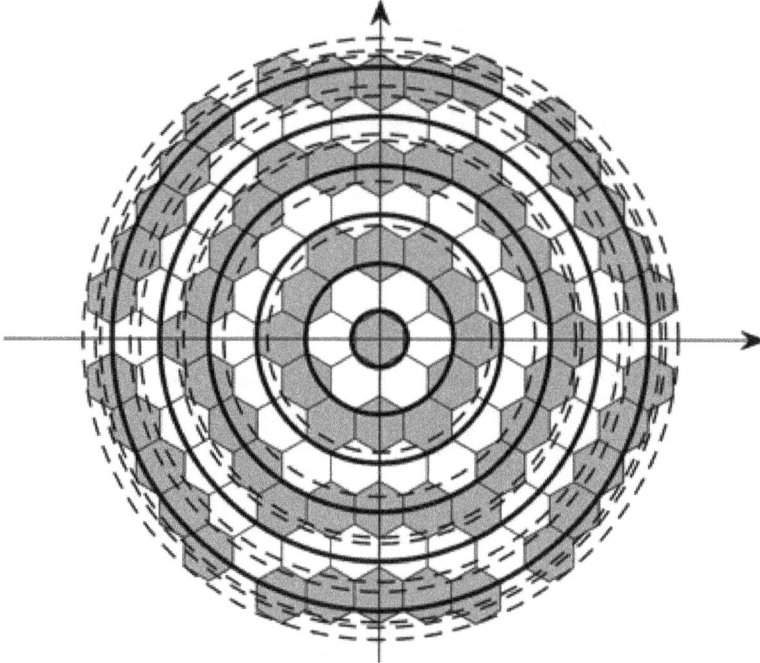

Figure 15. *A large hexagonally packed power mat.*

Let us index the seven transmitters from 1 to 7 and the single receiver have an index of 8. In order to describe the strongly coupled system, a set of differential equations based on CMT is given by [22].

$$\dot{a}_8(t) = (j\omega_0 - \Gamma_8 - \Gamma_L)a_8(t) + \sum_{j=1}^{7} j\kappa_{ij}a_j(t)$$

$$\dot{a}_i(t) = (j\omega_0 - \Gamma_i)a_i(t) + \sum_{\substack{j=1 \\ j \neq i}}^{8} j\kappa_{ij}a_j(t) + f_i(t), \quad i = 1, ..., 7 \tag{11}$$

Or, if written in matrix form

$$\begin{bmatrix} \dot{a}_1(t) \\ \dot{a}_2(t) \\ \vdots \\ \dot{a}_7(t) \\ \dot{a}_8(t) \end{bmatrix} = \begin{bmatrix} (j\omega_0 - \Gamma_1) & j\kappa_{12} & \cdots & j\kappa_{17} & j\kappa_{18} \\ j\kappa_{21} & (j\omega_0 - \Gamma_2) & \cdots & j\kappa_{27} & j\kappa_{28} \\ \vdots & \vdots & \ddots & \vdots & \vdots \\ j\kappa_{71} & j\kappa_{72} & \cdots & (j\omega_0 - \Gamma_7) & j\kappa_{78} \\ j\kappa_{81} & j\kappa_{82} & \cdots & j\kappa_{87} & (j\omega_0 - \Gamma_8 - \Gamma_L) \end{bmatrix} \begin{bmatrix} a_1(t) \\ a_2(t) \\ \vdots \\ a_7(t) \\ a_8(t) \end{bmatrix} + \begin{bmatrix} f(t) \\ f(t) \\ \vdots \\ f(t) \\ 0 \end{bmatrix} \tag{12}$$

where $a_i(t)$, $i = 1, 2, \cdots, 7$, and $a_8(t)$ are, respectively, the first eigenmodes of the transmitter and receiver resonators corresponding to the natural frequency ω_0, Γs are the intrinsic loss rates of resonators due to absorption and radiation, Γ_L represents the rate of energy going into the load, κs are pairwise coupling coefficients between resonators, and f is are the inputs to the transmitter resonators. In our case, all f is are the same, i.e., $f_1 = f_2 = \cdots = f_7 = f$. Note that a_is are also known as positive frequency components in terms of CMT. Although a_i (generally complex-valued) does not represent a voltage or current directly, the energy contained in each resonator can be represented as $|a_i|^2$, and the power output of the system is $2\Gamma_L |a_8|^2$. Using the CMT concept, the goal of obtaining a uniform power output becomes finding a constant $|a_8|$ within the WPT space.

To make Eq. (12) more concise, we write it into the following form:

$$\dot{\mathbf{a}} = \mathbf{A}\mathbf{a} + \mathbf{f} \tag{13}$$

where the vectors and matrices are in correspondence with Eq. (12).

If the WPT system is driven by a sinusoidal input, e.g., $\mathbf{f}(t) = Fe^{j\omega_0 t}[1, 1, \cdots 1, 0]^T$, the positive frequency components have the form of $\mathbf{a}(t) = \mathbf{a}e^{j\omega_0 t}$ at the steady state. Substituting this into Eq. (13), we can solve for $\mathbf{a}(t)$

$$\mathbf{a}(t) = -\mathbf{B}^{-1}\mathbf{f}(t) \tag{14}$$

where

$$\mathbf{B} = \begin{bmatrix} -\Gamma_1 & j\kappa_{12} & \cdots & j\kappa_{17} & j\kappa_{18} \\ j\kappa_{21} & -\Gamma_2 & \cdots & j\kappa_{27} & j\kappa_{28} \\ \vdots & \vdots & \ddots & \vdots & \vdots \\ j\kappa_{71} & j\kappa_{72} & \cdots & -\Gamma_7 & j\kappa_{78} \\ j\kappa_{81} & j\kappa_{82} & \cdots & j\kappa_{87} & -\Gamma_8 - \Gamma_L \end{bmatrix}$$

In case where $\mathbf{B}$ is not invertible, its pseudo inverse can be used instead. Thus, given Γ_i, κ_{ij}, and f i, we can compute $a_i(t)$ analytically by Eq. (14). The CMT approach provides a powerful analytical tool for the multi-resonator WPT system. For example, it has been utilized to maximize the efficiency of power transfer and investigate the relay effect by inserting one or more resonators between the transmitter and receiver [26]. Using CMT, we have studied the dynamics of the system involving an array of resonators [27]. Although the previous studies have shown that CMT well characterizes the temporal behavior of the WPT system, it has clear limitations when the system parameter changes. For example, when the receiving resonator moves over the HPT mat, the coupling coefficients κ_{8i} ($i = 1, 2, \cdots, 7$) change, and the variations of the system behavior are difficult to determine analytically. In order to study the motion effect of the receiving resonator and answer the critical question whether the receiver resonator can harvest sufficient amount of power at different locations over the HPT mat, we performed numerical simulation and conducted an experimental test.

Simulation of mat-based WPT system

For a clear illustration of the design principle of the mat-based WPT system, we simulated a single HPT cell consisting of seven PSCs, as limited by the computational complexity. This simulation does not cause a loss of generality because the results of multiple cells can be obtained simply by superposition of single cell results. In most cases, changes in the position of a device lead to a variation in mutual inductance which results from a change in the magnetic field distribution. Although some unevenness in the distribution is unavoidable, we expect this distribution to be nearly uniform with enhanced misalignment tolerability for WPT applications involving moving targets.

We utilize the concentric model to approximate the coil where the total magnetic field is a superposition of the fields of individual loops in the coil. Assuming that a loop with a radius of a is centered at the origin carrying a current I. Based on the Biot-Savart law, the x-, y- and z-components of the magnetic field at point r (x, y, z) are given by [28].

$$B_x = \frac{Cxz}{2\alpha^2\beta\rho^2}\left[(a^2+r^2)E(k^2)-\alpha^2K(k^2)\right]$$

$$B_y = \frac{Cyz}{2\alpha^2\beta\rho^2}\left[(a^2+r^2)E(k^2)-\alpha^2K(k^2)\right] = \frac{y}{x}B_x \qquad (15)$$

$$B_z = \frac{C}{2\alpha^2\beta}\left[(a^2-r^2)E(k^2)-\alpha^2K(k^2)\right]$$

where $\rho^2 = x^2 + y^2$, $r^2 = x^2 + y^2 + z^2$, $\alpha^2 + a^2 + r^2 - 2a\rho$, $\beta^2 = a^2 + r^2 + 2a\rho$, $k^2 = 1 - \alpha^2/\beta^2$, $C = u_0 I/\pi$, and $K(.)$ and $E(.)$ are the complete elliptic integrals of the first and second k inds, respectively. For easier calculation, without loss of generality, the current is chosen so that $C = 1$. With the magnetic field of a single loop, we can apply the superposition rule to get the magnetic field generated by a multi-loop circular coil. Since the receiver coil is always in parallel with the transmitter coil (which is installed below the animal cage floor), the fluxes in the receiver coil are contributed by the z-component of the magnetic field. Therefore, we will focus on the z-component of the magnetic field in our analysis.

With these simplifications, we can calculate the magnetic field of the seven- transmitter powering platform. **Figure 16** shows a magnetic field with different distances between the transmitter and the surface (evaluating plane) where the field was evaluated. It can be seen that the separation between the evaluating plane and the transmitter enables the variation of the z-component of magnetic field. When the distance is increasing, the flatness of the magnetic field also improves at first. However, when the height is too large, the magnetic field distribution becomes similar to that of a larger spiral coil, which affects both the flatness and magnitude of the magnetic field. On the other hand, when the height is too small, although the peak magnitude is larger, the z-component is inversed at the gap between resonators, which implies a large fluctuation of the magnetic field.

In order to investigate the effect of mutual inductance between turns in the spiral coil, we also performed a simulation study on the 7-coil mat using a commercial finite element (FE) software

HFSS (Ansys Corp., Pittsburgh, PA). **Figure 17** shows the 3D model of the HPT mat used in the simulation, where each PSC was 20 cm in outer diameter, 1 cm in conductive trace width, and 1 cm in trace spacing. The input power was set at 1 W. As stated previously, the goal of the power mat design was to obtain a nearly uniform magnetic field within an extended region to support WPT for moving targets, rather than optimizing PTE (the animal cage is powered from a regular AC socket).

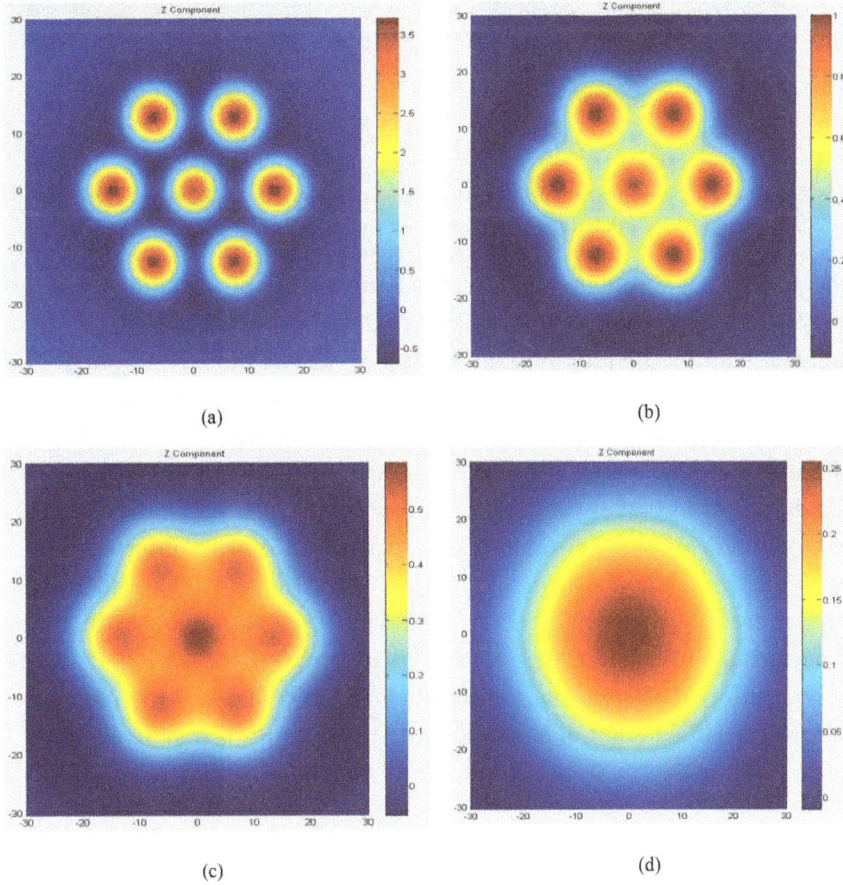

(a)   (b)

(c)   (d)

Figure 16. *Z-component of the magnetic field distribution with different separations between the evaluation plane and the transmitter consisting of seven coils. (a) 3.25 cm, (b) 7.25 cm, (c) 10.25 cm, and (d) 20.05 cm.*

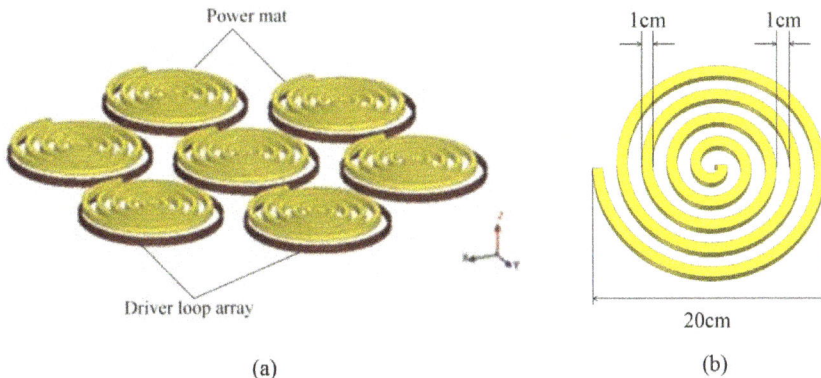

(a)   (b)

Figure 17. *HFSS simulation. (a) 3D model of the transmitter mat; (b) Dimensions of each PSC.*

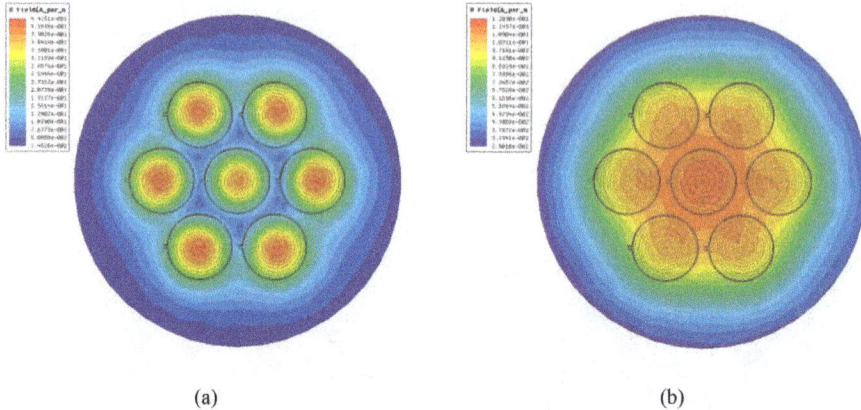

(a)   (b)

Figure 18. *Distribution of the z-component of the magnetic field in a plane at (a) 8 cm and (b) 20 cm above the HPT mat at the resonant frequency of 85.2 MHz.*

We excited the seven PSCs simultaneously using a common RF power source. Energy was injected into the driver coil array to maintain resonance in the presence of losses and energy drawn from the magnetic field by the receiver coil. **Figure 18** shows the z-component distribution of the magnetic field at 8 and 20 cm distances, respectively, above the power mat (i.e., the X-Y plane). Color indicates the magnitude of the magnetic field in the z-direction. It can be seen that, at z = 8 cm (**Figure 18a**), the magnitude of the magnetic field was the highest (peak) at the center of each coil, and the lowest (valley) at the junction of three coils. When the distance to the HPT mat increased to 20 cm, a smoother magnetic field distribution was observed, but approaching to the field generated by a large spiral coil (**Figure 18b**). In order to evaluate the evenness of distribution quantitatively, the coefficient of variation (COV) was utilized which was defined as the standard deviation of the field values divided by the mean. Thus, a smaller value of the COV indicates a more uniform distribution. **Figure 19** shows the COVs of the magnetic field in the z-direction above the HPT mat at distances from 5 to 40 cm. It can be observed that the COV achieves a value <10% when the distance is larger than the size of the transmitter coil.

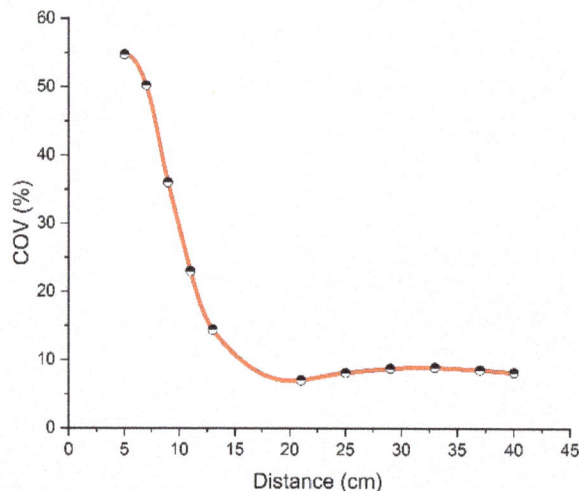

Figure 19. *Variation in coefficient of variation (COV) of vertical field distribution as a function of distance above the HPT mat at the resonant frequency of 85.2 MHz.*

Receiver coil design

In practical applications, it is almost always beneficial to reduce the size of an implantable device, whereas the WPT system requires a match of the resonant frequencies of the primary and secondary coils. We have designed a new structure of the implantable coil with a miniaturized size while having enough turns to match the resonant frequency of the primary coil. **Figure 20** illustrates our design in which coils serve as both implant exterior housing and power receiving elements. It consists of two planar sub-coils and one helical sub-coil. The sub-coils are combined into a single coil within a shallow box assembly. By choosing different geometric designs for the sub-coils, different shaped boxes can be obtained. In practical implementation, the exterior of the box must be covered by a biocompatible material for biological safety.

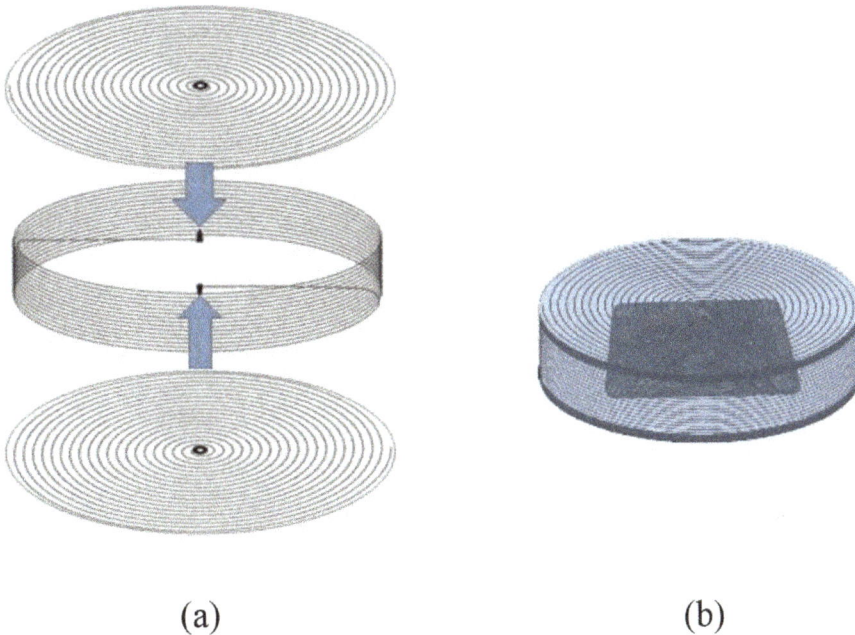

(a)   (b)

Figure 20. *(a) Three sub-coils winded in proper directions are combined and connected to form a single coil shown in (b).*

Experiment tests

In order to study the performance of the mat-based WPT system experimentally, we constructed a prototype mat-based system shown in **Figure 21**, where a transmitter consists of seven circular spiral coils arranged in a hexagonal form. Each PSC was 13.2 cm in diameter, 2.9 mm in trace width, and 1.6 mm in trace spacing. On the reverse side of each PSC, several conductor strips were utilized to form distributed capacitances with respect to the coil on the front side. By changing the numbers of these strips, the resonant frequencies of the PSCs became adjustable. The frequencies were adjusted to 29.453 ± 0.072 MHz with a Q-factor of approximately 100. The prototype of a receiver coil includes three coils and is 25 mm in diameter, 7 mm in height, and 3.39 g in weight. The resonant frequency and Q factor were measured to be 29.075 MHz and 61, respectively.

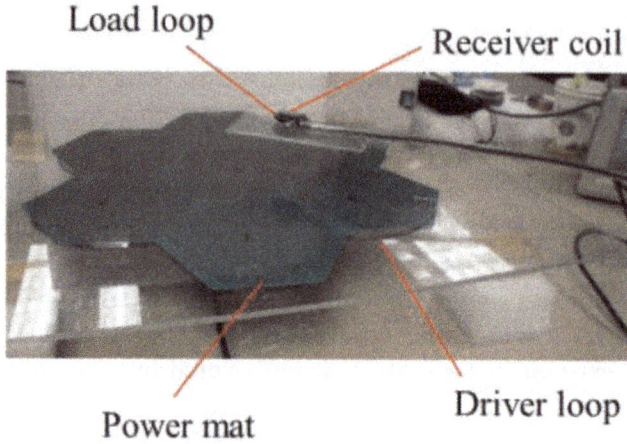

Figure 21. *Experimental platform for measuring magnetic field distribution.*

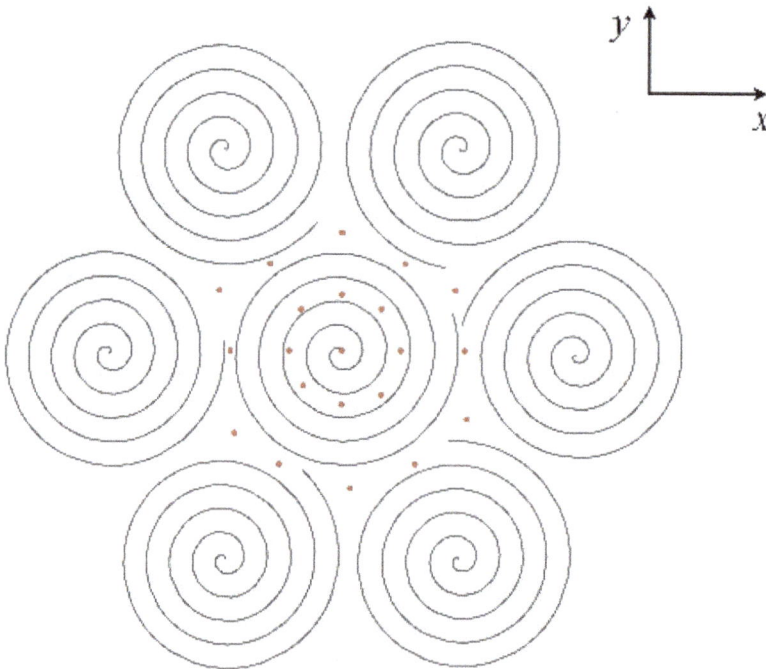

Figure 22. *Sketch of a seven-resonator mat and the test points.*

By measuring the induced voltage in the small receiver coil, the variation of the vertical magnetic field is given by

$$V = -\frac{\partial \varphi}{\partial t} = -N \frac{\partial \overline{B}_z S}{\partial t} \propto B_z \tag{16}$$

When the measuring coil is sufficiently small and the system is driven by sinusoidal input, the induced voltage V in the measuring coil is proportional to the local Bz. As our objective function cancels out the constants relating these two values, we can directly evaluate the cost function using V instead of Bz and compare with the calculation results. As shown in **Figure 22**, these measurement locations were chosen because the magnetic field distribution at any interior PSC in a regular mat can be approximated by the central PSC in each single seven-coil cell (**Figure 15**).

In order to compare the measured voltage with the calculated Bz, we need to normalize both to the same scale, as they are proportional to each other. **Figure 23** shows that the measured data matches the calculated ones very well, except for that the measured data tends to be larger than the calculated ones, which is because the receiver can capture a small portion of horizontal field in additional to the field in the vertical direction. When the measured data are normalized to the center value where the horizontal component is almost zero, the other positions will have larger field than the expected one.

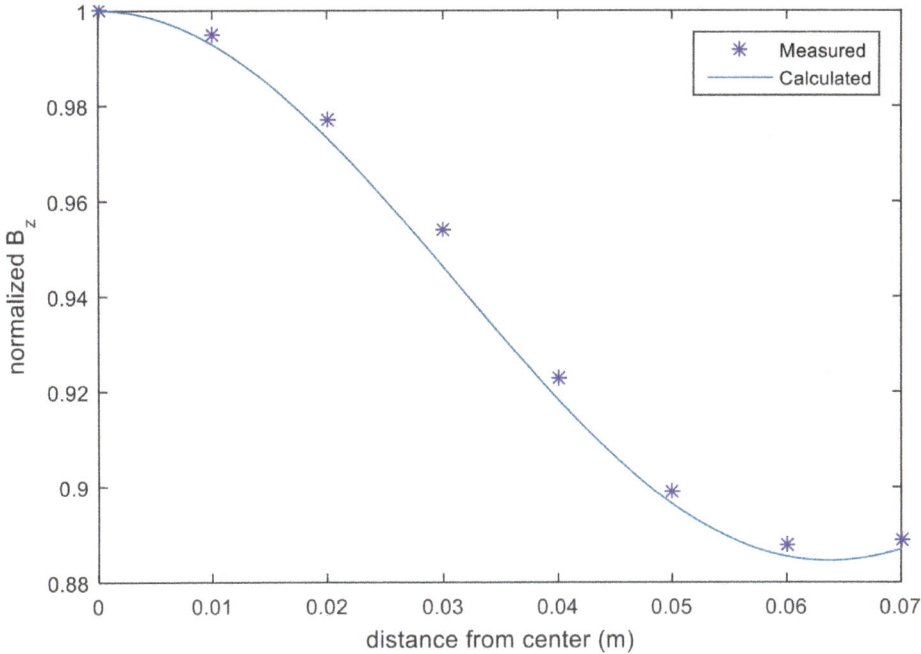

Figure 23. *Comparison of measured and calculated vertical magnetic field over the prototype mat.*

In our WPT system design (**Figure 13**), the separation between the primary and secondary coils includes a distance between the mat and the floor. This distance can be adjusted to achieve both a high WPT performance and an acceptably uniform magnetic field distribution. At different separations, we measured the peak-to-peak values of induced RF voltages in the load coil when the system was powered by a sinusoidal wave at the resonant frequency (approximately 26.6 MHz). Our experiments show that, for our particular system design, the separations of approximately 10 cm and 8 cm between the primary and secondary coils provide a good compromise between performance and magnetic field distribution. In order to visualize this distribution, we interpolated the 21 measured values and plotted the results in **Figure 24**. At a separation of 10 cm, the measured 21 voltage values were in the range between 1.12 and 1.64 V, whereas the mean and standard deviation were 1.23 and 0.11 V, respectively. The relatively small standard deviation indicates a nearly uniform magnetic field distribution, as observed in **Figure 24**. The results indicate that a flat magnetic field can be achieved by our power mat design and that this design is effective for WPT to biomedical implants in freely moving animals.

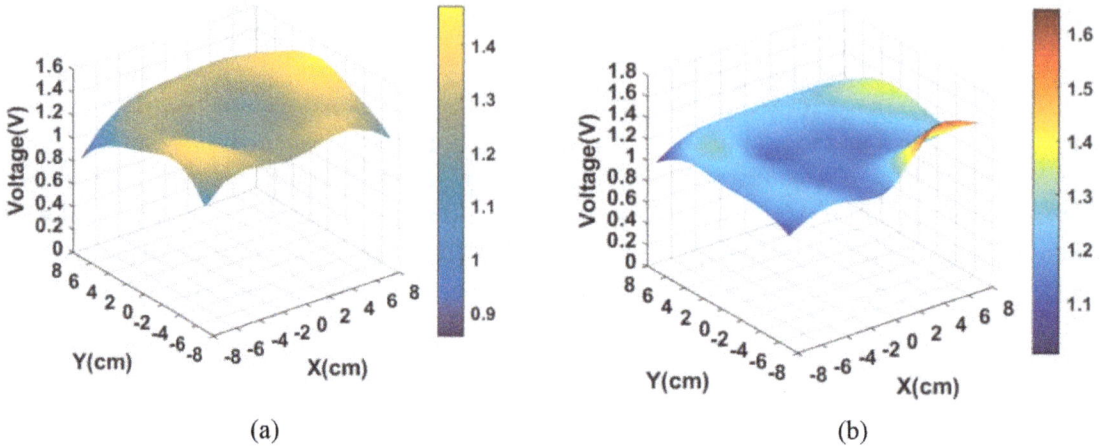

Figure 24. *Voltage distribution computed by interpolating the 21 measured voltage values across the load coil at a constant primary and secondary coil separation of (a) 8 cm and (b) 10 cm.*

Conclusion

We have presented a DH coil as the WPT receiver in implantable medical devices. This new coil has several attractive properties that it can be made conveniently at high precision on a flexible PCB along with other electronic components, forming a complete flexible sheet. This sheet, after being hermetically sealed, can be wrapped around a tubular biological structure, such as a blood vessel or a nerve bundle, to perform diagnostic, monitoring and therapeutic functions. The DH coil has been mathematically analyzed, and expressions for both mutual inductance and self-inductance have been derived. We have found that the DH coil provides a higher coupling factor than the conventional solenoid coil when a lateral or angular misalignment exists. In addition, the DH coil achieves the largest coupling factor and energy transfer efficiency when the axis of the DH coil is in parallel with the plane of the planar spiral transmitter coil. Our computer simulation and experiments under lateral and angular misalignments have been conducted and their results have verified our analytical results.

In order to support biomedical studies using the animal model, we have designed a new power mat, enabling wireless power delivery to miniaturized moving targets. The power mat contains a single or multiple transmitter cells and each cell consists of seven hexagonally packed PSCs. We have conducted theoretical, computational and experimental studies on the special WPT system to meet the challenge of distributing the electromagnetic field evenly over the power mat. We have analyzed the HPT cell using the CMT. Formulas have been derived relating the received power to the inputs and system parameters. Then, we utilize computer simulation to study the evenness of the magnetic field distribution over the power mat at different distances between the power mat and the floor of the animal cage. Finally, we constructed a prototype system, measured its magnetic field distribution and verified that our design has met the challenge. We have also presented a new design of the receiver coil consisting of three serially connected sub-coils. This new design of the receiver coil allows it to capture the most magnetic flux produced by the transmitter, facilitates a match of resonant frequencies of the transmitter and receiver, and reduces the volume of the implant.

Author details

Qi Xu[1,4], Tianfeng Wang[2,4], Shitong Mao[3], Wenyan Jia[3], Zhi-Hong Mao[3,5] and Mingui Sun[3,4,5]*

1 School of Artificial Intelligence and Automation, Huazhong University of Science and Technology, Wuhan, China

2 Department of Electrical Engineering, Shanghai Jiao Tong University, Shanghai, China

3 Department of Electrical and Computer Engineering, University of Pittsburgh, Pittsburgh, USA

4 Department of Neurosurgery, University of Pittsburgh, Pittsburgh, USA 5 Department of Bioengineering, University of Pittsburgh, Pittsburgh, USA

*Address all correspondence to: drsun@pitt.edu

References

[1] Ma A, Poon ASY. Midfield wireless power transfer for bioelectronics. IEEE Circuits and Systems Magazine. 2015; 15(2):54-60. DOI: 10.1109/MCAS.2015.2418999

[2] Jow U, Ghovanloo M. Design and optimization of printed spiral coils for efficient transcutaneous inductive power transmission. IEEE Transactions on Biomedical Circuits and Systems. 2007;1(3):193-202. DOI: 10.1109/ TBCAS.2007.913130

[3] RamRakhyani AK, Mirabbasi S, Mu C. Design and optimization of resonance-based efficient wireless power delivery systems for biomedical implants. IEEE Transactions on Biomedical Circuits and Systems. 2011; 5(1):48-63. DOI: 10.1109/TBCAS.2010. 2072782

[4] Zhong WX, Zhang C, Xun L, Hui SYR. A methodology for making a three-coil wireless power transfer system more energy efficient than a two-coil counterpart for extended transfer distance. IEEE Transactions on Power Electronics. 2015;30(2):933-942. DOI: 10.1109/TPEL.2014.2312020

[5] Zarifi T, Maunder A, Moez K, Mousavi P. Tunable open ended planar spiral coil for wireless power transmission. In: 2015 IEEE International Symposium on Antennas and Propagation & USNC/URSI National Radio Science Meeting, Canada; 19-24 July 2015

[6] Seung Hee J, Zhigang W. Stretchable wireless power transfer with a liquid alloy coil. In: 28th IEEE International Conference on Micro Electro Mechanical Systems (MEMS); 18-22 January 2015; Portugal; 2015

[7] Hasan N, Yilmaz T, Zane R, Pantic Z. Multi-objective particle swarm optimization applied to the design of wireless power transfer systems. In: Wireless Power Transfer Conference (WPTC); USA: IEEE; 2015. 13-15 May 2015

[8] Yakovlev A, Kim S, Poon A. Implantable biomedical devices: Wireless powering and communication. IEEE Communications Magazine. 2012; 50(4):152-159. DOI: 10.1109/ MCOM.2012.6178849

[9] Ho JS, Yeh AJ, Neofytou E, Kim S, Tanabe Y, Patlolla B, et al. Wireless power transfer to deep-tissue microimplants. Proceedings of the National Academy of Sciences. 2014; 111(22):7974-7979. DOI: 10.1073/pnas. 1403002111

[10] Available from: http://www.impla ntable-device. com/category/technologie s/wireless-power-transmission/

[11] Available from: http://news.b ostonscientific.com/news-releases?ite m/58598

[12] Ahn D, Ghovanloo M. Optimal design of wireless power transmission links for millimeter-sized biomedical implants. IEEE Transactions on Biomedical Circuits and Systems. 2016; 10(1):125-137. DOI: 10.1109/TBCAS. 2014.2370794

[13] Kyungmin N, Heedon J, Hyunggun M, Bien F. Tracking optimal efficiency of magnetic resonance wireless power transfer system for biomedical capsule endoscopy. IEEE Transactions on Microwave Theory and Techniques. 2015;63(1):295-304. DOI: 10.1109/TMTT.2014.2365475

[14] Loeb GE, Richmond FJ, Baker LL. The BION devices: Injectable interfaces with peripheral nerves and muscles. Neurosurgical Focus. 2006;20(5):1-9. DOI: 10.3171/foc.2006.20.5.3

[15] Kimchi T, Xu J, Dulac C. A functional circuit underlying male sexual behaviour in the female mouse brain. Nature. 2007;448:1009-1014

[16] Xu Q, Hu D, Duan B, He J. A fully implantable stimulator with wireless power and data transmission for experimental investigation of epidural spinal cord stimulation. IEEE Transactions on Neural Systems and Rehabilitation Engineering. 2015;23(4): 683-692. DOI: 10.1109/TNSRE.2015. 2396574

[17] OpenStax College. Anatomy and Physiology. Rice University, Houston: OpenStax College; 2013

[18] Mao S, Wang H, Mao ZH, Sun M. A miniature implantable coil that can be wrapped around a tubular organ within the human body. AIP Advances. 2018; 8(5):056629. DOI: 10.1063/1.5007258

[19] Mao S, Wang H, Mao ZH, Sun M. A double-helix and cross-patterned solenoid used as a wirelessly powered receiver for medical implants. AIP Advances. 2018;8(5):056603. DOI: 10.1063/1.5007236

[20] Yin N, Xu G, Yang Q, Zhao J, Yang X, Jin J, et al. Analysis of wireless energy transmission for implantable device based on coupled magnetic resonance. IEEE Transactions on Magnetics. 2012;48(2):723-726. DOI: 10.1109/TMAG.2011.2174341

[21] Mayordomo I, Dräger T, Spies P, Bernhard J, Pflaum A. An overview of technical challenges and advances of inductive wireless power transmission. Proceedings of the IEEE. 2013;101(6): 1302-1311. DOI: 10.1109/JPROC.2013.2243691

[22] Kurs A, Karalis A, Moffatt R, Joannopoulos JD, Fisher P, Soljacic M. Wireless power transfer via strongly coupled magnetic resonances. Science. 2007;317(5834):83-86. DOI: 10.1126/science.1143254

[23] Xu Q, Wang H, Gao Z, Mao Z, He J, Sun M. A novel mat-based system for position-varying wireless power transfer to biomedical implants. IEEE Transactions on Magnetics. 2013;49(8): 4774-4779. DOI: 10.1109/TMAG.2013. 2245335

[24] Xu Q, Gao Z, Wang H, He J, Mao Z, Sun M. Batteries not included: A mat-based wireless power transfer system for implantable medical devices as a moving target. IEEE Microwave Magazine. 2013;14(2):63-72. DOI: 10.1109/MMM.2012.2234640

[25] Graham RL, Lubachevsky BD, Nurmela KJ, Östergård PRJ. Dense packings of congruent circles in a circle. Discrete Mathematics. 1998;181(1-3): 139-154. DOI: 10.1016/S0012-365X(97) 00050-2

[26] Zhang F, Hackworth SA, Fu W, Li C, Mao Z, Sun M. Relay effect of wireless power transfer using strongly coupled magnetic resonances. IEEE Transactions on Magnetics. 2011;47(5): 1478-1481. DOI: 10.1109/ TMAG.2010. 2087010

[27] Zhang F, Liu J, Mao Z, Sun M. Mid-range wireless power transfer and its application to body sensor networks. Open Journal of Applied Sciences. 2012; 2(1):35-46. DOI: 10.4236/ojapps.2012. 21004

[28] Simpson J, Lane J, Immer J, Youngquist R. Simple analytic expressions for the magnetic field of a circular current loop. NASA Technical Document Collection. Document ID: 20010038494; 2001

Permissions

List of Contributors

Dac-Binh Ha
Duy Tan University, Da Nang, Vietnam

Jai P. Agrawal
Purdue University Northwest, Hammond, IN, USA

Bin-Jie Hu, Zhi-Wu Lin and Peng Liao
School of Electronic and Information Engineering, South China University of Technology, Guangzhou, China

Anming Dong
Shandong Provincial Key Laboratory of Wireless Communication Technologies and School of Control Science of Engineering, Shandong University, Jinan, China
School of Computer Science and Technology, Qilu University of Technology (Shandong Academy of Sciences), Jinan, China
Computer Science Center (National Supercomputer Center in Jinan), Shandong Provincial Key Laboratory of Computer Networks, Qilu University of Technology (Shandong Academy of Sciences), Jinan, China

Haixia Zhang
Shandong Provincial Key Laboratory of Wireless Communication Technologies and School of Control Science of Engineering, Shandong University, Jinan, China

Mohamed Aboualalaa and Hala Elsadek
Electronics Research Institute, Cairo, Egypt

Ahmed M.A. Sabaawi
College of Electronics Engineering, Ninevah University, Mosul, Iraq

Oras Ahmed Al-Ani
College of Electrical Engineering Techniques, Middle Technical University, Baghdad, Iraq

Marcos R. Pino and Rafael G. Ayestarán
Department of Electrical, Electronics, Computers and Systems Engineering, University of Oviedo, Gijón, Spain

Paolo Nepa and Giuliano Manara
Department of Information Engineering, University of Pisa, Pisa, Italy

Abdul Quddious and Marco A. Antoniades
Department of Electrical and Computer Engineering, University of Cyprus, Nicosia, Cyprus

Photos Vryonides and Symeon Nikolaou
Frederick University and Frederick Research Center, Nicosia, Cyprus

Antonio Lazaro, Marti Boada, Ramon Villarino and David Girbau
Department of Electronic, Electric and Automatic Control Engineering, Rovira i Virgili University, Tarragona, Spain

Lai Ly Pon, Sharul Kamal Abdul Rahim and Chee Yen Leow
Wireless Communication Center, School of Electrical Engineering, Faculty of Engineering, Universiti Teknologi Malaysia, Johor Bahru, Malaysia

Mohamed Himdi
Institute of Electronics and Telecommunication of Rennes, University of Rennes, Rennes, France

Qi Xu
School of Artificial Intelligence and Automation, Huazhong University of Science and Technology, Wuhan, China
Department of Neurosurgery, University of Pittsburgh, Pittsburgh, USA

Tianfeng Wang
Department of Electrical Engineering, Shanghai Jiao Tong University, Shanghai, China
Department of Neurosurgery, University of Pittsburgh, Pittsburgh, USA

Shitong Mao and Wenyan Jia
Department of Electrical and Computer Engineering, University of Pittsburgh, Pittsburgh, USA

Zhi-Hong Mao
Department of Electrical and Computer Engineering, University of Pittsburgh, Pittsburgh, USA
Department of Bioengineering, University of Pittsburgh, Pittsburgh, USA

Mingui Sun
Department of Electrical and Computer Engineering, University of Pittsburgh, Pittsburgh, USA
Department of Neurosurgery, University of Pittsburgh, Pittsburgh, USA
Department of Bioengineering, University of Pittsburgh, Pittsburgh, USA

Index

www.ingramcontent.com/pod-product-compliance
Lightning Source LLC
Chambersburg PA
CBHW061959190326
41458CB00009B/2915